T0258546

Nanoparticles: Toxic and Drug Nanoparticles

Nanoparticles: Toxic and Drug Nanoparticles

Edited by **Mindy Adams**

New York

Published by NY Research Press,
23 West, 55th Street, Suite 816,
New York, NY 10019, USA
www.nyresearchpress.com

Nanoparticles: Toxic and Drug Nanoparticles
Edited by Mindy Adams

International Standard Book Number: 978-1-63238-339-6 (Hardback)

Printed in the United States of America.

Contents

Preface

The purpose of the book is to provide a glimpse into the dynamics and to present opinions and studies of some of the scientists engaged in the development of new ideas in the field from very different standpoints. This book will prove useful to students and researchers owing to its high content quality.

The properties of materials change depending upon their size. This book explores the field of nano-scale particles and their grounding methods. Nanoparticle is a universal challenge for today's technology and future researches. Nanoparticles encompass approximately all kinds of sciences and manufacturing technologies. The characteristics of these particles are overcoming current scientific hurdles and have passed the restrictions of conventional science. This is why nanoparticles have been classified for use in various fields. The book discusses significant topics related to toxic and drug nanoparticles. This book will be beneficial for students and experts interested in this field.

At the end, I would like to appreciate all the efforts made by the authors in completing their chapters professionally. I express my deepest gratitude to all of them for contributing to this book by sharing their valuable works. A special thanks to my family and friends for their constant support in this journey.

Editor

Section 1

Toxic Nanoparticles

In vivo Toxicity Studies of Pristine Carbon Nanotubes: A Review

Jelena Kolosnjaj-Tabi, Henri Szwarc and Fathi Moussa

UMR CNRS 8612 and LETIAM, EA 4041, University of Paris Sud
France

1. Introduction

Discovered in 1991(Iijima, 1991), carbon nanotubes (CNTs) have attracted considerable attention in many fields of science and technology because of their unique structural, mechanical, and electronic properties. Their potential seemed paramount in the fields of materials science, including conductive and high-strength composites, energy storage and energy conversion devices, sensors, field emission displays and radiation sources, hydrogen storage, nanometer-sized semiconductor devices, probes and interconnects (Dresselhaus et al., 2004). Thus their studies progressed rapidly during the last two decades.

The chemistry of CNTs grew out from the efforts to open and fill the tubes (Ruoff et al., 1993; Sloan et al., 1998 as cited in Harris, 2009) and, indeed, to functionalize their sidewalls (J. Chen et al., 1998; Y. Chen et al., 1998). The first approach was driven partly for the study of matter in confined spaces and partly in order to use the tubes as templates for nanowires (Harris, 2009).

On the other hand, CNT functionalization was needed to disperse ("to solubilize") the tubes in aqueous media (Hirsh, 2002). Then, functionalization with biomolecules has become in vogue and many research groups have begun to investigate biological uses of these new types of nanostructures. According to some authors, CNTs could be possibly used as biosensors (Balavoine et al., 1999; Bekyarova et al., 2005; Lin et al., 2004; Richard et al., 2003; J. Wang et al., 2003; S. Wang et al., 2003), substrates for neuronal growth (Hu et al., 2004; Mattson et al., 2000; Lovat et al., 2005), supports for adhesion of liposaccharides to mimic cell membranes (X. Chen et al., 2004), delivery systems (Bianco et al., 2005), medical imaging agents (Ashcroft et al., 2007; Hartman et al., 2008; Sitharaman et al., 2005) and radiotherapeutics (Hartman et al., 2007a). Such potential uses remind those proposed for fullerenes and derivatives in the nineties (Jensen et al., 1996).

The growing use and mass production of CNTs raised concerns about their safety and environmental impact soon after the announcement of the national nanotechnology initiative by the US president in 2000 (http://www.nano.gov/). First toxicity studies addressed their safety at workplace (Oberdorster et al., 2005). Since then, the investigations of their toxicity first *in vitro* and then *in vivo* were reported in countless publications but their results remain contradictory (Kolosnjaj et al., 2007). As mentioned by most authors, this is due to several factors including surface defects, sizes, and degree of aggregation of the tested material, and exposure protocols (Oberdorster et al., 2005, Kolosnjaj et al., 2007).

In this chapter we will try to present an uptake of the current knowledge on CNT toxicity as a function of the route of administration. Because of the great number of papers devoted to this subject, we are quite aware that we will miss quoting a number of works and we apologize to forgotten colleagues.

For a better understanding of biological behaviour of CNTs we will first describe their main general characteristics.

2. General characteristics

Carbon nanotubes are mainly composed of sp^2 bonds, similar to graphite, and are categorized as single-walled nanotubes (SWNTs) and multi-walled nanotubes (MWNTs).

2.1 Structure

2.1.1 Single-walled carbon nanotubes (SWNTs)

Single-walled carbon nanotubes, composed by a rolled monolayered graphene sheet (that might be end-capped by half a C60 molecule), exist in a variety of structures corresponding to the many ways a sheet of graphite can be wrapped into a seamless tube. Each structure type has a specific diameter and chirality, or wrapping angle (α). The "armchair" structures (Fig.1 a), with $\alpha = 30°$, have metallic character. The "zigzag" tubes (Fig.1 b), for which $\alpha = 0°$, can be either semi-metallic or semiconducting, depending on the specific diameter. Nanotubes with chiral angles intermediate between 0 and 30° (Fig.1 c) include both semimetals and semiconductors. The terms "armchair" and "zigzag" refer to the pattern of carbon–carbon bonds along a tube's circumference (Dresselhaus et al., 2004).

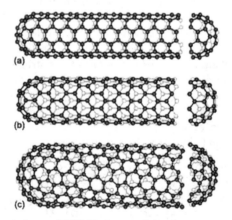

Fig. 1. Schematic representations of SWNTs in a variety of structures: (a) "armchair" (b) "zigzag" stricture (c) "intermediate" structure.

The diameter of the tubes generally varies from 0.4 to 20 nm, while the length usually reaches several micrometers. The tubes are often entangled and the ropes (Fig. 2) of SWNTs are held together by van der Waals forces (Popov et al., 2004).

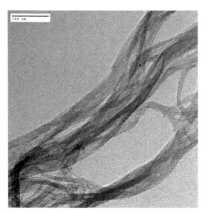

Fig. 2. Transmission electron micrograph of ropes of SWNTs

2.1.2 Multi-walled carbon nanotubes (MWNTs)

Multi-walled carbon nanotubes possess several graphitic concentric layers (Fig.3), made either by a single rolled graphene layer (resembling a scroll of parchment) or, more commonly, encased within one another (as Russian nesting dolls). The distance between each layer of graphene in a MWNT is about 0.34 nm (Iijima, 1991).

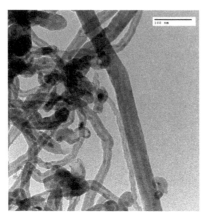

Fig. 3. Transmission electron micrograph of a bundle made of individual entangled MWNTs.

2.2 Impurities

Carbon nanotube powder often contains up to 30% metal (mainly iron and nickel) catalyst particles, as we can see on figure 2 (note the little round electron-dense particles present on nanotube ropes), and some amorphous carbon. Carbon nanotubes should be purified prior to their administration, in order to avoid any metal catalyst-related toxicity (Valko et al., 2005).

Several techniques of purification have been reviewed elsewhere (Sinha & Yeow, 2005) and the most commonly used technique appears to be the oxidation using strong acid treatments, which allows solubilisation and removal of a large part of the metallic impurities. Nevertheless, this methodology has an impact on the tubes. Strong acid conditions cut the tubes in shorter pieces and generate carboxylic functions at the tips and around the sidewalls where the curvatures of the tubes present a higher strain (Ziegler et al., 2005).

2.3 Dispersibility

Pristine (chemically unmodified) CNTs are insoluble and hardly dispersible in water. In order to disperse CNTs several groups proposed covalent and non-covalent functionalization, which will be described in this chapter in terms of *in vivo* interactions.

3. Interfacing pristine carbon nanotubes with living organisms

The interaction of CNTs with cells has been described by many research groups and has been reviewed by several authors, all converging to the conclusions that the toxicity of CNTs *in vivo* depends on the type of CNTs, including the method of production, impurities and purification process (which might affect the sidewalls), length, aggregation state, surface coating, and chemical modification. Moreover, special care is needed for the choice of reagents used to evaluate the viability of the cells (Kroll et al., 2009), as these nanoparticles may interact with assays and even dispersion agents, potentially resulting in a secondary rather than primary toxicity (Casey et al., 2007).

These physical and chemical characteristics are, indeed, important in *in vivo* studies. However, we should also consider other phenomena that may occur in complex living systems, such as interaction of CNTs with several different cells at a time, biotransformation and innate foreign body reactions, etc.

Organism-CNT interactions were first described in *in vivo* toxicity studies, which were performed in order to assess the exposure risks to CNTs at workplace. Airborne CNTs might represent a danger to people handling these materials on daily basis either by crossing their skin barrier or entering and residing in their lungs. However, in recent years, potential applications of CNTs in the biomedical field intensified the research on the *in vivo* behaviour of these materials and increased the number of studies devoted to the evaluation of their potential toxicity after non-pulmonary routes of administration.

3.1 Pilot studies on carbon nanotubes in regards to workplace safety

In 2001 the potential of CNTs to induce skin irritation was evaluated by performing two routine dermatological tests (Huczko & Lange, 2001). Initially, 40 volunteers with allergy susceptibilities were exposed for 96 h to a patch test consisting in a filter paper saturated with a water suspension of unrefined CNTs synthesized by the arc discharge process. Secondly, a modified Draize rabbit eye test using a water suspension of unrefined CNTs was conducted with four albino rabbits monitored for 72 h after exposure. Both tests showed no irritation in comparison to a CNT-free soot control and it was concluded that "no special precautions have to be taken while handling these nanostructures" (Huczko & Lange, 2001).

In a second two-part study, other investigations have been made to seek for exposure routes and toxicity of SWNTs (Maynard et al., 2004). The study was undertaken to evaluate the physical nature of the aerosol formed from SWNTs during mechanical agitation. This was complemented by a field study in which airborne and dermal exposure to SWNTs was evaluated while handling unrefined material. Although laboratory studies indicated that unrefined SWNT material could release fine particles into air under sufficient agitation, concentrations generated while handling material in the field were very low. Estimates of the airborne concentrations of nanotube materials generated during handling suggest that concentrations were lower than 53 $\mu g/m^3$ in all instances. In another way, glove deposits of SWNTs during handling were estimated at between 0.2 mg and 6 mg per hand (Maynard et al., 2004).

3.2 Respiratory exposure: Pulmonary toxicity

Carbon nanotubes are very light and could become airborne and potentially reach the lungs; therefore the earliest *in vivo* studies tried to assess their pulmonary toxicity (Lam et al., 2004; Warheit et al., 2004).

For this purpose three kinds of SWNTs were studied, namely raw and purified HiPco and CarboLex CNTs. The first material is rich in iron impurities and the last one contains nickel and yttrium impurities. The particles were dispersed by brief shearing (2 min in a small glass homogenizing tube) and subsequent sonication (0.5 min) in heat-inactivated mouse serum. Mice were then intra-tracheally instilled with 0, 0.1, or 0.5 mg of CNT or carbon black or quartz particles used as negative and positive control, respectively. Seven and 90 days after this single treatment the animals were sacrificed for histopathological examination of the lungs. All CNT treatments induced dose-dependent epithelioid granulomas and, in some cases, interstitial inflammation in the animals euthanized after 7 days (Lam et al., 2004). These lesions persisted and were more pronounced in the group euthanized after 90 days. The lungs of some animals also revealed peri-bronchial inflammation and necrosis that had extended into the alveolar septa. The lungs of mice treated with carbon black were normal, whereas those treated with high-dose quartz revealed mild to moderate inflammation. These results show that, under these conditions and on an equal- weight basis, if carbon nanotubes reach the lungs, they are much more toxic than carbon black and can be more toxic than quartz, which is considered a serious occupational health hazard in chronic inhalation exposures (Lam et al., 2004).

In a similar way, a parallel pulmonary toxicity assessment of pristine SWNTs was described (Warheit et al., 2004). The aim of the study was to evaluate the acute lung toxicity of intratracheally instilled SWNTs in rats. The applied CNTs were produced by laser ablation and contained about 30 to 40% amorphous carbon (by weight) and 5% each of nickel and cobalt. The lungs of rats were instilled either with 1 or 5 mg/kg of the following control or particle types: SWNTs, quartz particles (positive control), carbonyl iron particles (negative control), and the vehicle - phosphate buffered saline (PBS) and 1% Tween 80, or graphite particles (Warheit et al., 2004). Following exposure, the lungs of treated rats were assessed using bronchoalveolar fluid biomarkers and cell proliferation methods, as well as by histopathological examination of lung tissue at 24 h, 1 week, 1 month, and 3 months post-instillation. Exposures to high-dose (5 mg/kg) of SWNT produced mortality in approximately 15% of the instilled rats within 24 h post-instillation. This mortality resulted from mechanical blockage of the upper airways by the instilled particulate SWNT. In the surviving animals, SWNT produced temporary inflammatory and cell injury effects. Results from the lung histopathology indicated that pulmonary exposures to SWNT in rats produced a non-dose-dependent series of multifocal granulomas, which were evidence of a foreign tissue body reaction. However, they were non-uniform in distribution and not progressive beyond one month of post-exposure. The observation of SWNT-induced multifocal granulomas was inconsistent with the following: lack of lung toxicity by assessing lavage parameters, lack of lung toxicity by measuring cell proliferation parameters, apparent lack of a dose response relationship, non-uniform distribution of lesions, the paradigm of dust-related lung toxicity effects, and possible regression of effects over time. The observation of granulomas, in the absence of adverse effects measured by pulmonary endpoints was surprising, and did not follow the normal inflammogenic/fibrotic pattern produced by fibrogenic dusts, such as quartz, asbestos, and silicon carbide whiskers (Warheit et al., 2004).

While the first authors (Lam et al., 2004) concluded that SWNT were more toxic than quartz nanoparticles and crystalline silica particles, the second ones (Warheit et al., 2004) observed only a transient pulmonary inflammation and granuloma formation after SWNT exposure, contrarily to sustained lung inflammation, cytotoxicity, enhanced lung cell proliferation, foamy macrophage accumulation and lung fibrosis after exposure to quartz particles. The differences between these findings may be related in part to species differences (mouse vs. rat), but are more likely due to the differences in the experimental designs of the two studies (Warheit, 2006).

Respiratory toxicity of MWNTs has been also evaluated after intra-tracheal administration of MWNTs or ground MWNTs suspended and sonicated in sterile 0.9 % saline containing 1 % of Tween 80, at doses of 0.5, 2.0 or 5.0 mg, corresponding to approximately 2.2 mg/kg, 8.9 mg/kg and 22.2 mg/kg body-weight (bw) to Sprague-Dawley rats (Muller et al. 2005). The applied CNTs were still present in the lungs after 60 days (80% and 40% of the lowest dose) and both induced inflammatory and fibrotic reactions (Muller et al. 2005). At 2 months, pulmonary lesions induced by MWNTs were characterized by the formation of collagen-rich granulomas protruding in the bronchial lumen, in association with alveolitis in the surrounding tissues. These lesions were caused by the accumulation of large MWNT agglomerates in the airways. Ground CNTs were better dispersed in the lung parenchyma and also induced inflammatory and fibrotic responses. Both MWNTs and ground MWNTs stimulated the production of TNF-α in the lung of treated animals (Muller et al. 2005).

The physiological relevance of intra-tracheal instillation of CNTs is debatable since physiologically inspired particles would probably encounter several barriers in the upper respiratory tract before reaching the trachea and the lungs. Nevertheless, purified SWNTs elicited inflammation, fibrosis and granulomas formation in C57BL/6 mice even when administered by pharyngeal aspiration (Shvedova et al., 2005). The nanotubes used in this study were produced by HiPco and where further purified by acidic treatment. The analysis also proved that CNTs accounted for more than 99% of carbon. The animals were treated with either SWNT (0, 10, 20, 40 µg/mouse) or two reference materials (ultrafine carbon black or SiO2 at 40 µg/mouse). The animals were sacrificed at 1, 3, 7, 28, and 60 days following exposures. A rapid progressive fibrosis found in mice exhibited two distinct morphologies: 1- SWNT-induced granulomas mainly associated with hypertrophied epithelial cells surrounding dense micrometer-scale SWNT aggregates and 2- diffuse interstitial fibrosis and alveolar wall thickening likely associated with dispersed SWNTs. These differences in fibrosis morphology were attributed to the distinct particle morphologies of compact aggregates and dispersed SWNT structures. Importantly, deposition of collagen and elastin was also observed in both granulomatous regions as well as in the areas distant from granulomas. Increased numbers of alveolar type II (AT- II) cells, the progenitor cells that replicate following alveolar type I (AT-I) cell death, were also observed as a response to SWNT administration. Moreover, functional respiratory deficiencies and decreased bacterial clearance (Listeria monocytogenes) were found in mice treated with SWNT (Shvedova et al., 2005).

In a mechanistically oriented study, the physicochemical determinants of the MWNTs' toxicity mechanism were investigated (Muller et al., 2008). In this experiment the toxicity of MWNTs was evaluated after the tubes were heated at 600°C (which allowed loss of oxygenated carbon functionalities and reduction of oxidized metals) or at 2400°C (which annealed structural defects and eliminated metals) or after the MWNTs heated at 2400°C were grinded (introduction of structural defects in a metal-deprived framework). The MWNTs were suspended in 1% Tween 80 and physiological saline and administered intra-tracheally (2 mg/rat). The results show that pulmonary toxicity (and genotoxicity of MWNTs, determined *in vitro*) were reduced upon heating but restored upon grinding, indicating that the intrinsic toxicity of the tubes was mainly mediated by the presence of defective sites in their carbon framework (Muller et al., 2008).

Finally, in order to check the hypothesis linking lung toxicity to CNT aggregates (Mutlu, Budinger et al., 2010), the authors instilled intratracheally unpurified aggregated or highly dispersed SWNTs in 1% Pluronic F 108NF to mice. As-produced HiPco SWNTs were either dispersed in PBS or highly dispersed in Pluronic solution in a dose of 40 µg, which was chosen to match or exceed those previously reported to cause pulmonary fibrosis in mice (Mutlu, Budinger et al., 2010; Shvedova et al., 2005). According to the authors, lung inflammation induced by SWNTs is minimal compared to that induced by urban particulate matter or asbestos fibers (used as positive control). Aggregated SWNTs in PBS caused areas of chronic inflammation, while highly dispersed SWNTs do not cause any inflammation or fibrosis. Moreover, nanoscale dispersed SWNTs are taken up by alveolar macrophages and cleared from the lung over time (Mutlu, Budinger et al., 2010). Besides, by administering unpurified CNTs, the authors (Mutlu, Budinger et al., 2010) avoided a potential effect due to surface defects of tube sidewalls (Ziegler et al., 2005), which might contribute to an increase in collagen deposition (Mercer et al., 2008).

In a study where rats were instilled with 0.04, 0.2, or 1 mg/kg of individually dispersed MWNTs in Tween 80 (Kobayashi et al., 2010), it has been observed that pulmonary inflammatory responses occur only in the lungs of the group treated with the highest dose. Moreover, the authors did not find any evidence of chronic inflammation, such as angiogenesis or fibrosis, induced by MWNT instillation (Kobayashi et al., 2010). Light microscopic examination indicated that MWNT aggregates deposited in the lungs were phagocytized by alveolar macrophages and were accumulated in the lungs until 6-month post-exposure. These aggregates were located in the alveolar or interstitial macrophages, but individual MWNTs were not present in the cells of the interstitial tissue (Kobayashi et al., 2010). However, in the light micrograph panels, provided by the authors (Kobayashi et al., 2010) MWNTs seem to extend outside the macrophages in several directions, which is commonly referred to as incomplete or frustrated phagocytosis that is known to be a pro-inflammatory condition (Balkwill & Mantovani, 2001). According to the authors, the MWNTs used in this study were less than 20 μm long, but after measuring the tubes inside the alveolar macrophages, they concluded that median length was approximately 1.5 μm, although some tubes were measuring up to 6 μm (Kobayashi et al., 2010).

In contrast, a recent study reported that highly dispersed MWNTs could, depending on the way of administration (i.e. intratracheal instillation or inhalation) and dose, produce pulmonary lesions (Morimoto et al., 2011). The MWNTs that were used in this study were ground in a fructose mold - the fructose was rinsed afterwards with water and hydrogen peroxide. According to the authors this process slightly oxidized the tubes, which were subsequently dispersed in a 0.05% Triton X-100 solution (Morimoto et al., 2011). Triton X-100 is often used in cell biology to digest the cell membrane and cytoplasm to access the cell nucleus (http://fr.wikipedia.org/wiki/Triton_X-100). In the experiment of intratracheal instillation, two single doses (0.2 mg or 1 mg, 1.1 μm of mean particle length) were administered to rats and the study was conducted up to 6 months. While only a transient infiltration of inflammatory cells was observed for 0.2 mg treated animals, the high dose caused small granulomatous lesions and transient collagen depositions. In parallel, the authors conducted an inhalation study of dispersed MWNTs in a daily average mass of 0.37 ± 0.18 mg/m^3 (Morimoto et al., 2011). The rats were exposed to aerosol particles for 6h per day, 5 days a week for 4 weeks. At the end of the experiment, the dispersed MWNTs with the average length of 1.1 μm caused only a minimal transient inflammation, which did not cause neutrophil infiltration into alveolar space. Moreover granulomatous lesions or collagen depositions were not observed (Morimoto et al., 2011).

In conclusion, the studies performed thus far indicate that due to van der Waals interactions individual SWNTs are prone to form large aggregates, in air or in aqueous solutions, which can be more than one hundred micrometers in diameter. While accidental industrial exposure is the most probable risk and might have a serious impact at workplace, toxicity was not observed after the intratracheal instillation of nanoscale dispersed SWNTs at a dose of 1.6 mg/kg (Mutlu, Budinger et al., 2010), or nanoscale dispersed MWNTs at a dose of 0.66 mg/kg (Morimoto et al., 2011). It is worthy to note that these doses would be the equivalent to a single instilled dose of approximately 112 g or 46 g, respectively, in an average weighting human.

3.2.1 Respiratory exposure and health risks

The analogy between CNTs and asbestos fibres was pointed out in the late nineties (Service, 1998). The term asbestos refers to a variety of fibrous silicates, which were exploited commercially in past for their desirable physical properties, such as sound absorption, average tensile strength, and resistance to fire, heat, electrical and chemical damage (http://en.wikipedia.org/wiki/Asbestos). Asbestos fibres have a high aspect ratio and they are characterized by high chemical stability in physiological environment (Kane & Hurt, 2008). The pathologies related to asbestos exposure, especially lung fibrosis (asbestosis) and lung cancer (mesothelioma) that most often originates in the pleura, the outer lining of the lungs, have caused a major worldwide occupational health disaster (Donaldson & Poland, 2009) and founded reasonable fear of these airborne fibres.

Each lung is invested by a membrane, the pleura, which is arranged in the form of a closed invaginated sac. The membrane lining on the lung with its 'visceral' mesothelial layer is the visceral pleura and the membrane attached to the chest wall, covered by a continuous 'parietal' mesothelial cell layer is the parietal pleura. The pleural mesothelium is the primary mesothelial target for inhaled fibres (Donaldson et al, 2010). The space between the visceral and the parietal pleura contains the pleural fluid and a population of pleural macrophages. Pleural liquid is derived mainly from capillaries in the parietal pleura and is principally removed by lymphatic stomata in the parietal pleura (Lai-Fook, 2004), which drains the pleural fluid to the lymph nodes. This turnover is important for clearance of particles and fibres that reach the pleural space (Donaldson et al, 2010). While the exact mechanism of fibre deposition in the pleural mesothelium remains unclear, research indicates that retention of biopersistent fibres at the parietal pleura initiates mesothelial injury and inflammation that, overtime, lead to mesothelioma. When the fibre diameter is small, the fibre will align with the flow and deeply penetrate the lungs. The fibres are more or less cleared by macrophages, depending on their length. If the fibers are too long they cannot be entirely phagocytized. This unachieved – 'frustrated' phagocytosis is pro-inflammatory condition, characterized by the release of inflammatory mediators into the environment. These mediators may recruit other cells (for example collagen synthesizing fibroblasts) or cause DNA damage and mutations to proliferating cells, which may in term cause tumour development (Balkwill & Mantovani, 2001). Carbon nanotubes are fiber-shaped, however, for what concerns SWNTs, they are flexible and bendable, and often entangled. These particle-sized tangles would not obey the fibre toxicity paradigm because of their non-fibrous geometry (Donaldson et al, 2010; Kane & Hurt, 2008). Multi-walled carbon nanotubes, on the other hand, are much stiffer and generally less entangled; therefore, if long enough, they might present a risk (Donaldson et al, 2010).

A study performed with nickel containing milled pluronic-suspended MWNTs (Ryman-Rasmussen et al., 2009) with a length ranging between 100 nm and 10 µm showed that the nanotubes are observed inside the sub-pleural tissue macrophages after a single 6 h inhalation exposure of 30 mg/m³. Fibrotic lesions, which increase 2 and 6 weeks after exposure, remain focal and regional. This effect did not occur after exposure to a dose of 1 mg/m³, which according to the authors corresponds to 0.2 mg/kg. While the authors did not find MWNTs-loaded macrophages inside the pleura, they did notice an increased number of pleural mononuclear cells.

A study published about the same time reported that MWNTs could reach the pleura after pharyngeal aspiration (Porter et al., 2010). The inflammation extended from lungs to pleura in half of the MWNT-exposed mice. At 56 days post exposure, MWNTs penetrated the pleura in two out of four mice treated with the highest MWNT dose. The inflammation induced by the nanotubes was transient at low doses but persistent through day 56 at a dose of 40 µg.

In another study performed by the same group (Mercer et al., 2010), the authors reported an initial high density of penetrations into the sub-pleural tissue and the intra-pleural space one day following aspiration of MWNTs (80 µg per mice). The kinetics of penetration decreased due to the clearance by alveolar macrophages by day 7 and reached steady state levels in the sub-pleural tissue and intra-pleural space from days 28 to 56. The majority of the tubes (62% of the dose) resided in alveolar macrophages, while 0.6% of tubes, reached the visceral pleura region (sub-pleura and intra-pleural space).

As it has been already emphasized (Donaldson et al, 2010), the question that we should ask with regard to any fibre in relation to mesothelioma is not "Do fibres reach the pleura?" but "Are fibres retained in the parietal pleura", which is the site of origin of pleural mesothelioma.

When long and short CNTs (as well as long and short asbestos fibres) were injected directly into the pleural space, the authors found (Murphy et al., 2011) evidence of length-related inflammation, with no significant inflammation when short tubes (fibres) were administered. While no short samples were visible at day 1 or day 7, the mesothelium, that was thicker on day 1, returned to its normal thickness by day 7. While short tubes and fibres cleared from the pleura through stomatal openings, long tubes and fibres remained inside the pleura near stomata, where they persisted and caused inflammation and progressive fibrosis.

3.3 Effects of carbon nanotubes after intra-peritoneal administration

While pulmonary toxicity studies clearly indicate that inhalation of CNTs aggregates represents a possible occupational health hazard, the toxicity of CNTs after *in vivo* administration through bio-medically relevant routes is still a matter of debate. Among different routes of administration, the intra-peritoneal way has several advantages, firstly it offers the possibility to administer larger doses of suspended nanoparticles and secondly, the peritoneal cavity has a recognized particle-clearance mechanism. Particles leaving the peritoneal cavity pass *via* the retrosternal route through stomata (pore like structures) to the parathymic (mediastinal) nodes to the upper terminal thoracic duct or right lymphatic duct (Abu-Hijleh et al., 1995). Moreover, the peritoneal cavity and its mesothelium-covered viscera were recognized as a convenient substitute for pleural cavity mesothelium in fibre toxicity studies (Donaldson et al., 2010).

The first study of *in vivo* toxicity of CNTs after intra-peritoneal administration was conducted in our laboratory in collaboration with the Department of Chemistry of Rice University (Hartmann et al., 2007b). We compared the acute toxicity of full-length and ultra-short CNTs suspended in a Tween 80 aqueous solution, under the same conditions we used since 1996 to study the acute and sub-acute toxicity of [60]fullerene in mice (Moussa, F. et al, 1996). Our preliminary results showed that irrespective of the length of the administered CNT material, CNT aggregates induced a granulomatous response inside the organs like that which occurs in lungs after inhalation or intra-tracheal instillation (Hartmann et al. 2007).

One year later, in a comparative study of MWNTs and asbestos fibres, it was reported that MWNTs exhibit a length-dependent pathogenic behaviour (Poland et al., 2008) including granuloma formation and inflammation. In order to assess the role of fibre length, samples of long and short asbestos fibres and MWNTs with length ranges less than 5 µm, less than 20 µm- referred to as short, tangled MWNTs; and long tubes of the mean of 13 µm (24% of them were larger than 15 µm) and maximum 56 µm have been suspended in bovine serum albumin and saline and administered intra-peritoneally to mice in a dose of 50 µg per mouse. The MWNTs samples differed in the source, preparation and purification method (the short ones being purified by acid treatment). At day 7 after injection, the authors reported that only the samples containing long fibres (asbestos or MWNTs) caused significant polymorphonuclear leukocyte infiltration, protein exudation and granulomas. However, the mesothelial lining on the pleural side of the diaphragm was normal in every case (Poland et al., 2008) and the inflammation decreased by day 7. Short fibres of any kind did not cause significant inflammation, neither at day 1 nor at day 7 after administration, except for one mouse out of three in the group treated with short tangled MWNTs of the length < 20 µm. The overall conclusion of the study was that short MWNTs do not mimic the behaviour of long asbestos, but that their data cannot preclude the possibility that short MWNTs may be by some other mechanism that was not assessed in this study. Long MWNTs produced inflammation, foreign body giant cells and granulomas that were qualitatively and quantitatively similar to the foreign body inflammatory response caused by long asbestos. However, for the specimen treated with shorter MWNTs that did exhibit granuloma, the authors concluded that it was maybe due to the fact that the sample they injected was contaminated with long fibres, caused by some other unidentified component specific for the precise MWNTs sample, or the granulomas could have arisen spontaneously by chance (Poland et al., 2008).

In our laboratory (Kolosnjaj-Tabi et al., 2010) we administered Tween-suspended ultra-short (20-80 nm long) and full length SWNTs in a dose up to 1000 mg/kg. Our results indicated that regardless of the administered dose (50-1000 mg/kg b.w.), length, or surface state of the administered material, large aggregates of CNTs (>10 µm) irremediably induce granuloma formation (Fig. 4).

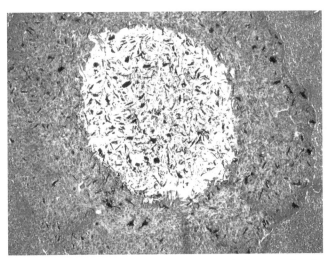

Fig. 4. Light micrograph after hematoxylin-eosin staining of a liver section from mice i.p. injected with a single dose of ultra-short SWNTs at 90 days post-administration showing a US-tube-laden granuloma. (Magnification = 10).

The administered doses were high, yet necessary to ascertain a sufficient circulating dose of administered material. The bolus dose was responsible, in set terms, for granulomas that were formed after aggregation of intra-peritoneally administered tubes. Smaller agglomerates (< 300 nm), on the other hand, did not induce granuloma formation nor did they cause any major life-threatening condition under our experimental conditions. A large portion of well-dispersed CNTs was eliminated through the kidney and the bile ducts. However, the aggregated part of the administered dose was not cleared from the body and persisted inside cells 5 months after administration (Fig. 5).

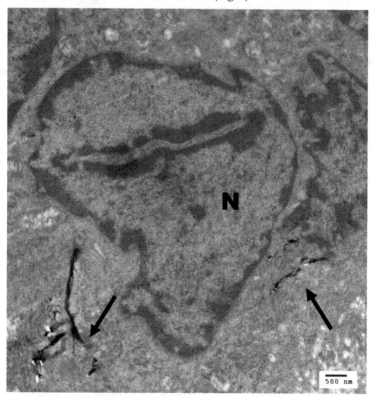

Fig. 5. Transmission electron micrograph of bundles of SWNTs in a Kuppfer cell, found in the liver of a mouse 5 months after treatment

The persistence of the nanotube aggregates inside the cells was probably due to the slow disaggregation and slow elimination of larger aggregates.

3.3.1 Carbon nanotubes and mesothelioma

The ability of MWNNTs to induce mesothelioma, a deadly cancer, in experimental models in rodents is still a matter of debate. This cancer is a highly specific response to bio-persistent fibres and may occur in the pleura (outer lining of the lungs and internal chest wall), peritoneum (the lining of the abdominal cavity), pericardium (a sac that surrounds the heart), or tunica vaginalis (a sac that surrounds the testis) (Moore et al., 2008).

In 2008, a Japanese team observed for the first time that MWNTs could induce mesothelioma after intra-peritoneal administration to p53 heterozygous mice (Takagi et al., 2008a). This type of mice are, however, more sensitive and have a shorter tumour onset rate than standard wild-type rodents. Indeed, different groups contested this study (Donaldson et al., 2008, Ichihara et al., 2008), but Takagi et al. explained and justified their experimental choices letting other studies to confirm mesothelium threats (Takagi et al., 2008b, c). The authors administered intra-peritoneally 3 mg of CNTs per mouse suspended in 1% Tween 80 and 0.5% methyl cellulose- aqueous solution. While the samples of crocidolite asbestos were evenly dispersed, MWNTs contained aggregates and fibres. The mice were monitored until one of the groups (MWNTs treated group) reached 100% mortality, which happened on day 180. The mice treated with MWNTs exhibited moderate to severe fibrous peritoneal adhesions, slight ascites, fibrous peritoneal thickening and a high incidence of macroscopic peritoneal tumours. Similar, but less severe findings were noted in the asbestos group. As the authors emphasized, it is important to limit the significance of this study to the monitoring of biological activity of a compartment of MWNTs longer that 5 μm. There is no information that this study method would be sensitive to pure nanometer-sized particles within the same timeframe (Takagi et al. 2008a).

An analogous study was made, in time, on wild-type rats (Muller et al., 2009) in a two years exposure period, where up to 20 mg of MWNTs with and without sidewall defects (induced by grounding of the raw material) suspended in phosphate buffer were intraperitoneally injected to a large number of rats. After 24 months, crocidolite induced mesothelioma in 34.6% animals whereas mesothelioma occurred only in 3.8% in the vehicle-treated rats. MWNTs with or without structural defects did not induce significant mesothelioma in this study, as mesothelioma occurrence was detected in up to 6% of MWNT-treated rats, which is in line with the spontaneous incidences of mesothelioma in rats (up to 6%). The incidence of tumours other than mesothelioma was not significantly increased across the groups (Bignon et al., 1995).

In contrast to what was observed after intra-peritoneal administration to wild type rats (Muller et al., 2009), MWNTs injected in a single intra-scrotal dose to rats, induced mesothelioma 37 to 40 weeks after treatment (Sakamoto et al., 2009). In the latter experiment, MWNTs were suspended in a 2% carboxymethyl cellulose aqueous solution and administered to rats at a dose of 0.24 mg/animal. According to the authors this dose corresponds to the maximal value recommended by the guideline for man-made mineral fibers (Bernstein & Riego Sintes, 1999, as cited in Sakamoto et al., 2009).

In conclusion, these findings also show that under some experimental conditions, MWNTs may induce mesothelioma formation. Thus, further investigations are urgently needed.

3.4 Effects of carbon nanotubes after sub-cutaneous administration

Subcutaneous implantations of clusters of MWNTs of different lengths (220 nm and 825 nm) in rats showed that the degree of inflammatory response around the shorter MWNTs was slighter than that observed around the longer ones, thus indicating that macrophages could envelop the shorter nanotubes more readily than the longer ones (Sato et al., 2005). However, no severe inflammatory response such as necrosis, degeneration or neutrophils infiltration was observed around both types of MWNTs.

These biological responses have also been described after subcutaneous implantations in mice of 2 mg/animal of SWNTs, two different types of MWNTs (20 and 80 nm of average diameter) and cup-stacked carbon nanotubes (CSNTs made with stacked truncated cones), for up to 3 months (Koyama et al., 2006). The nanotubes used in this study were purified by thermal treatment process during which the metal particles evaporate. After 1, 2, 3 weeks, 1 month, 2 months and 3 months post-implantation, the animals were sacrificed, blood was collected for CD4+ and CD8+ T-cells counting by flow-cytometry and tissue of skin (including muscle layers) were collected for histo-pathological examination. All mice survived, and no large changes in their weights were observed during the experimental period. After one week of implantation, only SWNTs activated the major histocompatibility complex (MHC) class I pathway of antigen-antibody response system (higher CD4+/CD8+ value), leading to the appearance of an oedematous aspect. After two weeks, significantly high values of CD4+ without changes in CD8+ signified the activation of MHC class II for all samples. The authors noted that antigenic mismatch becomes less evident with time, notably one month post-implantation, indicating an establishment of granuloma formation. Furthermore, the toxicological response of CNTs was absolutely lower than that of asbestos.

3.5 Effects of carbon nanotubes after intra-venous administration

3.5.1 Effects of SWNTs

The first report on the effects of SWNTs after intravenous administration to mice (3 mg/kg) monitored over four months indicated normal blood chemistries and normal histological examinations (Schipper et al., 2008). The animals did not show significant inflammatory lesions and SWNTs accumulated in liver and spleen as evidenced by Raman spectroscopy. The CNTs used in this study were highly dispersed with polyethylene glycol (PEG). CNT aggregates were eliminated with ultracentrifugation before administration to the animals.

Another study used highly dispersed pristine HiPco SWNTs with different PEGylated phospholipids (Liu et al., 2008). Big bundles and impurities were removed by centrifugation, and individually suspended tubes or small bundles were administered intravenously at a dose of approximately 20 µg or 100 µg per mouse. Blood, tissue and organ distribution and elimination in urines and faeces were evaluated by Raman spectroscopy, by assessment of the tangential graphite-like phonon mode (G band). Administered SWNTs accumulated mainly in liver and spleen, however the quantity decreased over a 3-month period. The authors concluded that SWNTs were mainly eliminated by the biliary pathway; only a small portion of short tubes (< 50 nm in length) was eliminated in the urines. Finally, the authors did not report any obvious sign of toxicity.

Other authors also reported low toxicity of SWNTs in mice over a 3-month period (Yang et al., 2008). Purified SWNTs were suspended in 1% Tween 80 aqueous solution and sonicated for 30 minutes prior administration to animals at various doses from 40 µg to 1 mg per mouse. Some of the serum biochemical parameters (ALT, AST and LDH) were higher in SWNT-treated animals compared to the control group 90 days post-exposure, indicating that induced hepatic injury and tissue breakdown were dose-dependent. The long-term accumulation of aggregated SWNTs was evidenced in histological sections of lungs, livers and spleens and was confirmed by Raman spectroscopy and transmission electron microscopy in organ lysates. However, no fibrosis was detected in the organs.

Embriotoxicity was recently reported as an effect caused by intravenous administration of SWNTs in mice (30 µg/mouse) (Pietroiusti et al., 2011). The authors used pristine, oxidized or ultra-oxidized by acid treatment SWNTs. Cobalt was the only impurity that was released in the medium in which the tubes were dispersed (DMEM cell culture medium with fetal bovine serum). Before the end of gestation, the animals were sacrificed, and uteri, placentas, and foetuses were examined. A high percentage of early miscarriages and foetal malformations were observed in females exposed to oxidized SWNTs, while lower percentages were found in animals exposed to the pristine material. The lowest effective dose was 100 ng/mouse. Extensive vascular lesions and increased production of reactive oxygen species (ROS) were detected in placentas of malformed but not of normally developed foetuses. Increased ROS levels were likewise detected in malformed foetuses. No increased ROS production or evident morphological alterations were observed in maternal tissues (Pietroiusti et al., 2011).

3.5.2 Effects of MWNTs

MWNTs dispersed in mouse-serum (10 min of sonication) and injected to mice at a dose of 200 or 400 µg per mouse showed no severe acute response, 24 hours after administration (Lacerda et al., 2008). However, mice treated with aggregates of pristine MWNTs exhibited subdued behavior, hunched posture, and signs of respiratory distress. While serum biochemistry data did not show significant increase, optical microscopy revealed aggregate accumulation, mostly in livers and lung vessels, which were probably responsible for respiratory distress. Nevertheless, no tissue degeneration, inflammation, necrosis or fibrosis occurred 24 hours after injection.

The effect of Tween-suspended pristine MWNTs and PBS-suspended acid oxidized MWNTs up to 2 months were investigated after administration to rats at a dose level of 10 or 60 mg/kg (Ji et al., 2009). The authors reported an impact on body weight gain of the highest dose. Severe inflammatory cell infiltration in the portal region, cellular necrosis and focal necrosis were seen at a dose of 60 mg/kg in the MWNT-treated group 15 and 60 days following the treatment. Moreover, severe mitochondrial swelling, bile canaliculi expansion, mitochondrial destruction, loss and lysis of mitochondrial crest were also observed. In the acid treated (oxidized) MWNTs group only slight inflammatory cell infiltration was observed after 2 months. A slight increase of AST activity used as biochemical marker of liver injury was also reported in treated animals, but it did not increase more than twofold. Interestingly, 329 genes were up-regulated and 31 genes that were down-regulated more than twofold in MWNT-treated mice and 1139 genes were up-regulated and 505 genes were down-regulated over twofold in the mice treated with oxidized MWNTs.

In order to avoid mechanical blockage by the administered material, it is of capital importance to administer only individually suspended, short CNTs. Thus, further studies with individually suspended CNTs have to be made in order to confirm the direct effect of these materials after administration by the intravenous route.

3.6 Effects of carbon nanotubes after oral administration

Oral administration of 1000 mg/kg of body weight of SWNTs to mice (Kolosnjaj-Tabi et al.) resulted in neither animal death nor behavioral abnormalities. Compound-colored stool was found 24 h after gavage in all treated groups. Two weeks after treatment, regardless of the length or of the iron content, the nanotube materials did not induce any abnormalities after pathological examination, indicating that under these conditions, the lowest lethal dose (LDLo) is greater than 1000 mg/kg b.w. in Swiss mice.

The potential effects of MWNTs after oral administration were also investigated on pregnant dams and embryo-fetal development in rats (Lim et al., 2011). MWNTs were administered to pregnant rats by gavage at 0, 40, 200, and 1,000 mg/kg/day. All dams were subjected to Cesarean section on day 20 of gestation, and the foetuses were examined for morphological abnormalities. A decrease in thymus weight was observed in the high dose group in a dose-dependent manner. However, maternal body weight, food consumption, and oxidant-antioxidant balance in the liver were not affected by treatment with MWNTs. No treatment-related differences in gestation index, foetal deaths, foetal and placental weights, or sex ratio were observed between the groups. Morphological examinations of the foetuses demonstrated no significant difference in incidences of abnormalities between the groups.

Intriguingly, it has also been reported that CNTs may be involved in oxidative stress with oxidative damage of DNA in the colon mucosa, liver, and lung of rats after oral administration of SWNTs in a dose of 0.064 or 0.64 mg/kg b.w. suspended in saline solution or corn oil (Folkmann et al., 2009). Suspensions of particles in saline solution or corn oil yielded a similar extent of genotoxicity. However, corn oil per se generated more genotoxicity than the particles (Folkmann et al., 2009).

4. Conclusion

Considered together, these diverging results highlight the difficulties in evaluating the toxicity of CNT materials. While the toxicity is certainly governed by the state of aggregation, length and stiffness of CNTs, other parameters might be involved. Most of them probably depend of the method of production of the sample, the method of purification and the method of preparation of the tested formulations. As the reactivity and the general behaviour of CNTs in biological media are not completely understood, assessing the safety of these nanoparticles should also include a careful selection of appropriate experimental methods. Thus, more studies are needed in order to determine the safety of CNTs. For the time being, precaution is necessary notably in case of CNT-exposure at workplace.

5. References

Abu-Hijleh, M.F., Habbal, O.A. & Moqattash, S.T. (1995). The role of the diaphragm in lymphatic absorption from the peritoneal cavity. *Journal of Anatomy, 18*, 453-467.

Ashcroft, J. M., Hartman, K. B., Kissell, K. R., Mackeyev, Y., Pheasant, S., Young, S., . . . Wilson, L. J. (2007). Single Molecule I2@ US Tube Nanocapsules: A New X ray Contrast Agent Design. *Advanced Materials, 19*(4), 573-576.

Balavoine, F., Schultz, P., Richard, C., Mallouh, V., Ebbesen, T. W., & Mioskowski, C. (1999). Helical crystallization of proteins on carbon nanotubes: a first step towards the development of new biosensors. *Angewandte Chemie International Edition, 38*(13/14), 1912-1915.

Balkwill, F., & Mantovani, A. (2001). Inflammation and cancer: back to Virchow? *The Lancet, 357*(9255), 539-545.

Bekyarova, E., Ni, Y., Malarkey, E. B., Montana, V., McWilliams, J. L., Haddon, R. C., & Parpura, V. (2005). Applications of carbon nanotubes in biotechnology and biomedicine. *Journal of biomedical nanotechnology, 1*(1), 3.

Bernstein, D., & Riego-Sintes, J. (1999). Methods for the determination of the hazardous properties for human health of man made mineral fibers (MMMF). Vol. EUR 18748 EN, April. 93.

Bianco, A., Kostarelos, K., & Prato, M. (2005). Applications of carbon nanotubes in drug delivery. *Current Opinion in Chemical Biology, 9*(6), 674-679.

Bignon, J., Brochard, P., Brown, R., Davis, J., Vu, V., Gibbs, G., . . . Sebastien, P. (1995). Assessment of the toxicity of man-made fibres. A final report of a workshop held in Paris, France 3-4 February 1994. *Annals of Occupational Hygiene, 39*(1), 89-106.

Casey, A., Herzog, E., Davoren, M., Lyng, F., Byrne, H., & Chambers, G. (2007). Spectroscopic analysis confirms the interactions between single walled carbon nanotubes and various dyes commonly used to assess cytotoxicity. *Carbon, 45*(7), 1425-1432.

Chen, J., Hamon, M. A., Hu, H., Chen, Y., Rao, A. M., Eklund, P. C., & Haddon, R. C. (1998). Solution properties of single-walled carbon nanotubes. *Science, 282*(5386), 95.

Chen, X., Lee, G. S., Zettl, A., & Bertozzi, C. R. (2004). Biomimetic engineering of carbon nanotubes by using cell surface mucin mimics. *Angewandte Chemie International Edition, 43*(45), 6111-6116.

Chen, Y., Haddon, R., Fang, S., Rao, A., Eklund, P., Lee, W., . . . Chavan, A. (1998). Chemical attachment of organic functional groups to single-walled carbon nanotube material. *J. Mater. Res, 13*(9), 2423-2431.

Donaldson, K., Murphy, F. A., Duffin, R., & Poland, C. A. (2010). Asbestos, carbon nanotubes and the pleural mesothelium: a review of the hypothesis regarding the role of long fibre retention in the parietal pleura, inflammation and mesothelioma. *Part Fibre Toxicol, 7*(5).

Donaldson, K., & Poland, C. A. (2009). Nanotoxicology: new insights into nanotubes. *Nature Nanotechnology, 4*(11), 708-710.

Donaldson, K., Stone, V., Seaton, A., Tran, L., Aitken, R., & Poland, C. (2008). Re: Induction of mesothelioma in p53+/-mouse by intraperitoneal application of multi-wall carbon nanotube. *The Journal of toxicological sciences, 33*(3), 385.

Dresselhaus, M., Dresselhaus, G., Charlier, J., & Hernandez, E. (2004). Electronic, thermal and mechanical properties of carbon nanotubes. *Philosophical Transactions of the Royal Society of London. Series A: Mathematical, Physical and Engineering Sciences, 362*(1823), 2065.

Folkmann, J. K., Risom, L., Jacobsen, N. R., Wallin, H., Loft, S., & M⁻ller, P. (2009). Oxidatively damaged DNA in rats exposed by oral gavage to C60 fullerenes and single-walled carbon nanotubes. *Environmental health perspectives, 117*(5), 703.

Harris, PJF. (2009). *Carbon nanotube science Synthesis, properties and applications* (1st Edition), Cambridge University Press, ISBN 9780521828956, Cambridge

Hartman, K. B., Hamlin, D. K., Wilbur, D. S., & Wilson, L. J. (2007). 211AtCl@ US Tube Nanocapsules: A New Concept in Radiotherapeutic Agent Design. *Small, 3*(9), 1496-1499.

Hartman, K., Kolosnjaj, J., Gharbi, N., Boudjemaa, S. Wilson, L. J. and Moussa, F. (2007). Comparative In vivo Toxicity Assessment of Singlewalled Carbon Nanotubes in Mice. The 211th Meeting of The Electrochemical Society, Chicago, USA, May 6-10

Hartman, K. B., Laus, S., Bolskar, R. D., Muthupillai, R., Helm, L., Toth, E., . . . Wilson, L. J. (2008). Gadonanotubes as ultrasensitive pH-smart probes for magnetic resonance imaging. *Nano letters, 8*(2), 415-419.

Hirsch, A. (2002). Functionalization of single walled carbon nanotubes. *Angewandte Chemie International Edition, 41*(11), 1853-1859.

Hu, H., Ni, Y., Montana, V., Haddon, R. C., & Parpura, V. (2004). Chemically functionalized carbon nanotubes as substrates for neuronal growth. *Nano letters, 4*(3), 507-511.

Huczko, A., & Lange, H. (2001). Carbon nanotubes: experimental evidence for a null risk of skin irritation and allergy. *Fullerene Science and Technology, 9*(2), 247-250.

Ichihara, G., Castranova, V., Tanioka, A., & Miyazawa, K. (2008). Letter to the editor. *The Journal of toxicological sciences, 33*(3), 381.

Iijima, S. (1991). Helical microtubules of graphitic carbon. *Nature, 354*(6348), 56-58.

Jensen, A. W., Wilson, S. R., & Schuster, D. I. (1996). Biological applications of fullerenes. *Bioorganic & medicinal chemistry, 4*(6), 767-779.

Ji, Z., Zhang, D., Li, L., Shen, X., Deng, X., Dong, L., . . . Liu, Y. (2009). The hepatotoxicity of multi-walled carbon nanotubes in mice. *Nanotechnology, 20*, 445101.

Kane, A. B., & Hurt, R. H. (2008). Nanotoxicology: The asbestos analogy revisited. *Nature Nanotechnology, 3*(7), 378-379.

Kobayashi, N., Naya, M., Ema, M., Endoh, S., Maru, J., Mizuno, K., & Nakanishi, J. (2010). Biological response and morphological assessment of individually dispersed multi-wall carbon nanotubes in the lung after intratracheal instillation in rats. *Toxicology , 276*(3), 143-153.

Kolosnjaj, J., Szwarc, H., & Moussa, F. (2007). Toxicity studies of carbon nanotubes. *Bio-Applications of Nanoparticles,* 181-204.

Kolosnjaj-Tabi, J., Hartman, K. B., Boudjemaa, S., Ananta, J. S., Morgant, G., Szwarc, H., . . . Moussa, F. (2010). In vivo behavior of large doses of ultrashort and full-length single-walled carbon nanotubes after oral and intraperitoneal administration to Swiss mice. *Acs Nano, 4*(3), 1481-1492.

Koyama, S., Endo, M., Kim, Y. A., Hayashi, T., Yanagisawa, T., Osaka, K., . . . Kuroiwa, N. (2006). Role of systemic T-cells and histopathological aspects after subcutaneous implantation of various carbon nanotubes in mice. *Carbon, 44*(6), 1079-1092.

Kroll, A., Pillukat, M. H., Hahn, D., & Schnekenburger, J. (2009). Current in vitro methods in nanoparticle risk assessment: Limitations and challenges. *European Journal of Pharmaceutics and Biopharmaceutics, 72*(2), 370-377.

Lacerda, L., Ali-Boucetta, H., Herrero, M. A., Pastorin, G., Bianco, A., Prato, M., & Kostarelos, K. (2008). Tissue histology and physiology following intravenous administration of different types of functionalized multiwalled carbon nanotubes. *Nanomedicine, 3*(2), 149-161.

Lai-Fook, S. J. (2004). Pleural mechanics and fluid exchange. *Physiological reviews, 84*(2), 385.

Lam, C. W., James, J. T., McCluskey, R., & Hunter, R. L. (2004). Pulmonary toxicity of single-wall carbon nanotubes in mice 7 and 90 days after intratracheal instillation. *Toxicological Sciences, 77*(1), 126.

Lim, J. H., Kim, S. H., Shin, I. S., Park, N. H., Moon, C., Kang, S. S., . . . Kim, J. C. (2011) Maternal exposure to multi wall carbon nanotubes does not induce embryo-fetal developmental toxicity in rats. *Birth Defects Research Part B: Developmental and Reproductive Toxicology. 92*(1),69-76.

Lin, Y., Taylor, S., Li, H., Fernando, K. A. S., Qu, L., Wang, W., . . . Sun, Y. P. (2004). Advances toward bioapplications of carbon nanotubes. *Journal of Materials Chemistry, 14*(4), 527-541.

Liu, Z., Davis, C., Cai, W., He, L., Chen, X., & Dai, H. (2008). Circulation and long-term fate of functionalized, biocompatible single-walled carbon nanotubes in mice probed by Raman spectroscopy. *Proceedings of the National Academy of Sciences, 105*(5), 1410.

Lovat, V., Pantarotto, D., Lagostena, L., Cacciari, B., Grandolfo, M., Righi, M., . . . Ballerini, L. (2005). Carbon nanotube substrates boost neuronal electrical signaling. *Nano letters, 5*(6), 1107-1110.

Mattson, M. P., Haddon, R. C., & Rao, A. M. (2000). Molecular functionalization of carbon nanotubes and use as substrates for neuronal growth. *Journal of Molecular Neuroscience, 14*(3), 175-182.

Maynard, A., Baron, P., Foley, M., Shvedova, A., Kisin, E., & Castranova, V. (2004). Exposure to carbon nanotube material: aerosol release during the handling of unrefined single-walled carbon nanotube material. *Journal of Toxicology and Environmental Health Part A, 67*(1), 87-107.

Mercer, R., Scabilloni, J., Wang, L., Kisin, E., Murray, A., Schwegler-Berry, D., . . . Castranova, V. (2008). Alteration of deposition pattern and pulmonary response as a result of improved dispersion of aspirated single-walled carbon nanotubes in a mouse model. *American Journal of Physiology-Lung Cellular and Molecular Physiology, 294*(1), L87.

Mercer, R. R., Hubbs, A. F., Scabilloni, J. F., Wang, L., Battelli, L. A., Schwegler-Berry, D., . . . Porter, D. W. (2010). Distribution and persistence of pleural penetrations by multi-walled carbon nanotubes. *Particle and Fibre Toxicology, 7*(1), 28.

Moore, A. J., Parker, R. J., & Wiggins, J. (2008). Malignant mesothelioma. *Orphanet J Rare Dis, 3*(1), 34.

Morimoto, Y., Hirohashi, M., Ogami, A., Oyabu, T., Myojo, T., Todoroki, M., . . . Lee, B. W. (2011). Pulmonary toxicity of well-dispersed multi-wall carbon nanotubes following inhalation and intratracheal instillation. *Nanotoxicology*(0), 1-15.

Moussa, F., Trivin, F., Céolin, R., Hadchouel, M., Sizaret, P.-Y., Greugny, V., Fabre, C., Rassat, A. and Szwarc, H. (1996). Early effects of C60 administration in Swiss Mice: a preliminary account for *in vivo* C60 toxicity. *Fullerenes Science & Technology, 4,* 21–29

Muller, J., Delos, M., Panin, N., Rabolli, V., Huaux, F., & Lison, D. (2009). Absence of carcinogenic response to multiwall carbon nanotubes in a 2-year bioassay in the peritoneal cavity of the rat. *Toxicological Sciences, 110*(2), 442.

Muller, J., Huaux, F., Fonseca, A., Nagy, J. B., Moreau, N., Delos, M., . . . Fenoglio, I. (2008). Structural defects play a major role in the acute lung toxicity of multiwall carbon nanotubes: Toxicological aspects. *Chemical research in toxicology, 21*(9), 1698-1705.

Muller, J., Huaux, F., Moreau, N., Misson, P., Heilier, J. F., Delos, M., . . . Lison, D. (2005). Respiratory toxicity of multi-wall carbon nanotubes. *Toxicology and Applied Pharmacology, 207*(3), 221-231.

Murphy, F. A., Poland, C. A., Duffin, R., Al-Jamal, K. T., Ali-Boucetta, H., Nunes, A., . . . Li, S. (2011). Length-Dependent Retention of Carbon Nanotubes in the Pleural Space of Mice Initiates Sustained Inflammation and Progressive Fibrosis on the Parietal Pleura. *The American Journal of Pathology, 178*(6), 2587-2600.

Mutlu, G. M., Budinger, G. R. S., Green, A. A., Urich, D., Soberanes, S., Chiarella, S. E., . . . Hersam, M. C. (2010). Biocompatible nanoscale dispersion of single-walled carbon nanotubes minimizes in vivo pulmonary toxicity. *Nano letters, 10*(5), 1664-1670.

Oberdorster, G., Maynard, A., Donaldson, K., Castranova, V., Fitzpatrick, J., Ausman, K., . . . Lai, D. (2005). ILSI Research Foundation/Risk Science Institute Nanomaterial Toxicity Screening Working Group. Principles for characterizing the potential human health effects from exposure to nanomaterials: Elements of a screening strategy. *Part Fibre Toxicol, 2*(8).

Pietroiusti, A., Massimiani, M., Fenoglio, I., Colonna, M., Valentini, F., Palleschi, G., . . . Bergamaschi, A. (2011) Low Doses of Pristine and Oxidized Single Wall Carbon Nanotubes Affect Mammalian Embryonic Development. *Acs Nano. 5*(6), 4624-4633.

Poland, C. A., Duffin, R., Kinloch, I., Maynard, A., Wallace, W. A. H., Seaton, A., . . . Donaldson, K. (2008). Carbon nanotubes introduced into the abdominal cavity of mice show asbestos-like pathogenicity in a pilot study. *Nature Nanotechnology, 3*(7), 423-428.

Popov, V. N. (2004). Carbon nanotubes: properties and application. *Materials Science and Engineering: R: Reports, 43*(3), 61-102.

Porter, D. W., Hubbs, A. F., Mercer, R. R., Wu, N., Wolfarth, M. G., Sriram, K., . . . Friend, S. (2010). Mouse pulmonary dose-and time course-responses induced by exposure to multi-walled carbon nanotubes. *Toxicology, 269*(2-3), 136-147.

Richard, C., Balavoine, F., Schultz, P., Ebbesen, T. W., & Mioskowski, C. (2003). Supramolecular self-assembly of lipid derivatives on carbon nanotubes. *Science, 300*(5620), 775.

Ruoff, R. S., Lorents, D. C., Chan, B., Malhotra, R., & Subramoney, S. (1993). Single crystal metals encapsulated in carbon nanoparticles. *Science, 259*(5093), 346.

Ryman-Rasmussen, J. P., Cesta, M. F., Brody, A. R., Shipley-Phillips, J. K., Everitt, J. I., Tewksbury, E. W., . . . Andersen, M. E. (2009). Inhaled carbon nanotubes reach the subpleural tissue in mice. *Nature Nanotechnology, 4*(11), 747-751.

Sakamoto, Y., Nakae, D., Fukumori, N., Tayama, K., Maekawa, A., Imai, K., . . . Ogata, A. (2009). Induction of mesothelioma by a single intrascrotal administration of multi-wall carbon nanotube in intact male Fischer 344 rats. *The Journal of toxicological sciences, 34*(1), 65-76.

Sato, Y., Yokoyama, A., Shibata, K., Akimoto, Y., Ogino, S., Nodasaka, Y., . . . Uo, M. (2005). Influence of length on cytotoxicity of multi-walled carbon nanotubes against human acute monocytic leukemia cell line THP-1 in vitro and subcutaneous tissue of rats in vivo. *Mol. BioSyst., 1*(2), 176-182.

Schipper, M. L., Nakayama-Ratchford, N., Davis, C. R., Kam, N. W. S., Chu, P., Liu, Z., . . . Gambhir, S. S. (2008). A pilot toxicology study of single-walled carbon nanotubes in a small sample of mice. *Nature Nanotechnology, 3*(4), 216-221.

Service, R. (1998). Chemistry: Nanotubes: The Next Asbestos? *Science, 281*(5379), 941-941.

Shvedova, A. A., Kisin, E. R., Mercer, R., Murray, A. R., Johnson, V. J., Potapovich, A. I., . . . Schwegler-Berry, D. (2005). Unusual inflammatory and fibrogenic pulmonary responses to single-walled carbon nanotubes in mice. *American Journal of Physiology-Lung Cellular and Molecular Physiology, 289*(5), L698.

Sinha, N., & Yeow, J. T. W. (2005). Carbon nanotubes for biomedical applications. *NanoBioscience, IEEE Transactions on, 4*(2), 180-195.

Sitharaman, B., Kissell, K. R., Hartman, K. B., Tran, L. A., Baikalov, A., Rusakova, I., . . . Chiu, W. (2005). Superparamagnetic gadonanotubes are high-performance MRI contrast agents. *Chem. Commun.*(31), 3915-3917.

Sloan, J., Hammer, J., Zwiefka-Sibley, M., & Green, M. L. H. (1998). The opening and filling of single walled carbon nanotubes (SWTs). *Chem. Commun.*(3), 347-348.

Takagi, A., Hirose, A., Nishimura, T., Fukumori, N., Ogata, A., Ohashi, N., Kitajima S. & Kanno, J. (2008). Induction of mesothelioma in p53+/-mouse by intraperitoneal application of multi-wall carbon nanotube. *The Journal of toxicological sciences, 33*(1), 105-116.

Takagi, A., Hirose, A., Nishimura, T., Fukumori, N., Ogata, A., Ohashi, N., Kitajima S. & Kanno, J. (2008b) Letter to the editor. *The Journal of toxicological sciences, 33*(3), 382-384.

Takagi, A., Hirose, A., Nishimura, T., Fukumori, N., Ogata, A., Ohashi, N., Kitajima S. & Kanno, J. (2008c) Letter to the editor. *The Journal of toxicological sciences, 33*(3), 386-388.

Valko, M., Morris, H., & Cronin, M. (2005). Metals, toxicity and oxidative stress. *Current Medicinal Chemistry, 12*(10), 1161-1208.

Wang, J., Musameh, M., & Lin, Y. (2003). Solubilization of carbon nanotubes by Nafion toward the preparation of amperometric biosensors. *Journal of the American Chemical Society, 125*(9), 2408-2409.

Wang, S., Humphreys, E. S., Chung, S. Y., Delduco, D. F., Lustig, S. R., Wang, H., . . . Chiang, Y. M. (2003). Peptides with selective affinity for carbon nanotubes. *Nature Materials, 2*(3), 196-200.

Warheit, D. (2006). What is currently known about the health risks related to carbon nanotube exposures? *Carbon, 44*(6), 1064-1069.

Warheit, D. B., Laurence, B., Reed, K. L., Roach, D., Reynolds, G., & Webb, T. (2004). Comparative pulmonary toxicity assessment of single-wall carbon nanotubes in rats. *Toxicological Sciences, 77*(1), 117.

Yang, S. T., Wang, X., Jia, G., Gu, Y., Wang, T., Nie, H., Liu, Y. (2008). Long-term accumulation and low toxicity of single-walled carbon nanotubes in intravenously exposed mice. *Toxicology letters, 181*(3), 182-189.

Ziegler, K. J., Gu, Z., Peng, H., Flor, E. L., Hauge, R. H., & Smalley, R. E. (2005). Controlled oxidative cutting of single-walled carbon nanotubes. *Journal of the American Chemical Society, 127*(5), 1541-1547.

Silver Nanoparticles

Hassan Korbekandi[1] and Siavash Iravani[2]
[1]Genetics and Molecular Biology Department, School of Medicine,
Isfahan University of Medical Sciences
[2]School of Pharmacy and Pharmaceutical Science,
Isfahan University of Medical Sciences
Iran

1. Introduction

Nanotechnology is an important field of modern research dealing with design, synthesis, and manipulation of particles structure ranging from approximately 1-100 nm. Tremendous growth in this emerging technology has opened novel fundamental and applied frontiers, including the synthesis of nanoscale materials and exploration or utilization of their exotic physicochemical and optoelectronic properties. Nanotechnology is rapidly gaining importance in a number of areas such as health care, cosmetics, food and feed, environmental health, mechanics, optics, biomedical sciences, chemical industries, electronics, space industries, drug-gene delivery, energy science, optoelectronics, catalysis, reorography, single electron transistors, light emitters, nonlinear optical devices, and photo-electrochemical applications (Colvin et al. 1994; Wang and Herron 1991; Schmid 1992; Hoffman et al. 1992; Hamilton and Baetzold 1979; Mansur et al. 1995).

Silver nanoparticles are of interest because of the unique properties (*e.g.,* size and shape depending optical, electrical, and magnetic properties) which can be incorporated into antimicrobial applications, biosensor materials, composite fibers, cryogenic superconducting materials, cosmetic products, and electronic components. Several physical and chemical methods have been used for synthesizing and stabilizing silver nanoparticles (Senapati 2005; Klaus-Joerger et al. 2001). The most popular chemical approaches, including chemical reduction using a variety of organic and inorganic reducing agents, electrochemical techniques, physicochemical reduction, and radiolysis are widely used for the synthesis of silver nanoparticles. Recently, nanoparticle synthesis is among the most interesting scientific areas of inquiry, and there is growing attention to produce nanoparticles using environmentally friendly methods (green chemistry). Green synthesis approaches include mixed-valence polyoxometalates, polysaccharides, Tollens, biological, and irradiation method which have advantages over conventional methods involving chemical agents associated with environmental toxicity. This chapter presents an overview of silver nanoparticle preparation by physical, chemical, and green synthesis approaches. The aim of this chapter is, therefore, to reflect on the current state and future prospects, especially the potentials and limitations of the above mentioned techniques for industries. Moreover, we discuss the applications of silver nanoparticles and their incorporation into other materials, the mechanistic aspects of the antimicrobial effects of silver nanoparticles.

2. Synthesis of silver nanoparticles

2.1 Physical approaches

Most important physical approaches include evaporation-condensation and laser ablation. Various metal nanoparticles such as silver, gold, lead sulfide, cadmium sulfide, and fullerene have previously been synthesized using the evaporation-condensation method. The absence of solvent contamination in the prepared thin films and the uniformity of nanoparticles distribution are the advantages of physical approaches in comparison with chemical processes. Physical synthesis of silver nanoparticles using a tube furnace at atmospheric pressure has some disadvantages, for example, tube furnace occupies a large space, consumes a great amount of energy while raising the environmental temperature around the source material, and requires a great deal of time to achieve thermal stability. Moreover, a typical tube furnace requires power consumption of more than several kilowatts and a preheating time of several tens of minutes to reach a stable operating temperature (Kruis et al. 2000; Magnusson et al. 1999). It was demonstrated that silver nanoparticles could be synthesized via a small ceramic heater with a local heating source (Jung et al. 2006). The evaporated vapor can cool at a suitable rapid rate, because the temperature gradient in the vicinity of the heater surface is very steep in comparison with that of a tube furnace. This makes possible the formation of small nanoparticles in high concentration. This physical method can be useful as a nanoparticle generator for long-term experiments for inhalation toxicity studies, and as a calibration device for nanoparticle measurement equipment (Jung et al. 2006).

Silver nanoparticles could be synthesized by laser ablation of metallic bulk materials in solution (Mafune et al. 2000; Mafune et al. 2001; Kabashin and Meunier 2003; Sylvestre et al. 2004; Dolgaev et al. 2002). The ablation efficiency and the characteristics of produced nano-silver particles depend upon many factors such as the wavelength of the laser impinging the metallic target, the duration of the laser pulses (in the femto-, pico- and nanosecond regime), the laser fluence, the ablation time duration and the effective liquid medium, with or without the presence of surfactants (Kim et al. 2005; Link et al. 2000; Tarasenko et al. 2006; Kawasaki and Nishimura 2006). One important advantage of laser ablation technique compared to other methods for production of metal colloids is the absence of chemical reagents in solutions. Therefore, pure and uncontaminated metal colloids for further applications can be prepared by this technique (Tsuji et al. 2002). Silver nanospheroids (20-50 nm) were prepared by laser ablation in water with femtosecond laser pulses at 800 nm (Tsuji et al. 2003). The formation efficiency and the size of colloidal particles were compared with those of colloidal particles prepared by nanosecond laser pulses. The results revealed the formation efficiency for femtosecond pulses was significantly lower than that for nanosecond pulses. The size of colloids prepared by femtosecond pulses were less dispersed than that of colloids prepared by nanosecond pulses. Furthermore, it was found that the ablation efficiency for femtosecond ablation in water was lower than that in air, while, in the case of nanosecond pulses, the ablation efficiency was similar in both water and air.

2.2 Chemical approaches

The most common approach for synthesis of silver nanoparticles is chemical reduction by organic and inorganic reducing agents. In general, different reducing agents such as sodium citrate, ascorbate, sodium borohydride (NaBH$_4$), elemental hydrogen, polyol process, Tollens reagent, N, N-dimethylformamide (DMF), and poly (ethylene glycol)-block copolymers are used for reduction of silver ions (Ag$^+$) in aqueous or non-aqueous solutions. The aforementioned reducing agents reduce silver ions (Ag$^+$) and lead to the formation of metallic silver (Ag^0), which is followed by agglomeration into oligomeric clusters. These clusters eventually lead to formation of metallic colloidal silver particles (Wiley et al. 2005; Evanoff and Chumanov 2004; Merga et al. 2007). It is important to use protective agents to stabilize dispersive nanoparticles during the course of metal nanoparticle preparation, and protect the nanoparticles that can be absorbed on or bind onto nanoparticle surfaces, avoiding their agglomeration (Oliveira et al. 2005). The presence of surfactants comprising functionalities (*e.g.*, thiols, amines, acids, and alcohols) for interactions with particle surfaces can stabilize particle growth, and protect particles from sedimentation, agglomeration, or losing their surface properties. Polymeric compounds such as poly(vinyl alcohol), poly(vinylpyrrolidone), poly(ethylene glycol), poly(methacrylic acid), and polymethylmethacrylate have been reported to be effective protective agents to stabilize nanoparticles. In one study, Oliveira et al. (Oliveira et al. 2005) prepared dodecanethiol-capped silver nanoparticles, based on Brust procedure (Brust and Kiely 2002), based on a phase transfer of an Au^{3+} complex from aqueous to organic phase in a two-phase liquid-liquid system, followed by a reduction with sodium borohydride in the presence of dodecanethiol as a stabilizing agent, binding onto the nanoparticles surfaces, thereby avoiding their aggregation and making them soluble in certain solvents. They reported that small changes in synthetic factors lead to dramatic modifications in nanoparticle structure, average size, size distribution width, stability and self-assembly patterns. Zhang et al. (2008) used a hyperbranched poly(methylene bisacrylamide aminoethyl piperazine) with terminal dimethylamine groups (HPAMAM-N(CH$_3$)$_2$) to produce colloids of silver. The amide moieties, piperazine rings, tertiary amine groups and the hyper-branched structure in HPAMAM-N(CH$_3$)$_2$ are important to its effective stabilizing and reducing abilities.

Uniform and size controllable silver nanoparticles can be synthesized using micro-emulsion techniques. The nanoparticles preparation in two-phase aqueous organic systems is based on the initial spatial separation of reactants (metal precursor and reducing agent) in two immiscible phases. The interface between the two liquids and the intensity of inter-phase transport between two phases, which is mediated by a quaternary alkyl-ammonium salt, affect the rate of interactions between metal precursors and reducing agents. Metal clusters formed at the interface are stabilized, due to their surface being coated with stabilizer molecules occurring in the non-polar aqueous medium, and transferred to the organic medium by the inter-phase transporter (Krutyakov et al. 2008). One of the major disadvantages of this method is the use of highly deleterious organic solvents. Thus large amounts of surfactant and organic solvent must be separated and removed from the final product. For instance, Zhang et al. (2007) used dodecane as an oily phase (a low deleterious and even nontoxic solvent), but there was no need to separate the prepared silver solution from the reaction mixture. On other hand, colloidal nanoparticles prepared in nonaqueous media for conductive inks are well-dispersed in a low vapor pressure organic solvent, to readily wet the surface of the polymeric substrate without any aggregation. These advantages can also be found in the applications of metal nanoparticles as catalysts to catalyze most organic reactions, which have been conducted in non-polar solvents. It is very important to transfer nanometal particles to different physicochemical environments in practical applications (Cozzoli et al. 2004).

A simple and effective method, UV-initiated photoreduction, has been reported for synthesis of silver nanoparticles in the presence of citrate, polyvinylpyrrolidone, poly(acrylic acid), and collagen. For instance, Huang and Yang produced silver nanoparticles via the photoreduction of silver nitrate in layered inorganic laponite clay suspensions which served as a stabilizing agent for the prevention of nanoparticles aggregation. The properties of the produced nanoparticles were studied as a function of UV irradiation time. Bimodal size distribution and relatively large silver nanoparticles were obtained when irradiated under UV for 3 h. Further irradiation disintegrated the silver nanoparticles into smaller sizes with a single distribution mode until a relatively stable size and size distribution was obtained (Huang and Yang 2008). Silver nanoparticles (nanosphere, nanowire, and dendrite) have been prepared by an ultraviolet irradiation photoreduction technique at room temperature using poly(vinylalcohol) (as protecting and stabilizing agent). Concentration of both poly(vinylalcohol) and silver nitrate played significant role in the growth of the nanorods and dendrites (Zhou et al. 1999). Sonoelectrochemistry technique utilizes ultrasonic power primarily to manipulate the material mechanically. The pulsed sonoelectrochemical synthetic method involves alternating sonic and electric pulses, and electrolyte composition plays a crucial role in shape formation (Socol et al. 2002). It was reported that silver nanospheres could be prepared by sonoelectrochemical reduction using a complexing agent, nitrilotriacetate to avoid aggregation (Socol et al. 2002).

Nano-sized silver particles with an average size of 8 nm were prepared by photoinduced reduction using poly(styrene sulfonate)/poly(allylamine hydrochloride) polyelectrolyte capsules as microreactors (Shchukin et al. 2003). Moreover, it was demonstrated that the photoinduced method could be used for converting silver nanospheres into triangular silver nanocrystals (nanoprisms) with desired edge lengths in the range of 30-120 nm (Jin et al. 2003). The particle growth process was controlled using dual-beam illumination of nanoparticles. Citrate and poly(styrene sulfonate) were used as stabilizing agents. In another study, silver nanoparticles were prepared through a very fast reduction of Ag^+ by α-aminoalkyl radicals generated from hydrogen abstraction toward an aliphatic amine by the excited triplet state of 2-substituted thioxanthone series ($TX-O-CH_2-COO^-$ and $TX-S-CH_2-COO^-$). The quantum yield of this prior reaction was tuned by a substituent effect on the thioxanthones, and led to a kinetic control of the conversion of Ag^+ to $Ag^{(0)}$ (Malval et al. 2010).

Electrochemical synthetic method can be used to synthesize silver nanoparticles. It is possible to control particle size by adjusting the electrolysis parameters and to improve homogeneity of silver nanoparticles by changing the composition of the electrolytic solutions. Polyphenylpyrrole-coated silver nanospheroids (3-20 nm) were synthesized by electrochemical reduction at the liquid/liquid interface. This nano-compound was prepared by transferring the silver metal ion from the aqueous phase to the organic phase, where it reacted with pyrrole monomer (Johans et al. 2002). In another study, monodisperse silver nanospheroids (1-18 nm) were synthesized by electrochemical reduction inside or outside zeolite crystals according to the silver exchange degree of the compact zeolite film modified electrodes (Zhang et al. 2002). Furthermore, spherical silver nanoparticles (10-20 nm) with narrow size distributions were conveniently synthesized in an aqueous solution by an electrochemical method (Ma et al. 2004). Poly N-vinylpyrrolidone was chosen as the stabilizer for the silver clusters in this study. Poly N-vinylpyrrolidone protects nanoparticles from agglomeration, significantly reduces silver deposition rate, and promotes silver nucleation and silver particle formation rate. Application of rotating platinum cathode effectively solves the technological difficulty of rapidly transferring metallic nanoparticles from the cathode vicinity to bulk solution, avoiding the occurrence of flocculates in vicinity of the cathode, and ensures monodispersity of particles. The addition of sodium dodecyl benzene sulfonate to the electrolyte improved the particle size and particle size distribution of the silver nanoparticles (Ma et al. 2004).

Silver nanoparticles can be synthesized by using a variety of irradiation methods. Laser irradiation of an aqueous solution of silver salt and surfactant can produce silver nanoparticles with a well defined shape and size distribution (Abid et al. 2002). Furthermore, the laser was used in a photo-sensitization synthetic method of making silver nanoparticles using benzophenone. Low laser powers at short irradiation times produced silver nanoparticles of approximately 20 nm, while an increased irradiation power produced nanoparticles of approximately 5 nm. Laser and mercury lamp can be used as light sources for the production of silver nanoparticles (Eutis et al. 2005). In visible light irradiation studies, the photo-sensitized growth of silver nanoparticles using thiophene (sensitizing dye) and silver nanoparticle formation by illumination of $Ag(NH_3)^+$ in ethanol has been accomplished (Sudeep and Kamat 2005; Zhang et al. 2003).

Microwave assisted synthesis is a promising method for the synthesis of silver nanoparticles. It was reported that silver nanoparticles could be synthesized by a microwave-assisted synthesis method employing carboxymethyl cellulose sodium as a reducing and stabilizing agent. The size of the resulting particles depended on the concentration of sodium carboxymethyl cellulose and silver nitrate. The produced nanoparticles were uniform and stable, and were stable at room temperature for 2 months without any visible change (Chen et al. 2008). The production of silver nanoparticles in the presence of Pt seeds, polyvinyl pyrrolidine and ethylene glycol was also reported (Navaladian et al. 2008). Additionally, starch has been employed as a template and reducing agent for the synthesis of silver nanoparticles with an average size of 12 nm, using a microwave-assisted synthetic method. Starch functions as a template, preventing the aggregation of the produced silver nanoparticles (Sreeram et al. 2008). Microwaves in combination with polyol process were applied in the synthesis of silver nanospheroids using ethylene glycol and poly N-vinylpyrrolidone as reducing and stabilizing agents, respectively (Komarneni et al. 2002). In a typical polyol process inorganic salt is reduced by the polyol (*e.g.*, ethylene glycol which serves as both a solvent and a reducing agent) at a high temperature. Yin et al. (Yin et al. 2004) reported that large-scale and size-controlled silver nanoparticles could be rapidly synthesized under microwave irradiation from an aqueous solution of silver nitrate and trisodium citrate in the presence of formaldehyde as a reducing agent. Size and size distribution of the produced silver nanoparticles are strongly dependent on the states of silver cations in the initial reaction solution. Silver nanoparticles with different shapes can be synthesized by microwave irradiation of a silver nitrate-ethylene-glycol-$H_2[PtCl_6]$-poly(vinylpyrrolidone) solution within 3 min (Tsuji et al. 2008). Moreover, the use of microwave irradiation to produce monodispersed silver nanoparticles using basic amino acids (as reducing agents) and soluble starch (as a protecting agent) has been reported (Hu et al. 2008). Radiolysis of silver ions in ethylene glycol, in order to synthesize silver nanoparticles, was also reported (Soroushian et al. 2005). Moreover, silver nanoparticles supported on silica aero-gel were produced using gamma radiolysis. The produced silver clusters were stable in the 2-9 pH range and started agglomeration at pH > 9 (Ramnami et al. 2007). Oligochitosan as a stabilizer can be used in a preparation of silver nanoparticles by gamma radiation. It was reported that stable silver nanoparticles (5-15 nm) were synthesized in a 1.8-9.0 pH range using this method (Long et al. 2007). Silver nanoparticles (4-5 nm) were also synthesized by γ-ray irradiation of acetic water solutions containing silver nitrate and chitosan (Cheng et al. 2007). In another study, silver nanospheroids (1-4 nm) were produced by γ-ray irradiation of a silver solution in optically transparent inorganic mesoporous silica. Reduction of silver ions within the matrix is brought about by hydrated electrons and hydroalkyl radicals generated during the radiolysis of a 2-propanol solution. The nanoparticles produced within the silica matrix were stable in the presence of oxygen for at least several months (Hornebecq et al. 2003). Moreover, silver nanoparticles (60-200 nm) have been produced by irradiating a solution, prepared by mixing silver nitrate and poly-vinyl-alcohol, with 6 MeV electrons (Bogle et al. 2006). A pulse radiolysis technique has been applied to study the reactions of inorganic and organic species in silver nanoparticle synthesis, to understand the factors controlling the shape and size of the nanoparticles synthesized by a common reduction method using citrate ions (as reducing and stabilizing agents) (Pillai and Kamat 2004), and to demonstrate the role of phenol derivatives in the formation of silver nanoparticles by the reduction of silver ions with dihydroxy benzene (Jacob et al. 2008). Dihydroxy benzene could be also used to reduce silver ions to synthesize stable silver nanoparticles (with an average size of 30 nm) in air-saturated aqueous solutions (Jacob et al. 2008).

In polysaccharide method, silver nanoparticles were prepared using water as an environmentally-friendly solvent and polysaccharides as capping/reducing agents. For instance, the synthesis of starch-silver nanoparticles was carried out with starch (as a capping agent) and β-D-glucose (as a reducing agent) in a gently heated system (Raveendran et al. 2003). The binding interactions between starch and produced silver nanoparticles were weak and could be reversible at higher temperatures, allowing for the separation of the synthesized nanoparticles. In dual polysaccharide function, silver nanoparticles were synthesized by the reduction of silver ions inside nanoscopic starch templates (Raveendran et al. 2003, 2005). The extensive network of hydrogen bands in templates provided surface passivation or protection against nanoparticle aggregation. Green synthesis of silver nanoparticles using negatively charged heparin (reducing/stabilizing agent and nucleation controller) was also reported by heating a solution of silver nitrate and heparin to 70 °C for approximately 8 h (Huang and Yang 2004). Transmission electron microscopy (TEM) micrographs demonstrated an increase in particle size of silver nanoparticles with increased concentrations of silver nitrate (as the substrate) and heparin. Moreover, changes in the heparin concentration influenced the morphology and size of silver nanoparticles. The synthesized silver nanoparticles were highly stable, and showed no signs of aggregation after two months (Huang and Yang 2004). In another study, stable silver nanoparticles (10-34 nm) were synthesized by autoclaving a solution of silver nitrate (as the substrate) and starch (as a capping/reducing agent) at 15 psi and 121 °C for 5 min (Vigneshwaran et al. 2006b). These nanoparticles were stable in solution for three months at approximately 25 °C. Smaller silver nanoparticles (≤10 nm) were synthesized by mixing two solutions of silver nitrate containing starch (as a capping agent), and NaOH solutions containing glucose (as a reducing agent) in a spinning disk reactor with a reaction time of less than 10 min (Tai et al. 2008).

Recently, a simple one-step process, Tollens method, has been used for the synthesis of silver nanoparticles with a controlled size. This green synthesis technique involves the reduction of $Ag(NH_3)_2^+$ (as a Tollens reagent) by an aldehyde (Yin et al. 2002). In the modified Tollens procedure, silver ions are reduced by saccharides in the presence of ammonia, yielding silver nanoparticle films (50-200 nm), silver hydrosols (20-50 nm) and silver nanoparticles of different shapes. In this method, the concentration of ammonia and the nature of the reducing agent play an important role in controlling size and morphology of the silver nanoparticles. It was revealed that the smallest particles were formed at the lowest ammonia concentration. Glucose and the lowest ammonia concentration (5 mM) resulted in the smallest average particle size of 57 nm with an intense maximum surface plasmon absorbance at 420 nm. Moreover, an increase in NH_3 from 0.005 M to 0.2 M resulted in a simultaneous increase in particle size and polydispersity (Kvítek et al. 2005). Silver nanoparticles with controllable sizes were synthesized by reduction of $[Ag(NH_3)_2]^+$ with glucose, galactose, maltose, and lactose (Panacek et al. 2006). The nanoparticle synthesis was carried out at various ammonia concentrations (0.005-0.20 M) and pH conditions (11.5-13.0), resulting in average particle sizes of 25-450 nm. The particle size was increased by increasing (NH_3), and the difference in the structure of the reducing agent (monosaccharides and disaccharides) and pH (particles obtained at pH 11.5 were smaller than those at pH 12.5) influenced the particle size. Polydispersity also decreased in response to decreases the pH. Produced silver nanoparticles were stabilized and protected by sodium dodecyl sulfate (SDS), polyoxyethylenesorbitane monooleate (Tween 80), and polyvinylpyrrolidone (PVP 360) (Kvitek et al. 2008; Soukupova et al. 2008).

Silver, gold, palladium, and platinum nanoparticles can be produced at room temperature, as a result of simply mixing the corresponding metal ions with reduced polyoxometalates which served as reducing and stabilizing agents. Polyoxometalates are soluble in water and have the capability of undergoing stepwise, multielectron redox reactions without disturbing their structure. It was demonstrated that silver nanoparticles were produced by illuminating a deaerated solution of polyoxometalate/S/Ag^+ (polyoxometalate: $[PW_{12}O_{40}]$ [3], $[SiW_{12}O_{40}]$ [4-]; S:propan-2-ol or 2,4-dichlorophenol) (Troupis et al. 2002). Furthermore, green chemistry-type one-step synthesis and stabilization of silver nanostructures with $Mo^V–Mo^{VI}$ mixed-valence polyoxometalates in water at room temperature has been reported (Zhang et al. 2007).

2.3 Biological approaches

In recent years, the development of efficient green chemistry methods employing natural reducing, capping, and stabilizing agents to prepare silver nanoparticles with desired morphology and size have become a major focus of researchers. Biological methods can be used to synthesize silver nanoparticles without the use of any harsh, toxic and expensive chemical substances (Ahmad et al. 2003a; Shankar et al. 2004; Ankamwar et al. 2005; Huang et al. 2007). The bioreduction of metal ions by combinations of biomolecules found in the extracts of certain organisms (*e.g.*, enzymes/proteins, amino acids, polysaccharides, and vitamins) is environmentally benign, yet chemically complex. Many studies have reported successful synthesis of silver nanoparticle using organisms (microorganisms and biological systems) (Korbekandi et al. 2009; Sastry et al. 2003; Iravani 2011). For instance, we demonstrated the bioreductive synthesis of silver nanoparticles using *F. oxysporum* (Figure 1). In this section, most of the organisms used in green synthesis of silver nanoparticles are shown.

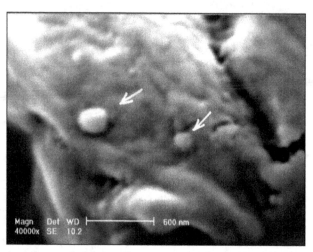

Fig. 1. SEM micrograph recorded from silver nanoparticles produced by reaction of AgNO$_3$ solution (1 mM) with *F. oxysporum* biomass.

2.3.1 Synthesis of silver nanoparticles by bacteria

It was reported that highly stable silver nanoparticles (40 nm) could be synthesized by bioreduction of aqueous silver ions with a culture supernatant of nonpathogenic bacterium, *Bacillus licheniformis* (Kalishwaralal et al. 2008b). Moreover, well-dispersed silver nanocrystals (50 nm) were synthesized using the bacterium *Bacillus licheniformis* (Kalishwaralal et al. 2008a). Saifuddin *et al.* (Saifuddin et al. 2009) have described a novel combinational synthesis approach for the formation of silver nanoparticles by using a combination of culture supernatant of *B. subtilis* and microwave irradiation in water. They reported the extracellular biosynthesis of monodispersed Ag nanoparticles (5-50 nm) using supernatants of *B. subtilis*, but in order to increase the rate of reaction and reduce the aggregation of the produced nanoparticles, they used microwave radiation which might provide uniform heating around the nanoparticles and could assist the digestive ripening of particles with no aggregation.

Silver nanocrystals of different compositions were successfully synthesized by *Pseudomonas stutzeri* AG259 (Klaus et al. 1999). The silver-resistant bacterial strain, *Pseudomonas stutzeri* AG259, isolated from a silver mine, accumulated silver nanoparticles intracellularly, along with some silver sulfide, ranging in size from 35 to 46 nm (Slawson et al. 1992). Larger particles were formed when *P. stutzeri* AG259 challenged with high concentrations of silver ions during culturing, resulted intracellular formation of silver nanoparticles, ranging in size from a few nm to 200 nm (Klaus-Joerger et al. 2001; Klaus et al. 1999). *P. stutzeri* AG259 detoxificated silver through its precipitation in the periplasmic space and its reduction to elemental silver with a variety of crystal typologies, such as hexagons and equilateral triangles, as well as three different types of particles: elemental crystalline silver, monoclinic silver sulfide acanthite (Ag_2S), and a further undetermined structure (Klaus et al. 1999). The periplasmic space limited the thickness of the crystals, but not their width, which could be rather large (100-200 nm) (Klaus-Joerger et al. 2001). In another study, rapid biosynthesis of metallic nanoparticles of silver using the reduction of aqueous Ag^+ ions by culture supernatants of *Klebsiella pneumonia*, *E. coli*, and *Enterobacter cloacae* (Enterobacteriacae) was reported (Shahverdi et al. 2007). The synthetic process was quite fast and silver nanoparticles were formed within 5 min of silver ions coming in contact with the cell filtrate. It seems that nitroreductase enzymes might be responsible for bioreduction of silver ions. It was also reported that visible-light emission could significantly increase synthesis of silver nanoparticles (1-6 nm) by culture supernatants of *K. pneumoniae* (Mokhtari et al. 2009). Monodispersed and stable silver nanoparticles were also successfully synthesized with bioreduction of [Ag $(NH_3)_2$] $^+$ using *Aeromonas* sp. SH10 and *Corynebacterium* sp. SH09 (Mouxing et al. 2006). It was speculated that [Ag $(NH_3)_2$] $^+$ first reacted with OH^- to form Ag_2O, which was then metabolized independently and reduced to silver nanoparticles by the biomass.

Lactobacillus strains, when exposed to silver ions, resulted in biosynthesis of nanoparticles within the bacterial cells (Nair and Pradeep 2002). It has been reported that exposure of lactic acid bacteria present in the whey of buttermilk to mixtures of silver ions could be used to grow nanoparticles of silver. The nucleation of silver nanoparticles occurred on the cell surface through sugars and enzymes in the cell wall, and then the metal nuclei were transported into the cell where they aggregated and grew to larger-sized particles.

2.3.2 Synthesis of silver nanoparticles by fungi

Silver nanoparticles (5-50 nm) could be synthesized extracellularly using *Fusarium oxysporum*, with no evidence of flocculation of the particles even a month after the reaction (Ahmad et al. 2003a). The long-term stability of the nanoparticle solution might be due to the stabilization of the silver particles by proteins. The morphology of nanoparticles was highly variable, with generally spherical and occasionally triangular shapes observed in the micrographs. Silver nanoparticles have been reported to interact strongly with proteins including cytochrome *c* (Cc). This protein could be self-assembled on citrate-reduced silver colloid surface (Macdonald and Smith 1996). Interestingly, adsorption of (Cc)-coated colloidal Au nanoparticles onto aggregated colloidal Ag resulted Ag: Cc: Au nanoparticle conjugate (Keating et al. 1998). In UV-vis spectra from the reaction mixture after 72 h, the presence of an absorption band at ca. 270 nm might be due to electronic excitations in tryptophan and tyrosine residues in the proteins. In *F. oxysporum*, the bioreduction of silver ions was attributed to an enzymatic process involving NADH-dependent reductase (Ahmad et al. 2003b). The exposure of silver ions to *F. oxysporum*, resulted in release of nitrate reductase and subsequent formation of highly stable silver nanoparticles in solution (Kumar et al. 2007). The secreted enzyme was found to be dependent on NADH cofactor. They mentioned high stability of nanoparticles in solution was due to capping of particles by release of capping proteins by *F. oxysporum*. Stability of the capping protein was found to be pH dependent. At higher pH values (>12), the nanoparticles in solution remained stable, while they aggregated at lower pH values (<2) as the protein was denatured. Kumar *et al.* (Kumar et al. 2007) have demonstrated enzymatic synthesis of silver nanoparticles with different chemical compositions, sizes and morphologies, using α-NADPH-dependent nitrate reductase purified from *F. oxysporum* and phytochelatin, *in vitro*. Silver ions were reduced in the presence of nitrate reductase, leading to formation of a stable silver hydrosol 10-25 nm in diameter and stabilized by the capping peptide. Use of a specific enzyme in *in vitro* synthesis of nanoparticles showed interesting advantages. This would eliminate the downstream processing required for the use of these nanoparticles in homogeneous catalysis and other applications such as non-linear optics. The biggest advantage of this protocol based on purified enzyme was the development of a new approach for green synthesis of nanomaterials over a range of chemical compositions and shapes without possible aggregation. Ingle *et al.* (Ingle et al. 2008) demonstrated the potential ability of *Fusarium acuminatum* Ell. and Ev. (USM-3793) cell extracts in biosynthesis of silver nanoparticles. The nanoparticles produced within 15-20 minutes and were spherical with a broad size distribution in the range of 5-40 nm with the average diameter of 13 nm. A nitrate-dependent reductase enzyme might act as the reducing agent. The white rot fungus, *Phanerochaete chrysosporium*, also reduced silver ions to form nano-silver particles (Vigneshwaran et al. 2006a). The most dominant morphology was pyramidal shape, in different sizes, but hexagonal structures were also observed. Possible involvement of proteins in synthesizing silver nanoparticles was observed in *Plectonema boryanum* UTEX 485 (a filamentous cyanobacterium) (Lengke et al. 2007).

Stable silver nanoparticles could be achieved by using *Aspergillus flavus* (Vigneshwaran et al. 2007). These nanoparticles were found to be stable in water for more than 3 months with no significant aggregation because of surface binding of stabilizing materials secreted by the fungus (Vigneshwaran et al. 2007). Extracellular biosynthesis of silver nanoparticles using *Aspergillus fumigatus* (a ubiquitous saprophytic mold) has also been investigated (Bhainsa and D'Souza 2006). The resulted TEM micrograph showed well-dispersed silver nanoparticles (5-25 nm) with variable shapes. Most of them were spherical in nature with some others having occasionally triangular shapes (Bhainsa and D'Souza 2006). Compared to intracellular biosynthesis of nanoparticles; extracellular synthesis could be developed as a simple and cheap method because of uncomplicated downstream processing and handling of biomasses.

The extracellular filtrate of *Cladosporium cladosporioides* biomass was used to synthesize silver nanoparticles (Balaji et al. 2009). It was suggested that proteins, organic acids and polysaccharides released by *C. cladosporioides* were responsible for formation of spherical crystalline silver nanoparticles. Kathiresan *et al.* (Kathiresan et al. 2009) have shown that when the culture filtrate of *Penicillium fellutanum* was incubated with silver ions and maintained under dark conditions, spherical silver nanoparticles could be produced. They also changed crucial factors such as pH, incubation time, temperature, silver nitrate concentration and sodium chloride to achieve the maximum nanoparticle production. The highest optical density at 430 nm was found at 24 h after the start of incubation time, 1 mM concentration of silver nitrate, pH 6.0, 5°C temperature and 0.3% sodium chloride. Fungi of *Penicillium* genus were used for green synthesis of silver nanoparticles (Sadowski et al.

2008). *Penicillium* sp. J3 isolated from soil was able to produce silver nanoparticles (Maliszewska et al. 2009). The bioreduction of silver ions occurred on the surface of the cells and proteins might have critical role in formation and stabilization of the synthesized nanoparticles.

Sanghi *et al.* (2009) have investigated the ability of *Coriolus versicolor* in formation of monodisperse spherical silver nanoparticles. Under alkaline conditions (pH 10) the time taken for production of silver nanoparticles was reduced from 72 h to 1 h. It was indicated that alkaline conditions might be involved in bioreduction of silver ions, water hydrolysis and interaction with protein functionalities. Findings of this study have shown that glucose was necessary for the reduction of silver nanoparticles, and S-H of the protein played an important role in the bioreduction.

2.3.3 Synthesis of silver nanoparticles by plants

Camellia sinensis (green tea) extract has been used as a reducing and stabilizing agent for the biosynthesis of silver nanoparticles in an aqueous solution in ambient conditions (Vilchis-Nestor et al. 2008). It was observed that when the amount of *C. sinensis* extract was increased, the resulted nanoparticles were slightly larger and more spherical. Phenolic acid-type biomolecules (*e.g.,* caffeine and theophylline) present in the *C. sinensis* extract seemed to be responsible for the formation and stabilization of silver nanoparticles. Black tea leaf extracts were also used in the production of silver nanoparticles (Begum et al. 2009). The nanoparticles were stable and had different shapes, such as spheres, trapezoids, prisms, and rods. Polyphenols and flavonoids seemed to be responsible for the biosynthesis of these nanoparticles.

Plant extracts from alfalfa (*Medicago sativa*), lemongrass (*Cymbopogon flexuosus*), and geranium (*Pelargonium graveolens*) have served as green reactants in silver nanoparticle synthesis. Harris *et al.* (2008) have investigated the limits (substrate metal concentration and time exposure) of uptake of metallic silver by two common metallophytes, *Brassica juncea* and *M. sativa*. They demonstrated that *B. juncea* and *M. Sativa* could be used in the phytosynthesis of silver nanoparticles. *B. juncea*, when exposed to an aqueous substrate containing 1,000 ppm silver nitrate for 72 h, accumulated up to 12.4 wt. % silver. *M. sativa* accumulated up to 13.6 wt. % silver when exposed to an aqueous substrate containing 10,000 ppm silver nitrate for 24 h. In the case of *M. sativa*, an increase in metal uptake was observed by increasing the exposure time and substrate concentration. In both cases, TEM analysis showed the presence of roughly spherical silver nanoparticles, with a mean size of 50 nm. Geranium leaf broth, when exposed to aqueous silver nitrate solution, resulted in enzymatic synthesis of stable crystalline silver nanoparticles, extracellularly (Shankar et al. 2003). Bioreduction of the metal ions was fairly rapid, occurred readily in solution, and resulted in a high density of stable silver nanoparticles (16-40 nm). The nanoparticles appeared to be assembled into open, quasilinear superstructures and were predominantly spherical in shape. It was believed that proteins, terpenoids and other bio-organic compounds in the geranium leaf broth participated in bioreduction of silver ions and in the stabilization of the nanoparticles via surface capping.

A high density of extremely stable silver nanoparticles (16-40 nm) was rapidly synthesized by challenging silver ions with *Datura metel* (*Solanaceae*) leaf extract (Kesharwani et al. 2009).

The leaf extracts of this plant contains biomolecules, including alkaloids, proteins/enzymes, amino acids, alcoholic compounds, and polysaccharides which could be used as reductant to react with silver ions, and therefore used as scaffolds to direct the formation of silver nanoparticles in the solution. Song and Kim (2008) elucidated the fact that *Pinus desiflora*, *Diospyros kaki*, *Ginko biloba*, *Magnolia kobus* and *Platanus orientalis* leaf broths synthesized stable · silver nanoparticles with average particle size ranging from 15 to 500 nm, extracellularly. In the case of *M. kobus* and *D. kaki* leaf broths, the synthesis rate and final conversion to silver nanoparticles was faster, when the reaction temperature was increased. But the average particle sizes produced by *D. kaki* leaf broth decreased from 50 nm to 16 nm, when temperature was increased from 25 °C to 95 °C. The researchers also illustrated that only 11 min was required for more than 90% conversion at the reaction temperature of 95 °C using *M. kobus* leaf broth (Song and Kim 2008). It was further demonstrated that leaf extracts from the aquatic medicinal plant, *Nelumbo nucifera* (Nymphaeaceae), was able to reduce silver ions and produce silver nanoparticles (with an average size of 45 nm) in different shapes (Santhoshkumar et al. 2010). The biosynthesized nanoparticles showed larvicidal activities against malaria (*Anopheles subpictus*) and filariasis (*Culex quinquefasciatus*) vectors. Silver nanoparticles were biosynthesized using *Sorbus aucuparia* leaf extract within 15 min. The nanoparticles were found to be stable for more than 3 months. The sorbate ion in the leaf extract of *S. aucuparia* encapsulated the nanoparticles and this action seemed to be responsible for their stability (Dubey et al. 2010).

Studying synthesis of silver nanoparticles with isolated/purified bio-organics may give better insight into the system mechanism. Glutathione (γ-Glu-Cys-Gly) as a reducing/capping agent can synthesize water-soluble and size tunable silver nanoparticles which easily bind to a model protein (bovine serum albumin) (Wu et al. 2008). Tryptophan residues of synthetic oligo-peptides at the C-terminus were identified as reducing agents producing silver nanoparticles (Si and Mandal 2007). Moreover, silver nanoparticles were synthesized by vitamin E in Langmuir-Blodgett technique, by bio-surfactants (sophorolipids) and by L-valine-based oligopeptides with chemical structures, Z-(L-Val)$_3$-OMe and Z-(L-Val)$_2$-L-Cys (S-Bzl)-OMe (Kasture et al. 2007; Mantion et al. 2008). The sulfur content in the Z-(L-Val)$_2$-L-Cys(S-Bzl)-OMe controls the shape and size of silver nanoparticles, which suggests interaction between silver ions and thioether moiety of the peptide (Mantion et al. 2008).

Pure natural constituents could be used to bioreduce and stabilize silver nanoparticles. Kasthuri *et al.* (Kasthuri et al. 2009) have shown the ability of apiin extracted from henna leaves to produce anisotropic gold and quasi-spherical silver nanoparticles. Secondary hydroxyl and carbonyl groups of apiin were responsible for the bioreduction of metal salts. In order to control size and shape of nanoparticles, they used different amounts of apiin (as the reducing agent). The nanoparticles were stable in water for 3 months, which could be attributed to surface binding of apiin to nanoparticles. Furthermore, geraniol as a volatile compound obtained from *P. graveolens* was used for biosynthesis of silver nanoparticles (1-10 nm). They also evaluated *in vitro* cytotoxicity of silver nanoparticles against Fibrosarcoma Wehi 164 at different concentrations (1-5 g ml^{-1}) (Safaepour et al. 2009). The presence of 5 g/ml of silver nanoparticles significantly inhibited the cell line's growth (up to 60 %).

The various synthetic and natural polymers such as poly(ethylene glycol), poly-(N-vinyl-2-pyrrolidone), starch, heparin, poly-cationic chitosan (1-4-linked 2-amino-2-deoxy-β-D-glucose), sodium alginate, and gum acacia have been reported as reducing and stabilizing agents for biosynthesis of silver nanoparticles. It was reported that monodisperse spherical silver nanoparticles (3 nm) could be synthesized using gum kondagogu (non-toxic polysaccharide derived as an exudate from the bark of *Cochlospermum gossypium*) (Kora et al. 2010). It was suggested that carboxylate and hydroxyl groups were involved in complexation and bioreduction of silver ions into nanoparticles. This method was compatible with green chemistry principles as the gum serves a matrix for both bioreduction and stabilization of the synthesized nanoparticles. Due to availability of low cost plant derived biopolymer, this method could be implemented for large-scale synthesis of highly stable monodispersed nanoparticles.

Spherical silver nanoparticles (40-50 nm) were produced using leaf extract of *Euphorbia hirta* (Elumalai et al. 2010). These nanoparticles had potential and effective antibacterial property against *Bacillus cereus* and *S. aureus*. *Acalypha indica* (Euphorbiaceae) leaf extracts have produced silver nanoparticles (20-30 nm) within 30 min (Krishnaraj et al. 2010). These nanoparticles had excellent antimicrobial activity against water borne pathogens, *E. coli* and *V. cholera* (Minimum Inhibitory Concentration (MIC) = 10 μg ml^{-1}). Moreover, silver nanoparticles (57 nm) were produced when 10 ml of *Moringa oleifera* leaf extract was mixed to 90 ml of 1 mM aqueous of $AgNO_3$ and was heated at 60-80 °C for 20 min. These nanoparticles had considerable antimicrobial activity against pathogenic microorganisms, including *Staphylococcus aureus*, *Candida tropicalis*, *Klebsiella pneumoniae*, and *Candida krusei* (Prasad and Elumalai 2011). It has been reported that cotton fibers loaded with biosynthesized silver nanoparticles (~20 nm) using natural extracts of *Eucalyptus citriodora* and *Ficus bengalensis* had excellent antibacterial activity against *E. coli*. These fibers had potential for utilization in burn/wound dressings as well as in the fabrication of antibacterial textiles and finishings (Ravindra et al. 2010). *Garcinia mangostana* leaf extract could be used as reducing agent in order to biosynthesize silver nanoparticles (35 nm). These nanoparticles had high effective antimicrobial activity against *E. coli* and *S. aureus* (Veerasamy et al. 2011). It was reported that *Ocimum sanctum* leaf extract could bioreduce silver ions into crystalline silver nanoparticles (4-30 nm) within 8 min of reaction time. These nanoparticles were stable due to the presence of proteins which may act as capping agents. *O. sanctum* leaves contain ascorbic acid which may play an important role in reduction of silver ions into metallic silver nanoparticles. These nanoparticles have shown strong antimicrobial activity against *E. coli* and *S. aureus* (Singhal et al. 2011). Green synthesis of silver nanoparticles using *Cacumen platycladi* extract was also investigated. Reducing sugars and flavonoids in the extract seemed to be mainly responsible for reduction of silver ions, and their reductive capability promoted at 90 °C, leading to formation of silver nanoparticles (18.4 ± 4.6 nm) with narrow size distribution. The produced nanoparticles had significant antibacterial activity against both gram negative and gram positive bacteria (*E. coli* and *S. aureus*) (Huang et al. 2011). In case of *Cinnamon zeylanicum*, the bark extract could be used in biosynthesis of cubic and hexagonal silver nanocrystals (31-40 nm) (Sathishkumar et al. 2009). The particle size distribution varied with variation in the amount of *C. zeylanicum* bark extract. The number of particles increased with increasing dosage due to the variation in the amount of reductive biomolecules. Small nanoparticles were formed at high pH. The shape of silver nanoparticles at high pH was more spherical in nature rather than ellipsoidal. Bactericidal effect of produced nano-crystalline silver particles was tested against *E. coli* strain. As a result, the various tested concentrations of 2, 5, 10, 25, and 50 mg

L[-1] produced inhibition of 10.9, 32.4, 55.8, 82, and 98.8 %, respectively. The minimum inhibitory concentration was found to be 50 mg L[-1].

3. Applications of silver nanoparticles and their incorporation into other materials

Nanoparticles are of great interest due to their extremely small size and large surface to volume ratio, which lead to both chemical and physical differences in their properties compared to bulk of the same chemical composition, such as mechanical, biological and sterical properties, catalytic activity, thermal and electrical conductivity, optical absorption and melting point (Daniel and Astruc 2004; Bogunia-Kubik and Sugisaka 2002; Zharov et al. 2005). Therefore, designing and production of materials with novel applications can be resulted by controlling shape and size at nanometer scale. Nanoparticles exhibit size and shape-dependent properties which are of interest for applications ranging from biosensing and catalysts to optics, antimicrobial activity, computer transistors, electrometers, chemical sensors, and wireless electronic logic and memory schemes. These particles also have many applications in different fields such as medical imaging, nano-composites, filters, drug delivery, and hyperthermia of tumors (Tan et al. 2006; Lee et al. 2008; Pissuwan et al. 2006). Silver nanoparticles have drawn the attention of researchers because of their extensive applications in areas such as integrated circuits (Kotthaus et al. 1997), sensors (Cao 2004), biolabelling (Cao 2004), filters (Cao 2004), antimicrobial deodorant fibres (Zhang and Wang 2003), cell electrodes (Klaus-Joerger et al. 2001), low-cost paper batteries (silver nano-wires) (Hu et al. 2009) and antimicrobials (Cho et al. 2005; Dura´n et al. 2007). Silver nanoparticles have been used extensively as antimicrobial agents in health industry, food storage, textile coatings and a number of environmental applications. Antimicrobial properties of silver nanoparticles caused the use of these nanometals in different fields of medicine, various industries, animal husbandry, packaging, accessories, cosmetics, health and military. Silver nanoparticles show potential antimicrobial effects against infectious organisms, including *Escherichia coli*, *Bacillus subtilis*, *Vibria cholera*, *Pseudomonas aeruginosa*, *Syphillis typhus*, and *S. aureus* (Cho et al. 2005; Dura´n et al. 2007). For instance, it was shown that silver nanoparticles mainly in the range of 1-10 nm attached to the surface of *E. coli* cell membrane, and disturbed its proper function such as respiration and permeability (Morones et al. 2005). It was observed that silver nanoparticles had a higher antibacterial effect on *B. subtilis* than on *E. coli*, suggesting a selective antimicrobial effect, possibly related to the structure of the bacterial membrane (Yoon et al. 2007). As antimicrobial agents, silver nanoparticles were applied in a wide range of applications from disinfecting medical devices and home appliances to water treatment (Jain and Pradeep 2005; Li et al. 2008). The synergetic activities of silver nanoparticles and antibiotics such as erythromycin, amoxicillin, penicillin G, clindamycin, and vancomycine against *E. coli* and *S. aureus* was reported (Shahverdi et al. 2007). Silver nanoparticles were also used in medical devices (*e.g.*, polymethylmetacrylate bone cement, surgical masks and implantable devices) to increase their antimicrobial activities (Figure 2). For instance, plastic catheters which were coated with silver nanoparticles using silver nitrate ($AgNO_3$), surfactant and N,N,N',N'-tetramethylethylenediamine (as a reducing agent), had significant antimicrobial effects against *E. coli*, Enterococcus, *S. aureus*, *C. albicans*, Staphylococci, and *P. aeroginosa* (Roe et al. 2008). These can be useful for reducing the risk of infectious complications in patients with indwelling catheters. Silver nanoparticles/clay was reported to have significant antimicrobial effects against dermal pathogens, including *S. aureus, P. aeruginosa,* and *Streptococcus pyrogens,* as well as the methicillin- and oxacillin resistant *S. aureus* (MRSA and ORSA) (Su et al. 2009).

Antifungal effects of silver nanoparticles against fungal pathogens are mostly unknown. However, it was reported that these nanoparticles had significant antifungal activities against *T. mentagrophytes* and Candida species (such as *C. albicans, C. tropicolis, C. glabrata, C. parapsilosis,* and *C. krusei*). Silver nanoparticles disrupt fungal envelope structure and lead to significant damage to fungal cells (Kim et al. 2008; Kim et al. 2009).

Sun et al demonstrated the cytoprotective activity of silver nanoparticles (5-20 nm) toward HIV-1 infected cells (Hut/CCR5). These nanoparticles inhibited the virus replication in Hut/CCR5 cells causing HIV-associated apoptosis (Sun et al. 2005). In addition, size-dependent antiviral activities of silver nanoparticles (exclusively within the range of 1-10 nm) against HIV-1 virus have been reported. Silver nanoparticles preferentially bind to the gp120 subunit of HIV-1 viral envelope glycoprotein, and this interaction caused the virus could not bind to the host cell (Elechiguerra et al. 2005). It seems that more investigations are needed to find the mechanisms of antiviral effects of silver nanoparticles.

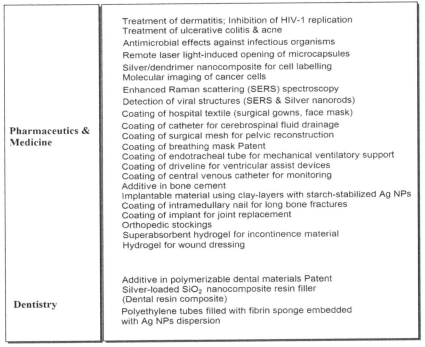

Pharmaceutics & Medicine	Treatment of dermatitis; Inhibition of HIV-1 replication
	Treatment of ulcerative colitis & acne
	Antimicrobial effects against infectious organisms
	Remote laser light-induced opening of microcapsules
	Silver/dendrimer nanocomposite for cell labelling
	Molecular imaging of cancer cells
	Enhanced Raman scattering (SERS) spectroscopy
	Detection of viral structures (SERS & Silver nanorods)
	Coating of hospital textile (surgical gowns, face mask)
	Coating of catheter for cerebrospinal fluid drainage
	Coating of surgical mesh for pelvic reconstruction
	Coating of breathing mask Patent
	Coating of endotracheal tube for mechanical ventilatory support
	Coating of driveline for ventricular assist devices
	Coating of central venous catheter for monitoring
	Additive in bone cement
	Implantable material using clay-layers with starch-stabilized Ag NPs
	Coating of intramedullary nail for long bone fractures
	Coating of implant for joint replacement
	Orthopedic stockings
	Superabsorbent hydrogel for incontinence material
	Hydrogel for wound dressing
Dentistry	Additive in polymerizable dental materials Patent
	Silver-loaded SiO_2 nanocomposite resin filler (Dental resin composite)
	Polyethylene tubes filled with fibrin sponge embedded with Ag NPs dispersion

Fig. 2. Applications of silver nanoparticles in pharmaceutics, medicine, and dentistry

In general, therapeutic effects of silver particles (in suspension form) depend on important aspects, including particle size (surface area and energy), particle shape (catalytic activity), particle concentration (therapeutic index) and particle charge (oligodynamic quality) (Pal et al. 2007). Mechanisms of antimicrobial effects of silver nanoparticles are still not fully understood, but several studies have revealed that silver nanoparticles may attach to the negatively charged bacterial cell wall and rupture it, which leads to denaturation of protein and finally cell death (Figure 3). Fluorescent bacteria were used to study antibacterial effects of silver nanoparticles. Green fluorescent proteins were adapted to these investigations (Gogoi et al. 2006). It was found that silver nanoparticles attached to sulfur-containing proteins of bacteria, and caused death. Moreover, fluorescent measurements of cell-free supernatants showed the effect of silver nanoparticles on recombination of bacteria. The attachment of silver ions or nanoparticles to the cell wall caused accumulation of envelope protein precursors resulting in immediate dissipation of the proton motive force (Lin et al. 1998). Catalytic mechanism of silver nanoparticle composites and their damage to the cell by interaction with phosphorous- and sulfur-containing compounds such as DNA have been also investigated (Sharma et al. 2009). Furthermore, silver nanoparticles exhibited destabilization of the outer membrane and rupture of the plasma membrane, thereby causing depletion of intracellular ATP (Lok 2006). Another mechanism involves the association of silver with oxygen and its reaction with sulfhydryl groups on the cell wall to form R-S-S-R bonds, thereby blocking respiration and causing cell death (Kumar et al. 2004).

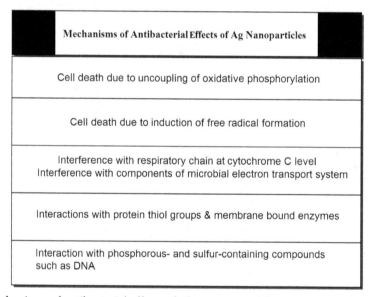

Fig. 3. Mechanisms of antibacterial effects of silver nanoparticles

Products made with silver nanoparticles have been approved by a range of accredited bodies such as US Food and Drug Administration (FDA), US Environmental Protection Agency (EPA), and Research Institute for Chemical Industry and FITI Testing (Zhong et al. 2007; Wei et al. 2007; Jia et al. 2008). Silver nanoparticles can be exploited in medicine and pharmacy for dental materials, burn treatments, coating stainless steel materials, textile fabrics, and sunscreen lotions (Duran et al. 2007). Silver nanoparticles are widely used in surface enhanced Raman scattering (SERS) spectroscopy (surface-sensitive technique which is known for its high sensitivity due to its surface plasmon effects lacks behind at uniform substrate formation) because of high local field enhancement factor. For example, it was demonstrated that SERS and silver nanorods quickly revealed the viral structures (Shanmukh et al. 2006). It can also differentiate between respiratory viruses, virus strains, and viruses containing gene deletions without manipulating the virus. Silver nanoparticles can also be used for modification of polymer based diffraction gratings which further can be used in grating light reflection spectroscopy for sensing (Šileikaitė et al. 2007; Abu Hatab et al. 2008). Furthermore, the electrochemical properties of silver nanoparticles incorporated them in nano-scale sensors which can offer faster response times and lower detection limits (Manno et al. 2008; Hahm and Lieber 2004). Manno et al. (Hahm and Lieber 2004) electrodeposited silver nanoparticles onto alumina plates gold micro-patterned electrode which showed a high sensitivity to hydrogen peroxide (H_2O_2). Silver nanoparticles were found to catalyze the chemiluminescence from luminal-hydrogen peroxide system with catalytic activity better than Au and Pt colloid (Guo et al. 2008). The catalytic properties of silver nanoparticles varied from their size and morphology. It was reported that when the silver colloid was injected, chemiluminoscence emission from the luminal-H_2O_2 system was greatly enhanced (Guo et al. 2008). Silver nanoparticles supported halloysite nanotubes (Ag/HNTs), with silver content of approximately 11%, was applied to catalyze reduction of 4-nitrophenol with $NaBH_4$ in alkaline aqueous solutions (Liu and Zhao 2009). Nano-clusters composed of 2-8 silver atoms could be the basis for a new type of optical data storage. Moreover, fluorescent emissions from the clusters could also potentially be used in biological labels and electroluminescent displays (Berciaud et al. 2005; Kossyrev et al. 2005). The optical properties of a metallic nanoparticle depend mainly on its surface plasmon resonance, where the plasmon refers to the collective oscillation of the free electrons within the metallic nanoparticle. The plasmon resonant peaks and line widths are sensitive to the size and shape of the nanoparticle, the metallic species and the surrounding medium. The optical properties of silver nanoparticles have been taken advantage of optical chemosensors and biosensors. For instance, triangular silver nanoparticles have been used in measurement of biological binding signal between antibodies and antigens (Zhu et al. 2009).

Metal nanoparticles embedded into polymer matrix can be used as sensors, materials with solvent switchable electronic properties, optical limiters and filters, and optical data storage. They can be applied for catalytic applications and antimicrobial coatings. Production of metal/polymer nanocomposites in poly(methyl methacrylate), N-isopropylacrylamide, diallyldimethylammonium chloride, poly(vinyl alcohol), poly(vinyl acetate), poly(vinyl carbazole), and polyimide films has been reported (Boyd et al. 2006; Sakamoto et al. 2007; Sakamoto et al. 2009). Nanocomposites of polymer/silver nanoparticles have attracted great attention because of their potential applications in the fields of catalysis, bioengineering, photonics, and electronics (Kiesow et al. 2003). Polymers are considered good host materials for metallic nanoparticles as well as other stabilizing agents, including citrates, organic solvents (*e.g.,* tetrahydrofuran and tetrahydrofuran/methanol), long chain alcohols, surfactants, and organometallics (Sarkar et al. 2005; Chen et al. 2001). Various chemical and physical methods were applied to prepare metal-polymer composites (*e.g.,* fabrication approach) (Kiesow et al. 2003; Rifai et al. 2006). Fabrication approach is often referred to as the evaporation method since the polymer solvent is evaporated from the reaction mixture after nanoparticle dispersion. However, this often leads to inhomogeneous distribution of the particles in the polymer. A successful production of nanoparticles is determined by the ability to produce particles with uniform distributions and long stability, given their tendency to rapidly agglomerate in aqueous solution (Rifai et al. 2006). Another approach is a system in which simultaneous polymerization and metal reduction occur. For instance, polychrome silver nanoparticles were obtained by microwave irradiation from a solution of silver nitrate in the presence of poly(*N*-vinyl-2-pyrrolidone) without any other reducing agent. The *in-situ* reduction of silver ions in poly(N-vinyl-2-pyrrolidone) by microwave irradiation produced particles with narrow size distribution (He et al. 2002; Mohan et al. 2007). Moreover, silver nanoparticles (approximately 5 nm) were prepared by mixing equal amounts of 0.5 wt % aqueous solutions of acacia (natural polymer) and silver nitrate. In this process, acacia polymeric chains promote the reduction and act as good stabilizers over 5 months (He et al. 2002; Mohan et al. 2007).

Conventional heating approach (simultaneous polymerization-reduction method) to polymerize acrylonitrile reduces silver ions resulting in homogeneous dispersal and narrow size distributions of the silver nanoparticles in silver-polyacrylonitrile composite powders (Zhang and Han 2003). Moreover, size-controlled synthesis of silver nano-complex was recently achieved in reduction of silver nitrate by a UV-irradiated argine-tungstonsilicate acid solution (Yang et al. 2007). Other various metal-polymer nano-composites have been prepared by these reduction methods, such as poly(vinyl alcohol)-Ag, Ag-polyacrylamide, Ag-acrylonitrile, Ag_2Se-polyvinyl alcohol, Ag-polyimide, Au-polyaniline, and Cu-poly(acrylic acid) (Ma et al. 2002; Akamatsu et al. 2003). Poly(vinyl alcohol) (a water soluble and biocompatible polymer with extremely low cytotoxicity) has a lot of biomedical applications. This biologically friendly polymer can be used as a stabilizer due to its optical clarity and classified into grades of partially (85-89%) and fully (97-99.5%) hydrolyzed polymers (Badr and Mahmoud 2006). This polymer was widely used in textile, paper industries, food packaging, pharmacy, and cosmetics. Introduction of nano-sized silver into poly(vinyl alcohol) provides antibacterial activity, which is highly desired in textiles used in medicine, wound dressing applications, clothing and household products (Hong et al. 2006). For instance, poly(vinyl alcohol) nanofibers containing silver nanoparticles were prepared by electro-spinning poly(vinyl alcohol)/silver nitrate aqueous solutions, followed by short heat treatment. The produced nanofibers containing silver nanoparticles showed strong antimicrobial activities (Hong et al. 2006). Various approaches such as solvent evaporation, electron and ultraviolet radiation, *in situ* chemical reduction, thermal annealing, and sono-chemical method have been applied to synthesize poly(vinyl alcohol)-silver nanoparticles (Sharma et al. 2009). In solvent evaporation methods, synthesis of poly(vinyl alcohol)-silver nanoparticles was achieved by first reducing silver salt with $NaBH_4$, followed by mechanical dispersion of the silver colloids into the dissolved polymer, and then the solvent was evaporated resulting in final structure. Electron and ultraviolet radiation approaches involve irradiation of a Ag^+ doped polymer film which gives poly(vinyl alcohol)-silver nano-composites (Sharma et al. 2009). In one study, silver nanoparticles have been prepared in poly(vinyl alcohol) matrix by thermal treatments. The annealing time and temperature vary morphologies of poly(vinyl alcohol)-silver nanoparticles (Clemenson et al. 2007). Moreover, hydrogels of poly(vinyl alcohol)- poly(N-vinyl pyrolidone containing silver nanoparticles were produced by repeated freezing–thawing treatments. These nanosilver-containing hydrogels had excellent antibacterial effects against *E. coli* and *S. aureus* (Yu et al. 2007).

Apatite-coated Ag/AgBr/titanium dioxide (TiO_2) was prepared by deposition of silver to generate electron–hole pairs by extending the excitation wavelength to the visible-light region, AgBr, and hydroxy apatite as photosensitive material and adsorption bioceramic, respectively (Elahifard et al. 2007). In another study, nanoparticle photocatalysts based on silver, carbon, and sulfur-doped titanium dioxide were synthesized by a modified sol-gel method (Hamal and Klabunde 2007). Silver/TiO_2 nano-composites and poly(vinyl alcohol)-capped colloidal silver-TiO_2 nano-composites were prepared. The synthesized nanocomposites were effective toward destroying Gram-negative bacteria, e.g. *Escherichia coli.* (Guin et al. 2007). These silver/TiO_2 surfaces had advantageous properties, including visible light photocatalysis, biological compatibility, and antimicrobial activities (Elahifard et al. 2007; Seery et al. 2007; Hamal and Klabunde 2007). For example, Ag/TiO_2 composite films were prepared by incorporating silver in the pores of titanium dioxide films with an impregnation method via photoreduction (a versatile and convenient process with advantage of space-selective fabrication). It was reported that these composite films are promising in application of photocatalysis, antimicrobial and self-clean technologies (Liu et al. 2008). In order to synthesize silver nanoparticles on titanium dioxide surfaces, aqueous reduction, photochemical (*in-situ* photochemical fabrication approach can be used for preparation of metal nanoparticles/polymer nano-composites), liquid phase deposition, and sol-gel methods can be applied (Zheng et al. 2008; Elahifard et al. 2007; Seery et al. 2007; Liu et al. 2008; Hamal and Klabunde 2007; Cozzoli et al. 2004). Silver nanoparticles with a narrow size distribution were prepared by simple aqueous reduction from silver ions in different molar ratios of titanium dioxide suspensions and $NaBH_4$ (reducing agent) (Nino-Martinez et al. 2008). One of the photochemical reduction methods involves loading silver nanoparticles (3-5 nm) onto the surface of titanium dioxide nano-tubes first using the liquid deposition approach followed by UV or laser irradiation (Li et al. 2008).

4. Toxicity of nano-silver particles

Nano-sized particles can pass through biological membranes and penetrate even very small capillaries throughout the body (*e.g.,* pass through blood-brain and blood testes barriers). Size, morphology, and surface area are recognized as important determinants for toxicity of nanoparticles (Ji et al. 2007). Information on the toxicological implication of silver nanoparticles is limited (Chen and Schluesener 2008). Toxicity of silver nanoparticles is mostly determined *in vitro* with particles ranging from approximately 1-100 nm. The *in vitro* and *in vivo* toxicity studies revealed that silver nanoparticles have the potential ability to cause chromosomal aberrations and DNA damage, to enter cells and cause cellular damage, and are capable of inducing proliferation arrest in cell lines of zebra fish (Ji et al. 2007; Asharani PV et al. 2007; Hussain SM et al. 2005). Limited health effects of the use of nano-silver particles in humans have been reported, such as argyria (a condition in which the skin becomes blue or bluish-grey colored because of improper exposure to chemical forms of element) which was appeared to occur only after intake of large amounts of colloidal silver particles (the suspension with nano-silver of different sizes). Several cross-sectional studies reported that argyria is the most frequent adverse outcome from exposure to silver nanoparticles. For instance, prolonged ingestion of colloidal silver can change the color of skin and cause blue-grey appearance on face (the symptoms of argyria) (Chang et al. 2006). Silver nanoparticles can bind to different tissues (bind to proteins and enzymes in mammalian cells) and cause toxic effects, such as adhesive interactions with cellular membrane and production of highly reactive and toxic radicals like reactive oxygen species (ROS), which can cause inflammation and show intensive toxic effects on mitochondrial function. Potential target organs for nano-silver toxicity may involve liver and immune system. Accumulation and histopathological effects were seen in livers from rats systemically exposed to silver nanoparticles (10-15 nm). Furthermore, increased liver enzymes were reported in Sprague-Dawley rats after Twenty-eight-day oral exposure to nano-silver particles (60 nm) (Kim et al. 2008). It was reported that silver-coated dressing acticoat caused raised liver enzymes and argyria-like symptoms in burn patients (Trop et al. 2006). The *in vitro* and *in vivo* studies revealed toxicity effects of silver nanoparticles on immune system, especially cytokine excretion. Application of 1% nano-silver cream (silver nanoparticles with size range of approximately 50 nm), inhibited DNB-induced allergic contact dermatitis, and accumulation in the spleen has also been reported (Bhol and Schechter 2007; Takenaka et al. 2001). It has been suggested that silver nanoparticles are especially effective at inhibiting inflammations and may be used to treat immunologic and inflammatory diseases (Shin et al. 2007). Oral administration of silver nanoparticles (60 nm) to rats induced some local inflammatory effects (Kim et al. 2008). Moreover, in one *in vitro* study, toxicity effects of nano-silver particles on erythrocytes was reported (Garner et al. 1994), while an increase in red blood cells was seen after oral administration of silver nanoparticles (60 nm) (Kim et al. 2008), but not after inhalation of silver nanoparticles (15 nm) (Ji et al. 2007). An *in vitro* study suggests that nano-silver particles may have cytotoxic effects on mammalian (mouse) spermatogonial germ line stem cells (Braydich-Stolle et al. 2005). Silver nanoparticles (10 µg ml[-1] and above concentration) showed dramatic changes such as necrosis and apoptosis of the cells and at 5-10 µg ml[-1] drastically reduced mitochondrial function and cell viability. It was reported that there is no evidence available to demonstrate that silver is a cause of neurotoxic damage even though silver deposits have been identified in the region of cutaneous nerves (Lansdown 2007). The respiratory system seemed relatively unaffected by exposure to nano-silver *in vivo* in a 28 days study (Ji et al. 2007). However, cytotoxic effect of nano-silver particles on alveolar macrophages and alveolar epithelial cells was demonstrated, *in vitro* (Soto et al. 2005; Soto et al. 2007; Park et al. 2007). In a 90 days inhalation study, decrease in lung functions, including tidal volume, minute volume and peak inspiration flow as well as inflammatory lesions in lung morphology and effects on inflammatory markers were reported (Sung et al. 2008). Kim et al. (2008) performed a bone marrow micronucleus test as a part of a 28-day oral administration study in order to investigate genotoxic effects or carcinogenicity of exposure to silver nanoparticles. No significant genotoxic potential of oral exposure to silver nanoparticles (60 nm) was found in this study. An increase in (mostly local) malignant tumors was found following chronic subcutaneous administration of high doses colloidal silver (Schmaehl and Steinhoff 1960).

5. Conclusion

Silver nanoparticles have attracted the attention of researchers because of their unique properties, and proven applicability in diverse areas such as medicine, catalysis, textile engineering, biotechnology, nanobiotechnology, bioengineering sciences, electronics, optics, and water treatment. Moreover, silver nanoparticles have significant inhibitory effects against microbial pathogens, and are widely used as antimicrobial agents in a diverse range of consumer products, including air sanitizer sprays, socks, pillows, slippers, respirators, wet wipes, cosmetics, detergents, soaps, shampoos, toothpastes, air and water filters, coatings of refrigerators, vacuum cleaners, bone cement, wound dressings, surgical dressings, washing machines, food storage packaging, and cell phones. The flexibility of silver nanoparticle synthetic methods and facile incorporation of silver nanoparticles into different media have interested researchers to further investigate the mechanistic aspects of antimicrobial, antiviral and anti-inflammatory effects of these nanoparticles.

In brief, there are limited well-controlled investigations on potential toxicities of nano-silver particles, and it seems that additional long-term studies (preferably using multiple particle sizes) are needed to better characterize and understand the risk of using these particles.

Various chemical, physical and biological synthetic methods have been developed to obtain silver nanoparticles of various shapes and sizes, including laser ablation, gamma irradiation, electron irradiation, chemical reduction, photochemical methods, microwave processing, and thermal decomposition of silver oxalate in water and in ethylene glycol, and biological synthetic methods. Most of these methods are still in the development stages and the problems experienced involve the stability and aggregation of nanoparticles, control of crystal growth, morphology, size and size distribution, and occasional difficulty in the management of the synthesis, as in the case of the radiolysis technique. Moreover, the separation of produced nanoparticles for further applications is still an important issue. By using different reducing agents and stabilizers, the particle size and morphology of silver nanoparticles have been controlled. Selection of solvent medium and selection of eco-friendly nontoxic reducing and stabilizing agents are the most important issues which must be considered in green synthesis of silver nanoparticles. In biological synthetic methods, it was shown that the silver nanoparticles produced by plants are more stable in comparison with those produced by other organisms. Plants (especially plant extracts) are able to reduce silver ions faster than fungi or bacteria. Furthermore, in order to use easy and safe green methods in scale-up and industrial production of well-dispersed silver nanoparticles, plant extracts are certainly better than plant biomass or living plants. However, better experimental procedures are needed for synthesis of well-characterized nanoparticles.

6. Acknowledgment

The authors appreciate the support of Isfahan University of Medical Sciences.

7. References

Abid, J.P.; Wark, A.W.; Brevet, P.F. & Girault H.H. (2002). Preparation of silver nanoparticles in solution from a silver salt by laser irradiation. *Chem Commun*, 792-793

Abu Hatab, N.A.; Oran, J.M. & Sepaniak, M.J. (2008). Surface-enhanced raman spectroscopy substrates created via electron beam lithography and nanotransfer printing. *American Chemical Society ACS NANO*, Vol.2, pp.377–385

Ahmad, A.; Mukherjee, P.; Senapati, S.; Mandal, D.; Khan, M.I.; Kumar, R. & Sastry, M. (2003a). Extracellular biosynthesis of silver nanoparticles using the fungus *Fusarium oxysporum*, *Colloids and Surfaces B: Biointerfaces*, Vol.28, pp.313-318

Ahmad, A.; Senapati, S.; Khan, M.I.; Kumar, R.; Ramani, R.; Srinivas, V. & Sastry, M.(2003b). Intracellular synthesis of gold nanoparticles by a novel alkalotolerant actinomycete, *Rhodococcus* species. *Nanotechnology*, Vol.14, pp.824-828

Akamatsu, K.; Ikeda, S. & Nawafune, H. (2003). Site-selective direct silver metallization on surface-modified polyimide layers. *Langmuir*, Vol.19, pp.10366–10371

Ankamwar, B.; Damle, C.; Ahmad, A. & Sastry, M. (2005). Biosynthesis of gold and silver nanoparticles using *Emblica officinalis* fruit extract, their phase transfer and transmetallation in an organic solution. *J Nanosci Nanotechnol*, Vol.5, pp.1665-1671

Asharani, P.V.; Nair, G.; Zhiyuan, H.; Manoor, P. & Valiyaveettil, S. (2007). Potential health impacts of silver nanoparticles. *Abstracts of Papers, 234th ACS National Meeting, Boston, MA, USA, August*:19-23, 2007. pp:TOXI-2099.

Badr, Y. & Mahmoud, M.A. (2006). Enhancement of the optical properties of poly vinyl alcohol by doping with silver nanoparticles. *J Appl Polym Sci*, Vol.99, pp.3608-3614

Balaji, D.S.; Basavaraja, S.; Deshpande, R.; Bedre Mahesh, D.; Prabhakar, B.K. & Venkataraman, A. (2009). Extracellular biosynthesis of functionalized silver nanoparticles by strains of *Cladosporium cladosporioides* fungus. *Colloids and Surfaces B: Biointerfaces*, Vol.68, pp.88-92

Begum, N.A.; Mondal, S.; Basu, S.; Laskar, R.A. & Mandal, D. (2009). Biogenic synthesis of Au and Ag nanoparticles using aqueous solutions of Black Tea leaf extracts. *Colloids and Surfaces B: Biointerfaces*, Vol.71, pp.113–118

Berciaud, S.; Cognet, L.; Tamarat, P. & Lounis, B. (2005). Observation of intrinsic size effects in the optical response of individual gold nanoparticles. *Nano Lett*, Vol.5,pp.515–518

Bhainsa, K.C. & D'Souza, S. (2006). Extracellular biosynthesis of silver nanoparticles using the fungus *Aspergillus fumigatus*. *Colloids and Surfaces B: Biointerfaces*, Vol.47, pp.160-164

Bhol, K.C. & Schechter, P.J. (2007). Effects of nanocrystalline silver (NPI 32101) in a rat model of ulcerative colitis. *Digestive Dis Sci*, Vol.52, pp.2732-2742

Bogle, K.A.; Dhole, S.D. & Bhoraskar, V.N. (2006). Silver nanoparticles: synthesis and size control by electron irradiation. *Nanotechnology*, Vol.17, pp.3204

Bogunia-Kubik, K. & Sugisaka, M. (2002). From molecular biology to nanotechnology and nanomedicine. *Biosystems*, Vol.65, pp.123-138

Boyd, D.A.; Greengard, L.; Brongersma, M.; El-Naggar, M.Y. & Goodwin, D. (2006). Plasmon-assisted chemical vapor deposition. *Nano Lett*, Vol.6, pp.2592-2597

Braydich-Stolle, L.; Hussain, S.; Schlager, J.J. & Hofmann, M.C. (2005). *In vitro* cytotoxicity of nanoparticles in mammalian germline stem cells. *Toxicol Sci*, Vol.88, pp.412-419

Brust, M. & Kiely, C. (2002). Some recent advances in nanostructure preparation from gold and silver particles: a short topical review. *Colloids Surf A: Phys Eng Aspects*, Vol.202, pp.175-186

Cao, G. (2004). *Nanostructures and nanomaterials: synthesis, properties and applications*. Imperial College Press, ISBN: 1-86094-4159, London

Chang, A.L.S.; Khosravi, V. & Egbert, B. (2006). A case of argyria after colloidal silver ingestion. *J Cutan Pathol*, Vol.33, pp.809-811

Chen, J.; Wang, K.; Xin, J. & Jin, Y. (2008). Microwave-assisted green synthesis of silver nanoparticles by carboxymethyl cellulose sodium and silver nitrate. *Mater Chem Phys*, Vol.108, pp.421-424

Chen, W.; Cai, W.; Zhang, L.; Wang, G. & Zhang, L. (2001). Sonochemical Processes and Formation of Gold Nanoparticles within Pores of Mesoporous Silica. *J Colloid Interface Sci*, Vol.238, pp.291-295

Chen, X. & Schluesener, H.J. (2008). Nano-silver: A nanoproduct in medical application. *Toxicol Lett*, Vol.176, pp.1-12

Cheng, P.; Song, L.; Liu, Y. & Fang, Y.E. (2007). Synthesis of silver nanoparticles by γ-ray irradiation in acetic water solution containing chitosan. *Radiat Phys Chem*, Vol.76, pp.1165-1168

Cho, K-H.; Park, J-E.; Osaka, T. & Park, S.G. (2005). The study of antimicrobial activity and preservative effects of nanosilver ingredient. *Electrochimica Acta*, Vol.51, pp.956-960

Clemenson, S.; David, L. & Espuche, E. (2007). Structure and morphology of nanocomposite films prepared from polyvinyl alcohol and silver nitrate: Influence of thermal treatment. *J Polym Sci Part A Polym Chem*, Vol.45, pp.2657-2672

Colvin, V.L.; Schlamp, M.C. & Alivisatos, A. (1994). Light emitting diodes made from cadmium selenide nanocrystals and a semiconducting polymer. *Nature*, Vol.370, pp.354-357

Cozzoli, P.; Comparelli, R.; Fanizza, E.; Curri, M.; Agostiano, A. & Laub, D. (2004). Photocatalytic synthesis of silver nanoparticles stabilized by TiO_2 nanorods: a semiconductor/metal nanocomposite in homogeneous nonpolar solution. *J Am Chem Soc*, Vol.126, pp.3868–3879

Daniel, M.C. & Astruc, D. (2004). Gold nanoparticles: assembly, supramolecular chemistry, quantum-size-related properties, and applications toward biology, catalysis, and nanotechnology. *Chem Rev*, Vol.104, pp.293-346

Dolgaev, S.I.; Simakin, A.V.; Voronov, V.V.; Shafeev, G.A. & Bozon-Verduraz, F. (2002). Nanoparticles produced by laser ablation of solids in liquid environment. *Appl Surf Sci*, Vol.186, pp.546-551

Dubey, S.P.; Lahtinen, M.; Särkkä, H. & Sillanpää, M. (2010). Bioprospective of *Sorbus aucuparia* leaf extract in development of silver and gold nanocolloids. *Colloids and Surfaces B: Biointerfaces*, Vol.80, pp.26-33

Dura´n, N.; Marcato, P.D.; De, S.; Gabriel, I.H.; Alves, O.L. & Esposito, E. (2007). Antibacterial effect of silver nanoparticles produced by fungal process on textile fabrics and their effluent treatment. *J Biomed Nanotechnol*, Vol.3, pp.203-208

Duran, N.; Marcato, D.P.; De Souza, H.I.; Alves, L.O. & Espsito, E. (2007). Antibacterial effect of silver nanoparticles produced by fungal process on textile fabrics and their effluent treatment. *J Biomedical Nanotechnology*, Vol.3, pp.203-208

Elahifard, M.R.; Rahimnejad, S.; Haghighi, S. & Gholami, M.R. (2007). Apatite-coated Ag/AgBr/TiO_2 visible-light photocatalyst for destruction of bacteria. *J Am Chem Soc*, Vol.129, pp.9552–9553

Elechiguerra, J.L.; Burt, J.L.; Morones, J.R.; Camacho-Bragado, A.; Gao, X.; Lara, H.H. & Yacaman, M.J. (2005). Interaction of silver nanoparticles with HIV-1. *J Nanobiotechnol*, Vol.3:6

Elumalai, E.K.; Prasad, T.N.V.K.V.; Hemachandran, J.; Viviyan Therasa, S.; Thirumalai, T. & David, E. (2010). Extracellular synthesis of silver nanoparticles using leaves of *Euphorbia hirta* and their antibacterial activities. *J Pharm Sci & Res*, Vol.2, pp.549-554

Eutis, S.; Krylova, G.; Eremenko, A.; Smirnova, N.; Schill, A.W. & El-Sayed, M. (2005). Growth and fragmentation of silver nanoparticles in their synthesis with a fs laser and CW light by photo-sensitization with benzophenone. *Photochem Photobiol Sci*, Vol.4, pp.154-159

Evanoff, Jr. & Chumanov, G. (2004). Size-controlled synthesis of nanoparticles. 2. measurement of extinction, scattering, and absorption cross sections. *J Phys Chem B*, Vol.108, pp.13957-13962

Garner, M.; Reglinski, J.; Smith, W.E. & Stewart, M.J. (1994). The interaction of colloidal metals with erythrocytes. *J Inorg Biochem*, Vol.56, pp.283-290

Gogoi, K.S.; Gopina, P.; Paul, A.; Ramesh, A.; Ghosh, S.S. & Chattopadhyay, A. (2006). Green fluorescent protein expressing *Escherichia coli* as a model system for investigating the antimicrobial activities of silver nanoparticles. *Langmuir*, Vol.22, pp.9322-9328

Guin, D.; Manorama, S.V.; Latha, J.N.L. & Singh, S. (2007). Photoreduction of silver on bare and colloidal TiO2 nanoparticles/nanotubes: synthesis, characterization, and tested for antibacterial outcome. *J Phys Chem C*, Vol.111, pp.13393-13397

Guo, J-Z.; Cui, H.; Zhou, W. & Wang, W. (2008). Ag nanoparticle-catalyzed chemiluminescent reaction between luminal and hydrogen peroxide. *J Photochem Photobiol A: Chem*, Vol.193, pp.89-96

Guo, J.; Cui, H.; Zhou, W. & Wang, W. (2008). Ag nanoparticle-catalyzed chemiluminescent reaction between luminol and hydrogen peroxide. *J Photochem Photobiol A: Chem*, Vol.193, pp.89-96

Hahm, J. & Lieber, C. (2004). Direct ultrasensitive electrical detection of DNA and DNA sequence variations using nanowire nanosensors. *Nano Lett*, Vol.4, pp.51-54

Hamal, D.B. & Klabunde, K.J. (2007). Synthesis, characterization, and visible light activity of new nanoparticle photocatalysts based on silver, carbon, and sulfur-doped TiO2. *J Colloid Interface Sci*, Vol.311, pp.514-522

Hamilton, J.F. & Baetzold, R. (1979). Catalysis by small metal clusters. *Science*, Vol.205, pp.1213-1220

Harris, A.T. & Bali, R. (2008). On the formation and extent of uptake of silver nanoparticles by live plants. *J Nanopart Res*, Vol.10, pp.691-695

He, R.; Qian, X.; Yin, J. & Zhu, Z. (2002). Preparation of polychrome silver nanoparticles in different solvents. *J Mater Chem*, Vol.12, pp.3783-3786

Hoffman, A.J.; Mills, G.; Yee, H. & Hoffmann, M. (1992). Q-sized cadmium sulfide: synthesis, characterization, and efficiency of photoinitiation of polymerization of several vinylic monomers. *J Phys Chem*, Vol.96, pp.5546-5552

Hong, K.H.; Park, J.L.; Sul, I.H.; Youk, J.H. & Kang, T.J. (2006). Preparation of antimicrobial poly(vinyl alcohol) nanofibers containing silver nanoparticles. *J Polym Sci Part B Polym Phys*, Vol.44, pp.2468-2474

Hornebecq, V.; Antonietti, M.; Cardinal, T. & Treguer-Delapierre, M. (2003). Stable silver nanoparticles immobilized in mesoporous silica. *Chemistry of Materials*, Vol.15, pp.1993-1999

Hu, B.; Wang, S-B.; Wang, K.; Zhang, M. & Yu, S.H. (2008). Microwave-assisted rapid facile "green" synthesis of uniform silver nanoparticles: Self-assembly into multilayered films and their optical properties. *J Phys Chem C*, Vol.112, pp.11169-11174

Hu, L.; Choi, J.W.; Yang, Y.; Jeong, S.; La Mantia, F.; Cui, L-F. & Cui, Y. (2009). Highly conductive paper for energy-storage devices. Proceedings of the National Academy of Sciences of the United States of America 106:21490-21494, S21490/21491-S21490/21413

Huang, H. & Yang, X. (2004). Synthesis of polysaccharide-stabilized gold and silver nanoparticles: A green method. *Carbohydr Res*, Vol.339, pp.2627-2631

Huang, H. & Yang, Y. (2008) Preparation of silver nanoparticles in inorganic clay suspensions. *Compos Sci Technol,* Vol.68, pp.2948-2953

Huang, J.; Li, Q.; Sun. D.; Lu, Y.; Su, Y.; Yang, X.; Wang, H.; Wang, Y.; Shao, W.; He, N.; Hong, J. & Chen, C. (2007). Biosynthesis of silver and gold nanoparticles by novel sundried *Cinnamomum camphora* leaf. *Nanotechnology,* Vol.18, pp.105,104

Huang, J.; Zhan, G.; Zheng, B.; Sun, D.; Lu, F.; Lin, Y.; Chen, H.; Zheng, Z.; Zheng, Y. & Li, Q. (2011). Biogenic silver nanoparticles by *Cacumen Platycladi* extract: synthesis, formation mechanism and antibacterial activity. *Ind Eng Chem Res,* Vol.50, pp.9095-9106

Hussain, S.M.; Hess, K.L.; Gearhart, J.M.; Geiss, K.T. & Schlager, J.J. (2005). *In vitro* toxicity of nanoparticles in BRL 3A rat liver cells. *Toxicol in vitro,* Vol.19, pp.975-983

Ingle, A.; Gade, A.; Pierrat, S.; Sönnichsen, C. & Mahendra, R. (2008). Mycosynthesis of silver nanoparticles using the fungus *Fusarium acuminatum* and its activity against some human pathogenic bacteria. *Current Nanoscience,* Vol.4, pp.141-144

Iravani, S. (2011). Green synthesis of metal nanoparticles using plants. *Green Chem,* Vol.13, pp. 2638-2650.

Jacob, J.A.; Mahal, H.S.; Biswas, N.; Mukerjee, T. & Kappor, S. (2008). Role of phenol derivatives in the formation of silver nanoparticles. *Langmuir,* Vol.24, pp.528–533

Jain, P. & Pradeep, T. (2005). Potential of silver nanoparticle-coated polyurethane foam as an antibacterial water filter. *Biotechnol Bioeng,* Vol.90, pp.59-63

Ji, J.H.; Jung, J.H.; Kim, S.S.; Yoon, J.U.; Park, J.D.; Choi, B.S.; Chung, Y.H.; Kwon, I.H.; Jeong, J.; Han, B.S.; Shin, J.H.; Sung, J.H.; Song, K.S. & Yu, I.J. (2007). Twenty-eight-day inhalation toxicity study of silver nanoparticles in Sprague-Dawley rats. *Inhal Toxicol,* Vol.19, pp.857-871

Jia, X.; Ma, X.; Wei, D.; Dong, J. & Qian, W. (2008). Direct formation of silver nanoparticles in cuttlebone-derived organic matrix for catalytic applications. *Colloids Surf A:Physicochem Eng Aspects,* Vol.330, pp.234-240

Jin, R.; Cao, Y.C.; Hao, E.; Metraux, G.S.; Schatz, G.C. & Mirkin, C. (2003). Controlling anisotropic nanoparticle growth through plasmon excitation. *Nature,* Vol.425, pp.487–490

Johans, C.; Clohessy, J.; Fantini, S.; Kontturi, K. & Cunnane, V.J. (2002). Electrosynthesis of polyphenylpyrrole coated silver particles at a liquid-liquid interface. *Electrochemistry Communications,* Vol.4, pp.227–230

Jung, J.; Oh, H.; Noh, H.; Ji, J. & Kim, S. (2006). Metal nanoparticle generation using a small ceramic heater with a local heating area. *J Aerosol Sci,* Vol.37, pp.1662-1670

Kabashin, A.V. & Meunier, M. (2003) Synthesis of colloidal nanoparticles during femtosecond laser ablation of gold in water. *J Appl Phys,* Vol.94, pp.7941-7943

Kalishwaralal, K.; Deepak, V.; Ramkumarpandian, S.; Bilal, M. & Sangiliyandi G. (2008a). Biosynthesis of silver nanocrystals by *Bacillus licheniformis. Colloids and Surfaces B: Biointerfaces,* Vol.65, pp.150-153

Kalishwaralal, K.; Deepak, V.; Ramkumarpandian, S.; Nellaiah, H. & Sangiliyandi, G. (2008b). Extracellular biosynthesis of silver nanoparticles by the culture supernatant of *Bacillus licheniformis. Mater Lett,* Vol.62, pp.4411-4413

Kasthuri, J.; Veerapandian, S. & Rajendiran, N. (2009). Biological synthesis of silver and gold nanoparticles using apiin as reducing agent. *Colloids and Surfaces B: Biointerfaces,* Vol.68, pp.55-60

Kasture, M.; Singh, S.; Patel, P.; Joy, P.A.; Prabhune, A.A.; Ramana, C.V. & Prasad, B.L.V. (2007). Multiutility sophorolipids as nanoparticle capping agents: synthesis of stable and water dispersible Co nanoparticles. *Langmuir*, Vol.23, pp.11409-11412

Kathiresan, K.; Manivannan, S.; Nabeel, M.A. & Dhivya, B. (2009). Studies on silver nanoparticles synthesized by a marine fungus, *Penicillium fellutanum* isolated from coastal mangrove sediment. *Colloids and Surfaces B: Biointerfaces*, Vol.71, pp.133-137

Kawasaki, M. & Nishimura, N. (2006). 1064-nm laser fragmentation of thin Au and Ag flakes in acetone for highly productive pathway to stable metal nanoparticles. *Appl Surf Sci*, Vol.253, pp.2208-2216

Keating, C.D.; Kovaleski, K.K. & Natan, M. (1998). Heightened electromagnetic fields between metal nanoparticles: surface enhanced Raman scattering from metal-Cytochrome c-metal sandwiches. *J Phys Chem B*, Vol.102, pp.9414

Kesharwani, J.; Yoon, K.Y.; Hwang, J. & Rai, M. (2009). Phytofabrication of silver nanoparticles by leaf extract of *Datura metel*: hypothetical mechanism involvedin synthesis. *Journal of Bionanoscience*, Vol.3, pp.1-6

Kiesow, A.; Morris, J.E.; Radehaus, C. & Heilmann, A. (2003). Switching behavior of plasma polymer films containing silver nanoparticles. *J Appl Lett*, Vol.94, pp.6988-6990

Kim, K.J.; Sung, W.S.; Moon, S.K.; Choi, J.S.; Kim, J.G. & Lee, D.G. (2008). Antifungal effect of silver nanoparticles on dermatophytes. *J Microbiol Biotechnol*, Vol.18, pp.1482–1484

Kim, K.J.; Sung, W.S.; Suh, B.K.; Moon, S.K.; Choi, J.S.; Kim, J.G. & Lee, D.G. (2009). Antifungal activity and mode of action of silver nano-particles on *Candida albicans*. *Biometals*, Vol.22, pp.235-242

Kim, S.; Yoo, B.; Chun, K.; Kang, W.; Choo, J.; Gong, M. & Joo, S. (2005). Catalytic effect of laser ablated Ni nanoparticles in the oxidative addition reaction for a coupling reagent of benzylchloride and bromoacetonitrile. *J Mol Catal A: Chem*, Vol.226, pp.231-234

Kim, Y.S.; Kim, J.S.; Cho, H.S.; Rha, D.S.; Kim, J.M.; Park, J.D.; Choi, B.S.; Lim, R.; Chang, H.K.; Chung, Y.H.; Kwon, I.H.; Jeong, J.; Han, B.S. & Yu, I.J. (2008). Twenty-eight-day oral toxicity, genotoxicity, and gender-related issue distribution of silver nanoparticles in Sprague-Dawley rats. *Inhal Toxicol*, Vol.20, pp.575-583

Klaus-Joerger, T.; Joerger, R.; Olsson, E. & Granqvist, C.G. (2001). Bacteria as workers in the living factory: metal-accumulating bacteria and their potential for materials science. *Trends in Biotechnology*, Vol.19, pp.15–20

Klaus, T.; Joerger, R.; Olsson, E. & Granqvist, C.Gr. (1999). Silver-based crystalline nanoparticles, microbially fabricated. *Proc Natl Acad Sci USA*, Vol.96, pp.13611-13614

Komarneni, S.; Li, D.; Newalkar, B.; Katsuki, H. & Bhalla, A.S. (2002). Microwave-Polyol Process for Pt and Ag Nanoparticles. *Langmuir*, Vol.18, pp.5959–5962

Kora, A.J.; Sashidhar, R.B. & Arunachalam, J. (2010). Gum kondagogu (*Cochlospermum gossypium*): A template for the green synthesis and stabilization of silver nanoparticles with antibacterial application. *Carbohydrate Polymers*, Vol.82, pp.670-679

Korbekandi, H.; Iravani, S. & Abbasi, S. (2009). Production of nanoparticles using organisms. *Critical Reviews in Biotechnology*, Vol.29, pp.279-306

Kossyrev, P.; Yin, A.; Cloutier, S.; Cardimona, D.; Huang, D.; Alsing, P. & Xu, J. (2005). Electric field tuning of plasmonic response of nanodot array in liquid crystal matrix. *Nano Lett*, Vol.5, pp.1978–1981

Kotthaus, S.; Gunther, B.H.; Hang, R. & Schafer, H. (1997). Study of isotropically conductive bondings filled with aggregates of nano-sited Ag-particles. *IEEE Trans Compon Packaging Technol*, Vol.20, pp.15-20

Krishnaraj, C.; Jagan, E.G.; Rajasekar, S.; Selvakumar, P.; Kalaichelvan, P.T. & Mohan, N. (2010). Synthesis of silver nanoparticles using *Acalypha indica* leaf extracts and its antibacterial activity against water borne pathogens. *Colloids and Surfaces B: Biointerfaces*, Vol.76, pp.50–56

Kruis, F.; Fissan, H. & Rellinghaus, B. (2000). Sintering and evaporation characteristics of gas-phase synthesis of size-selected PbS nanoparticles. *Mater Sci Eng B*, Vol.69, pp.329-324

Krutyakov, Y.; Olenin, A.; Kudrinskii, A.; Dzhurik, P. & Lisichkin, G. (2008). Aggregative stability and polydispersity of silver nanoparticles prepared using two-phase aqueous organic systems. *Nanotechnol Russia*, Vol.3, pp.303-310

Kumar, S.A.; Majid Kazemian, A.; Gosavi, S.W.; Sulabha, K.K.; Renu, P.;Ahmad A. & Khan M.I. (2007). Nitrate reductase-mediated synthesis of silver nanoparticles from AgNO$_3$. *Biotechnology Letters*, Vol.29, pp.439-445

Kumar, V.S.; Nagaraja, B.M.; Shashikala, V.; Padmasri, A.H.; Madhavendra, S.S. & Raju, B.D. (2004). Highly efficient Ag/C catalyst prepared by electro-chemical deposition method in controlling microorganisms in water. *J Mol Catal A*, Vol.223, pp.313–319

Kvitek, L.; Panacek, A.; Soukupova, J.; Kolar, M.; Vecerova, R.; Prucek, R.; Holecová, M. & Zbořil, R. (2008). Effect of surfactants and polymers on stability and antibacterial activity of silver nanoparticles (NPs). *J Phys Chem C*, Vol.112, pp.5825–5834

Kvítek, L.; Prucek, R.; Panáček, A.; Novotný, R.; Hrbác, J. & Zbořil, R. (2005). The influence of complexing agent concentration on particle size in the process of SERS active silver colloid synthesis. *J Mater Chem*, Vol.15, pp.1099-1105

Lansdown, A.B. (2007). Critical observations on the neurotoxicity of silver. *Crit Rev Toxicol*, Vol.37, pp.237-250

Lee, H.Y.; Li, Z.; Chen, K.; Hsu, A.R.; Xu, C.; Xie, J.; Sun, S.; Chen, X. (2008). PET/MRI dual-modality tumor imaging using arginine-glycine-aspartic (RGD)–conjugated radiolabeled iron oxide nanoparticles, *J Nucl Med*, Vol.49, pp.1371-1379

Lengke, M.F.; Fleet, M.E. & Southam, G. (2007). Biosynthesis of silver nanoparticles by filamentous cyanobacteria from a silver(I) nitrate complex. *Langmuir*, Vol.23, pp.2694–2699

Li, H.; Duan, X.; Liu, G. & Liu, X. (2008). Photochemical synthesis and characterization of Ag/TiO$_2$ nanotube composites. *J Mater Sci*, Vol.43, pp.1669-1676

Li, Q.; Mahendra, S.; Lyon, D.; Brunet, L.; Liga, M.; Li, D. & Alvarez, P. (2008). Antimicrobial nanomaterials for water disinfection and microbial control: Potential applications and implications. *Water Res*, Vol.42, pp.4591-4602

Lin, Y.E.; Vidic, R.D.; Stout, J.E.; McCartney, C.A. & Yu V.L. (1998). Inactivation of *Mycobacterium avium* by copper and silver ions. *Water Res*, Vol.32, pp.1997-2000

Link, S.; Burda, C.; Nikoobakht, B. & El-Sayed, M. (2000). Laser-induced shape changes of colloidal gold nanorods using femtosecond and nanosecond laser pulses. *J Phys Chem B*, Vol.104, pp.6152–6163

Liu, P. & Zhao, M. (2009). Silver nanoparticle supported on halloysite nanotubes catalyzed reduction of 4-nitrophenol (4-NP). *Appl Surf Sci*, Vol.255, pp.3989-3993

Liu, Y.; Wang, X.; Yang, F. & Yang, X. (2008). Excellent antimicrobial properties of mesoporous anatase TiO_2 and Ag/TiO_2 composite films. *Microporous Mesoporous Mater*, Vol.114, pp.431-439

Lok, C. (2006). Proteomic analysis of the mode of antibacterial action of silver nanoparticles. *J Proteome Res*, Vol.5, pp.916-924

Long, D.; Wu, G. & Chen, S. (2007). Preparation of oligochitosan stabilized silver nanoparticles by gamma irradiation. *Radiat Phys Chem*, Vol.76, pp.1126-1131

Ma, H.; Yin, B.; Wang, S.; Jiao, Y.; Pan, W.; Huang, S.; Chen, S. & Meng, F. (2004). Synthesis of silver and gold nanoparticles by a novel electrochemical method. *Chem Phys Chem*, Vol.24, pp.68-75

Ma, X-D.; Qian, X-F.; Yin, J. & Zhu, Z-K. (2002). Preparation and characterization of polyvinyl alcohol–selenide nanocomposites at room temperature. *J Mater Chem*, Vol.12, pp.663-666

Macdonald, I.D.G. & Smith, W. (1996). Orientation of Cytochrome c adsorbed on a citrate-reduced silver colloid surface. *Langmuir*, Vol.12, pp.706

Mafune, F.; Kohno, J.; Takeda, Y.; Kondow, T. & Sawabe, H. (2001). Formation of gold nanoparticles by laser ablation in aqueous solution of surfactant. *J Phys Chem B*, Vol.105, pp.5114-5120

Mafune, F.; Kohno, J.; Takeda, Y.; Kondow, T. & Sawabe, H. (2000). Structure and stability of silver nanoparticles in aqueous solution produced by laser ablation. *J Phys Chem B*, Vol.104, pp.8333-8337

Magnusson, M.; Deppert, K.; Malm, J.; Bovin, J. & Samuelson, L. (1999). Gold nanoparticles: production, reshaping, and thermal charging. *J Nanoparticle Res*, Vol.1, pp.243-251

Maliszewska, I.; Szewczyk, K. & Waszak, K. (2009). Biological synthesis of silver nanoparticles. *Journal of Physics: Conference Series*, Vol.146, pp.1-6

Malval, J-P.; Jin, M.; Balan, L.; Schneider, R.; Versace, D-L.; Chaumeil, H.; Defoin, A. & Soppera, O. (2010). Photoinduced size-controlled generation of silver nanoparticles coated with carboxylate-derivatized thioxanthones. *J Phys Chem C*, Vol.114, pp.10396-10402

Manno, D.; Filippo, E.; Giulio, M. & Serra, A. (2008). Synthesis and characterization of starch-stabilized Ag nanostructures for sensors applications. *J Non-Cryst Solids*, Vol.354, pp.5515-5520

Mansur, H.S.; Grieser, F.; Marychurch, M.S.; Biggs, S.; Urquhart, R.S. & Furlong, D. (1995). Photoelectrochemical properties of 'q-state' cds particles in arachidic acid langmuir-blodgett films. *J Chem Soc Faraday Trans*, Vol.91, pp.665-672

Mantion, A.; Guex, A.G.; Foelske, A.; Mirolo, L.; Fromn, K.M. & Painsi, M. (2008). Silver nanoparticle engineering via oligovaline organogels. *Soft Matter*, Vol.4, pp.606-617

Merga, G.; Wilson, R.; Lynn, G.; Milosavljevic, B. & Meisel, D. (2007). Redox catalysis on "naked" silver nanoparticles. *J Phys Chem C*, Vol.111, pp.12220–12226

Mohan, Y.M.; Raju, K.M.; Sambasivudu, K.; Singh, S. & Sreedhar, B. (2007). Preparation of acacia-stabilized silver nanoparticles: A green approach. *J Appl Polym Sci*, Vol.106, pp.3375-3381

Mokhtari, N.; Daneshpajouh, S.; Seyedbagheri, S.; Atashdehghan, R.; Abdi, K.; Sarkar, S.; Minaian, S.; Shahverdi, H.R. & Shahverdi, A.R. (2009). Biological synthesis of very small silver nanoparticles by culture supernatant of *Klebsiella pneumonia*: The effects of visible-light irradiation and the liquid mixing process. *Materials Research Bulletin*, Vol.44, pp.1415-1421

Morones, J.R.; Elechiguerra, L.J.; Camacho, A.; Holt, K.; Kouri, B.J.; Ramirez, T.J. & Yocaman, J.M. (2005). The bactericidal effect of silver nanoparticles. *Nanotechnology*, Vol.16, pp.2346-2353

FU, M.; Li Q.; Sun, D.; Lu, Y.; He, N.; Xu, D.; Wang, H. & Huang, J. (2006). Rapid preparation process of silver nanoparticles by bioreduction and their characterizations. *Chinese J Chem Eng*, Vol.14, pp.114-117

Nair, B. & Pradeep, T. (2002). Coalescence of nanoclusters and formation of submicron crystallites assisted by *Lactobacillus* strains. *Crystal Growth & Design*, Vol.2, pp.293-298

Navaladian, S.; Viswanathan, B.; Varadarajan, T.K. & Viswanath, R.P. (2008). Microwave-assisted rapid synthesis of anisotropic Ag nanoparticles by solid state transformation. *Nanotechnology*, Vol.19, pp.045603

Nino-Martinez, N.; Martinez-Castanon, G.A.; Aragon-Pina, A.; Martinez-Gutierrez, F.; Martinez-Mendoza, J.R. & Ruiz, F. (2008). Characterization of silver nanoparticles synthesized on titanium dioxide fine particles. *Nanotechnology*, Vol.19, pp.065711/065711-065711/065718

Oliveira, M.; Ugarte, D.; Zanchet, D. & Zarbin, A. (2005). Influence of synthetic parameters on the size, structure, and stability of dodecanethiol-stabilized silver nanoparticles. *J Colloid Interface Sci*, Vol.292, pp.429-435

Pal, S.; Tak, Y.K. & Song, J.M. (2007). Does the antibacterial activity of silver nanoparticles depend on the shape of the nanoparticle? a study of the gram-negative bacterium *Escherichia coli*. *Applied and Environmental Microbiology*, Vol.73, pp.1712–1720

Panacek, A.; Kvitek, L.; Prucek, R.; Kolar, M.; Vecerova, R.; Pizurova, N.; Sharma, V.K.; Nevěná, T. & Zbořil, R. (2006). Silver colloid nanoparticles: synthesis, characterization, and their antibacterial activity. *J Phys Chem B*, Vol.110, pp.16248–16253

Park, S.; Lee, Y.K.; Jung, M.; Kim, K.H.; Eun-Kyung Ahn, N.C.; Lim, Y. & Lee, K.H. (2007). Cellular toxicity of various inhalable metal nanoparticles on human alveolar epithelial cells. *Inhalat Toxicol*, Vol.19, pp.59-65

Pillai, Z.S. & Kamat, P.V. (2004). What factors control the size and shape of silver nanoparticles in the citrate ion reduction method?. *J Phys Chem B*, Vol.108, pp.945-951

Pissuwan, D.; Valenzuela, S.M. & Cortie, M.B. (2006). Therapeutic possibilities of plasmonically heated gold nanoparticles, *Trends Biotechnol*, Vol.24, pp.62-67

Prasad, T.N.V.K.V. & Elumalai, E. (2011). Biofabrication of Ag nanoparticles using *Moringa oleifera* leaf extract and their antimicrobial activity. *Asian Pacific Journal of Tropical Biomedicine*, Vol.1, pp.439-442

Ramnami, S.P.; Biswal, J. & Sabharwal, S. (2007). Synthesis of silver nanoparticles supported on silica aerogel using gamma radiolysis. *Radiat Phys Chem*, Vol.76, pp.1290-1294

Raveendran. P.; Fu, J. & Wallen, S.L. (2003). Completely "green" synthesis and stabilization of metal nanoparticles. *J Am Chem Soc*, Vol.125, pp.13940–13941

Raveendran, P.; Fu, J. & Wallen, S.L. (2005). A simple and "green" method for the synthesis of Au, Ag, and Au–Ag alloy nanoparticles, *Green Chem*, Vol.8, pp.34-38

Ravindra, S.; Murali Mohan, Y.; Narayana Reddy, N. & Raju, K.M. (2010). Fabrication of antibacterial cotton fibres loaded with silver nanoparticles via "Green Approach". *Colloids and Surfaces A: Physicochem Eng Aspects*, Vol.367, pp.31-40

Rifai, S.; Breen, C.A.; Solis, D.J. & Swager, T.M. (2006). Facile *in situ* silver nanoparticle formation in insulating porous polymer matrices. *Chem Mater,* Vol.18, pp.21-25

Roe, D.; Karandikar, B.; Bonn-Savage, N.; Gibbins, B. & Roullet, J.B. (2008). Antimicrobial surface functionalization of plastic catheters by silver nanoparticles. *J Antimicrob Chemother*, Vol.61, pp.869-876

Sadowski, Z.; Maliszewska, I.H.; Grochowalska, B.; Polowczyk, I. & Kozlecki, T. (2008). Synthesis of silver nanoparticles using microorganisms. *Materials Science-Poland*, Vol.26, pp.419-424

Safaepour, M.; Shahverdi, A.R.; Shahverdi, H.R.; Khorramizadeh, M.R. & Gohari, A.R. (2009). Green synthesis of small silver nanoparticles using geraniol and its cytotoxicity against Fibrosarcoma-Wehi 164. *Avicenna J Med Biotech*, Vol.1, pp.111-115

Saifuddin, N.; Wong, C.W. & Nur Yasumira, A.A. (2009). Rapid biosynthesis of silver nanoparticles using culture supernatant of bacteria with microwave irradiation. *E-Journal of Chemistry*, Vol.6, pp.61-70

Sakamoto, M.; Fujistuka, M. & Majima, T. (2009). Light as a construction tool of metal nanoparticles: synthesis and mechanism. *J Photochem and Photobiol C: Photochem Reviews*, Vol.10, pp.33-56

Sakamoto, M.; Tachikawa, T.; Fujitsuka, M. & Majima, T. (2007). Photochemical formation of Au/Cu bimetallic nanoparticles with different shapes and sizes in a PVA film. *Adv Funct Mater*, Vol.17, pp.857-862

Sanghi, R. & Verma P. (2009). Biomimetic synthesis and characterisation of protein capped silver nanoparticles. *Bioresource Technology*, Vol.100, pp.501-504

Santhoshkumar, T.; Rahuman, A.A.; Rajakumar, G.; Marimuthu, S.; Bagavan, A.; Jayaseelan, C.; Zahir, A.A.; Elango, G. & Kamaraj, C. (2010). Synthesis of silver nanoparticles using *Nelumbo nucifera* leaf extract and its larvicidal activity against malaria and filariasis vectors. *Parasitol Res*, Vol.108, pp.693-702

Sarkar, A.; Kapoor, S. & Mukherjee, T. (2005). Preparation, characterization, and surface modification of silver nanoparticles in formamide. *J Phys Chem*, Vol.109, pp.7698-7704

Sastry, M.; Ahmad, A.; Khan, M.I. & Kumar, R. (2003). Biosynthesis of metal nanoparticles using fungi and actinomycete. *Current Science*, Vol.85, pp.162-170

Sathishkumar, M.; Sneha, K.; Won, S.W.; Cho, C-W.; Kim, S. & Yun, Y.S. (2009). *Cinnamon zeylanicum* bark extract and powder mediated green synthesis of nano-crystalline silver particles and its bactericidal activity. *Colloids and Surfaces B: Biointerfaces*, Vol.73, pp.332-338

Schmaehl, D. & Steinhoff, D. (1960). Studies on cancer induction with colloidal silver and gold solutions in rats. *Z Krebsforsch*, Vol.63, pp.586-591

Schmid, G. (1992). Large clusters and colloids. Metals in the embryonic state. *Chem Rev*, Vol.92, pp.1709-1727

Seery, M.K.; George, R.; Floris, P. & Pillai, S.C. (2007). Silver doped titanium dioxide nanomaterials for enhanced visible light photocatalysis. *J Photochem Photobiol A Chem*, Vol.189, pp.258-263

Senapati, S. (2005). *Biosynthesis and immobilization of nanoparticles and their applications.* University of pune, India

Shahverdi, A.R.; Minaeian, S.; Shahverdi, H.R.; Jamalifar, H. & Nohi, A. (2007). Rapid synthesis of silver nanoparticles using culture supernatants of Enterobacteria: A novel biological approach. *Process Biochemistry,* Vol.42, pp.919-923

Shahverdi, R.A.; Fakhimi, A.; Shahverdi, H.R. & Minaian, S. (2007). Synthesis and effect of silver nanoparticles on the antibacterial activity of different antibiotics against *Staphyloccocus aureus* and *Escherichia coli. Nanomed: Nanotechnol Biol Med,* Vol.3, pp.168-171

Shankar, S.S.; Absar, A. & Murali, S. (2003). Geranium leaf assisted biosynthesis of silver nanoparticles. *Biotechnol Prog,* Vol.19, pp.1627-1631

Shankar, S.S.; Rai, A.; Ankamwar, B.; Singh, A.; Ahmad, A. & Sastry, M. (2004). Biological synthesis of triangular gold nanoprisms. *Nature Materials,* Vol.3, pp.482-488

Shanmukh, S.; Jones, L.; Driskell, J.; Zhao, Y.; Dluhy, R. & Tripp, R.A. (2006). Rapid and sensitive detection of respiratory virus molecular signatures using a silver nanorod array SERS substrate. *Nano Letters,* Vol.6, pp.2630-2636

Sharma, V.K.; Yngard, R.A. & Lin, Y. (2009). Silver nanoparticles: green synthesis and their antimicrobial activities. *Advances in Colloid and Interface Science,* Vol.145, pp.83-96

Shchukin, D.G.; Radtchenko, I.L. & Sukhorukov, G. (2003). Photoinduced reduction of silver inside microscale polyelectrolyte capsules. *Chem Phys Chem,* Vol.4, pp.1101–1103

Shin, S.H.; Ye, M.K.; Kim, H.S. & Kang, H.S. (2007). The effects of nanosilver on the proliferation and cytokine expression by peripheral blood mononuclear cells. *Int Immuno pharmacol,* Vol.7, pp.1813-1818

Si, S. & Mandal, T.K. (2007). Tryptophan-based peptides to synthesize gold and silver nanoparticles: a mechanistic and kinetic study. *Chem A Eur J,* Vol.13, pp.3160-3168

Šileikaitė, A.; Puišo, J.; Prosyčevas, I.; Guobienė, A.; Tamulevičius, S.; Tamulevičius, T. & Janušas, G. (2007). Polymer diffraction gratings modified with silver nanoparticles. *Materials Science (Medžiagotyra),* Vol.13, pp.273–277

Singhal, G.; Bhavesh, R.; Kasariya, K.; Sharma, A.R. & Singh, R.P. (2011). Biosynthesis of silver nanoparticles using *Ocimum sanctum* (Tulsi) leaf extract and screening its antimicrobial activity. *J Nanopart Res,* Vol.13, pp.2981-2988

Slawson, R.M.; Van, D.M.; Lee, H. & Trevor, J. (1992). Germanium and silver resistance, accumulation and toxicity in microorganisms. *Plasmid,* Vol.27, pp.73-79

Socol, Y.; Abramson, O.; Gedanken, A.; Meshorer, Y.; Berenstein, L. & Zaban, A. (2002). Suspensive electrode formation in pulsed sonoelectrochemical synthesis of silver nanoparticles. *Langmuir,* Vol.18, pp.4736–4740

Song, J.Y. & Kim, B. (2008). Rapid biological synthesis of silver nanoparticles using plant leaf extracts. *Bioprocess Biosyst Eng,* Vol.32, pp.79-84

Soroushian, B.; Lampre, I.; Belloni, J. & Mostafavi, M. (2005). Radiolysis of silver ion solutions in ethylene glycol: solvated electron and radical scavenging yields. *Radiat Phys Chem,* Vol.72, pp.111-118

Soto, K.; Garza, K.M. & Murr, L.E. (2007). Cytotoxic effects of aggregated nanomaterials. *Acta Biomater,* Vol.3, pp.351-358

Soto, K.F.; Carrasco, A.; Powell, T.G.; Garza, K.M. & Murr, L.E. (2005). Comparative *in vitro* cytotoxicity assessment of some manufactured nanoparticulate materials characterized by transmission electron microscopy. *J Nanopart Res*, Vol.7, pp.145-169

Soukupova, J.; Kvitek, L.; Panacek, A.; Nevecna, T. & Zboril, R. (2008). Comprehensive study on surfactant role on silver nanoparticles (NPs) prepared via modified Tollens process. *Mater Chem Phys*, Vol.111, pp.77-81

Sreeram, K.J.; M N. & Nair, B.U. (2008). Microwave assisted template synthesis of silver nanoparticles. *Bull Mater Sci*, Vol.31, pp.937-942

Su, H.L.; Chou, C.C.; Hung, D.J.; Lin, S.H.; Pao, I.C.; Lin, J.H.; Huang, F.L.; Dong, R.X. & Lin, J.J. (2009). The disruption of bacterial membrane integrity through ROS generation induced by nanohybrids of silver and clay. *Biomaterials*, Vol.30, pp.5979-5987

Sudeep, P.K. & Kamat, P.V. (2005). Photosensitized growth of silver nanoparticles under visible light irradiation: a mechanistic investigation. *Chem Mater*, Vol.17, pp.5404-5410

Sun, R.W.; Chen, R.; Chung, N.P.; Ho, C.M.; Lin, C.L. & Che, C.M. (2005). Silver nanoparticles fabricated in Hepes buffer exhibit cytoprotective activities toward HIV-1 infected cells. *Chem Commun*, Vol.40, pp.5059-5061

Sung, J.H.; Ji, J.H.; Yoon, J.U.; Kim, D.S.; Song, M.Y.; Jeong, J.; Han, B.S.; Han, J.H.; Chung, Y.H.; Kim, J.; Kim, T.S.; Chang, H.K.; Lee, E.J.; Lee, J,H. & Yu, I.J. (2008). Lung function changes in Sprague-Dawley rats after prolonged inhalation exposure to silver nanoparticles. *Inhal Toxicol*, Vol.20, pp.567-574

Sylvestre, J.P.; Kabashin, A.V.; Sacher, E.; Meunier, M. & Luong, J.H.T. (2004). Stabilization and size control of gold nanoparticles during laser ablation in aqueous cyclodextrins. *J Am Chem Soc*, Vol.126, pp.7176-7177

Tai, C.; Wang, Y-H. & Liu, H.S. (2008). A green process for preparing silver nanoparticles using spinning disk reactor. *AIChE J*, Vol.54, pp.445-452

Takenaka, S.; Karg, E.; Roth, C.; Schulz, H.; Ziesenis, A.; Heinzmann, U.; Schramel, P. & Heyder, J. (2001). Pulmonary and systemic distribution of inhaled ultrafine silver particles in rats. *Environ Health Perspect*, Vol.109, pp.547-551

Tan, M.; Wang, G.; Ye, Z. & Yuan, J. (2006). Synthesis and characterization of titania-based monodisperse fluorescent europium nanoparticles for biolabeling, *Journal of Luminescence*, Vol.117, pp.20-28

Tarasenko, N.; Butsen, A.; Nevar, E. & Savastenko, N. (2006). Synthesis of nanosized particles during laser ablation of gold in water. *Appl Surf Sci*, Vol.252, pp.4439-4444

Trop, M.; Novak, M.; Rodl, S.; Hellbom, B.; Kroell, W. & Goessler, W. (2006). Silver coated dressing acticoat caused raised liver enzymes and argyria-like symptoms in burn patient. *J Trauma Injury Infect Crit Care*, Vol.60, pp.648-652

Troupis, A.; Hiskia, A. & Papaconstantinou, E. (2002). Synthesis of metal nanoparticles by using polyoxometalates as photocatalysts and stabilizers. *Angew Chem Int Ed*, Vol.41, pp.1911-1914

Tsuji, M.; Matsumoto, K.; Jiang, P.; Matsuo, R.; Hikino, S.; Tang, X-L. & Nor Kamarudin, K.S. (2008). The Role of Adsorption Species in the Formation of Ag Nanostructures by a Microwave-Polyol Route. *Bull Chem Soc Jpn*, Vol.81, pp.393-400

Tsuji, T.; Iryo, K.; Watanabe, N. & Tsuji, M. (2002). Preparation of silver nanoparticles by laser ablation in solution: influence of laser wavelength on particle size. *Appl Surf Sci*, Vol.202, pp.80-85

Tsuji, T.; Kakita, T. & Tsuji, M. (2003). Preparation of nano-Size particle of silver with femtosecond laser ablation in water. *Applied Surface Science*, Vol.206, pp.314–320

Veerasamy, R.; Xin, T.Z.; Gunasagaran, S.; Xiang, T.F.W.; Yang, E.F.C.; Jeyakumar, N. & Dhanaraj. S.A. (2011). Biosynthesis of silver nanoparticles using mangosteen leaf extract and evaluation of their antimicrobial activities. *Journal of Saudi Chemical Society*, Vol.15, pp.113-120

Vigneshwaran, N.; Ashtaputre, N.M.; Varadarajan, P.V.; Nachane, R.P.; Paralikar, K.M. & Balasubramanya, R. (2007). Biological synthesis of silver nanoparticles using the fungus *Aspergillus flavus*. *Materials Letters*, Vol.61, pp.1413-1418

Vigneshwaran, N.; Kathe, A.A.; Varadarajan, P.V.; Nachane, R.P. & Balasubramanya, R. (2006a). Biomimetics of silver nanoparticles by white rot fungus, *Phaenerochaete chrysosporium*. *Colloids and Surfaces B: Biointerfaces*, Vol.53, pp.55-59

Vigneshwaran, N.; Nachane, R.P.; Balasubramanya, R.H. & Varadarajan, P.V. (2006b). A novel one-pot 'green' synthesis of stable silver nanoparticles using soluble starch. *Carbohydr Res*, Vol.341, pp.2012-2018

Vilchis-Nestor, A.R.; Sánchez-Mendieta, V.; Camacho-López, M.A.; Gómez-Espinosa, R.M.; Camacho-López, M.A. & Arenas-Alatorre, J. (2008). Solventless synthesis and optical properties of Au and Ag nanoparticles using *Camellia sinensis* extract. *Materials Letters*, Vol.62, pp.3103–3105

Wang, Y. & Herron, N. (1991). Nanometer-sized semiconductor clusters: materials synthesis, quantum size effects, and photophysical properties. *J Phys Chem*, Vol.95, pp.525-532

Wei, H.; Li, J.; Wang, Y. & Wang, E. (2007). Silver nanoparticles coated with adenine: preparation, self-assembly and application in surface-enhanced Raman scattering. *Nanotechnology*, Vol.18, pp.175610

Wiley, B.; Sun, Y.; Mayers, B. & Xi, Y. (2005). Shape-controlled synthesis of metal nanostructures: the case of silver. *Chem Eur J*, Vol.11, pp.454-463

Wu, Q.; Cao, H.; Luan, Q.; Zhang, J.; Wang, Z.; Warner, J-H. & Watt, A.A.R. (2008). Biomolecule-assisted synthesis of water-soluble silver nanoparticles and their biomedical applications. *Inorg Chem*, Vol.47, pp.5882–5888

Yang, L.; Shen, Y.; Xie, A. & Zhang, B. (2007). Facile size-controlled synthesis of silver nanoparticles in UV-irradiated tungstosilicate acid solution. *J Phys Chem C*, Vol.111, pp.5300–5308

Yin, H.; Yamamoto, T.; Wada, Y. & Yanagida, S. (2004). Large-scale and size-controlled synthesis of silver nanoparticles under microwave irradiation. *Materials Chemistry and Physics*, Vol.83, pp.66–70

Yin, Y.; Li, Z-Y.; Zhong, Z.; Gates, B. & Venkateswaran, S. (2002). Synthesis and characterization of stable aqueous dispersions of silver nanoparticles through the Tollens process. *J Mater Chem*, Vol.12, pp.522-527

Yoon, K.Y.; Byeon, J.H.; Park, J.H. & Hwang, J. (2007). Susceptibility constrants of *Escherichia coli* and *Bacillus subtilis* to silver and copper nanoparticles. *Science of the Total Environment*, Vol.373, pp.572-575

Yu, H.; Xu, X.; Chen, X.; Lu, T.; Zhang, P. & Jiang, X. (2007). Preparation and antibacterial effects of PVA-PVP hydrogels containing silver nanoparticles. *J Appl Polym Sci*, Vol.103, pp.125-133

Zhang, G.; Keita, B.; Dolbecq, A.; Mialane, P.; Secheresse, F.; F M. & Nadjo, L. (2007). Green chemistry-type one-step synthesis of silver nanostructures based on Mo^V–Mo^{VI} mixed-valence polyoxometalates. *Chem Mater,* Vol.19, pp.5821-5823

Zhang, L.; Yu, J.C.; Yip, H.Y.; Li, Q.; Kwong, K.W.; A-W X. & Wong, P.K. (2003). Ambient light reduction strategy to synthesize silver nanoparticles and silver-coated TiO$_2$ with enhanced photocatalytic and bactericidal activities. *Langmuir,* Vol.19, pp.10372–10380

Zhang, W.; Qiao, X. & Chen, J. (2007). Synthesis of nanosilver colloidal particles in water/oil microemulsion. *Colloids Surf A: Physicochem Eng Aspects,* Vol.299, pp.22-28

Zhang, W. & Wang, G. (2003). Research and development for antibacterial materials of silver nanoparticle. *New Chem Mater,* Vol.31, pp.42-44

Zhang, Y.; Chen, F.; Zhuang, J.; Tang, Y.; Wang, D.; Wang, Y.; Dong, A. & Ren, N. (2002). Synthesis of silver nanoparticles via electrochemical reduction on compact zeolite film modified electrodes. *Chemical Communications,* Vol.24, pp.2814–2815

Zhang, Y.; Peng, H.; Huang, W.; Zhou, Y. & Yan, D. (2008). Facile preparation and characterization of highly antimicrobial colloid Ag or Au nanoparticles. *J Colloid Interface Sci,* Vol.325, pp.371-376

Zhang, Z. & Han, M. (2003). One-step preparation of size-selected and well-dispersed silver nanocrystals in polyacrylonitrile by simultaneous reduction and polymerization. *J Mater Commun,* Vol.13, pp.641-643

Zharov, V.P.; Kim, J-W.; Curiel, D.T. & Everts, M. (2005). Self-assembling nanoclusters in living systems: application for integrated photothermal nanodiagnostics and nanotherapy. *Nanomedicine: Nanotechnology, Biology and Medicine,* Vol.1, pp.326-345

Zheng, J.; Hua, Y.; Xinjun, L. & Shanqing, Z. (2008). Enhanced photocatalytic activity of TiO$_2$ nano-structured thin film with a silver hierarchical configuration. *Appl Surf Sci,* Vol.254, pp.1630-1635

Zhong, L.; Hu, J.; Cui, Z.; Wan, L. & Song, W. (2007). *In-Situ* loading of noble metal nanoparticles on hydroxyl-group-rich titania precursor and their catalytic applications. *Chem Mater,* Vol.19, pp.4557–4562

Zhou, Y.; Yu, S.H.; Wang, C.Y.; Li, X.G.; Zhu, Y.R. & Chen, Z.Y. (1999). A novel ultraviolet irradiation photoreduction technique for the preparation of single-crystal Ag nanorods and Ag dendrites. *Advanced Materials,* Vol.11, pp.850–852

Zhu, S.; Du, C.L. & Fu, Y. (2009). Fabrication and characterization of rhombic silver nanoparticles for biosensing. *Optical Materials,* Vol.31, pp.769-774

3

Cytotoxicity of Tamoxifen-Loaded Solid Lipid Nanoparticles

Roghayeh Abbasalipourkabir, Aref Salehzadeh and Rasedee Abdullah
¹Hamedan University of Medical Science
²Universiti Putra Malaysia
¹Iran
²Malaysia

1. Introduction

Breast cancer is one the most important health concerns of the modern society (Ferlay *et al.*, 2007). Worldwide, it is estimated that over one million new cases of breast cancer are diagnosed every year, and more than 400 thousands will die from the breast cancer (Coughlin & Ekwueme, 2009) The life-time risk in women contracting breast cancers is estimated to be 1 in 8, which is the highest among all forms of cancers. (DevCan, 2004). Although the mortality rates from breast cancers have decreased in most developed countries because more frequent mammographic screening and extensive use of tamoxifen, it still remains the second highest in women (Clark, 2008). Breast cancer incidence rates were reported to have doubled or tripled in developing countries in the past 40 years (Anderson *et al.*, 2008; Porter, 2008). The main options for breast cancer treatment include surgery, radiation therapy and chemotherapy (Mirshahidi & Abraham, 2004). Surgical procedures usually lead to significant morbidity such as lymph edema, muscle wasting, neuropathy and chronic pain (Paci *et al.*, 1996). Radiation therapy is useful for cancer which is more localized, but it also carry a number of acute and chronic side- effects such as nausea, diarrhea, pain and fatigue (Ewesuedo & Ratain, 2003). Endocrine therapy may be used as a supplementary treatment. This method of therapy is applied to specific group of patients, e.g. women after menopause with hormone-responsive disease (Gradishar, 2005). In hormone-sensitive cancer patients receive chemotherapy with cytotoxic drugs. The cytotoxic drugs treat cancers by causing cell death or growth arrest. Effective cancer chemotherapy is able either to shrink a tumor or to help destroy cancer cells (Ewesuedo & Ratain, 2003). A number of obstacles such as drug toxicity, possible undesirable drug interactions and various forms of drug resistance have to be overcome to achieve effective chemotherapy (Cardosa *et al.*, 2009). Drug resistance is a general problem in the chemotherapy of several cancers including breast cancers (Wong *et al.*, 2006). Failures in treatment of cancers are common. Development of new drugs is also slow to progress. Among the reasons contributing to this are weak absorption, high rate of metabolism and elimination of drugs per oral administration resulting in less or variable concentrations in blood, poor drug solubility, unpredictable bioavailability of oral drugs due to food, and tissue toxicity (Sipos *et al.*, 1997). Thus alternative methods of drug administration like appropriate drug carrier system is needed to overcome this problem. Depending on the route of administration, the size of drug carriers may range from a few nanometers (colloidal carriers), to micrometers (microparticles) and to several millimeters (implants). Among these carriers, nanoparticles had shown great promise for parenteral application of chemotherapeutic drugs (Mehnert *et al.*, 2001). Targeting of unhealthy tissues and organs of the body is one of the important challenges of the drug delivery systems. Nanoparticles seem to show promise as a drug targeting systems supplying drug to target tissues at the right time (Kayser *et al.*, 2005). The main objective of new drug delivery systems is to improve the anti-tumor efficacy of drug and reduce their toxic effects on normal tissues. Nanoparticle is expected to be able to diminish toxicity of chemotherapy drug. Nanoparticles based on lipids that are solid at room temperature, namely solid lipid nanoparticle (SLN) using physiological well-tolerable lipids have potentially wide application (Siekmann & Westesen, 1992; Müller *et al.*, 1995; Müller *et al.*, 2000). The SLN is a drug delivery system that loads lipophilic or chemically unstable drugs (Fig. 1). Among the advantages of SLN are high potential for management of drug release and drug targeting, high stability for drug loading and high capacity for drug payload. This delivery system makes possible the encapsulation of lipophilic and hydrophilic drugs without the toxic effect of the carriers. This system also avoids the use

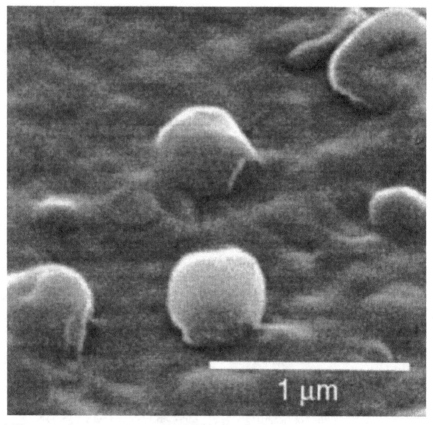

Fig. 1. Electron microscopy picture of solid lipid nanoparticles made from Compritol® stabilized with poloxamer 188, diameter 400nm (Adapted from Müller and Mäder 2000)

organic solvents, and has potential for large scale production. However, several disadvantages are associated with SLNs to include particle growth, particle aggregation, unpredictable gelation tendency, polymorphic transitions, burst drug release and inherently low incorporation capacities due to the crystalline structure of the solid lipid (Mehnert & Mäder, 2001).

The most common production technique of SLNs are high-pressure homogenization (HPH), high-shear homogenization combined with ultrasound, solvent emulsification/evaporation and microemulsion techniques. HPH is the predominant production method because it is easy to handle and scale-up. In this method, drug incorporation is achieved by dissolving or dispersing the drug in the melted lipid (He et al., 2007). The drug can be encapsulated in the matrix or attached to the particle surface. In spite of improved researches in production of high quality SLN, it is still not routinely used clinically.

2. Tamoxifen-encapsulated solid lipid nanoparticles

The chemical name of tamoxifen is trans-2-[4-(1,2-diphenyl-1-butenyl)phenoxyl] N,Ndimethylethylamine (Fig. 2). Tamoxifen, an antiestrogen molecule and strong hydrophobic drug (water solubility, 0.04 µg/mL at 37°C), is widely administered in breast cancer and high risk patients (McGregor & Jorda, 1998). Although tamoxifen was primarily used as a drug against hormone-dependent breast cancers (Wyld et al., 1998), it has also been used in the treatment of hormone-insensitive estrogen receptor-negative breast cancers (Jordan, 1994). Tamoxifen inhibits cell proliferation and induces apoptosis in breast cancer cells (MCF-7, MDA-MB231 and BT-20) (Mandlekar & Kong, 2001; Mandlekar et al., 2000). In spite of being high effective, tamoxifen has harmful dose-dependent long-term side-effects such as development of endometrial cancer (Brigger et al., 2001), hyperplasia, polyps, carcinoma, sarcoma (Peters-Engl et al., 1999; Cohen, 2004) vaginal hemorrhage, blazes and liquid retention in postmenopausal breast cancer patients (Mourits et al., 2001; Delima et al., 2003). Formulations with the encapsulation of low-dose tamoxifen in colloidal delivery systems have been effective. Tamoxifen has been formulated in nanospheres such as poly-ε-caprolactone nanoparticles (Chawla & Amiji, 2003) and long circulating Poly(MePEGcyanoacrylate-co-hexadecylcyanoacrylate) nanoparticles in the form of free base (Brigger, et al., 2001). Tamoxifen, as a nonsteroidal antiestrogen drug was recently encapsulated in SLNs and was shown to be effective on induced mammary tumor gland in Sprague-Dawley rats (Abbasalipourkabir et al., 2010,1) in parenteral administration. The SLN

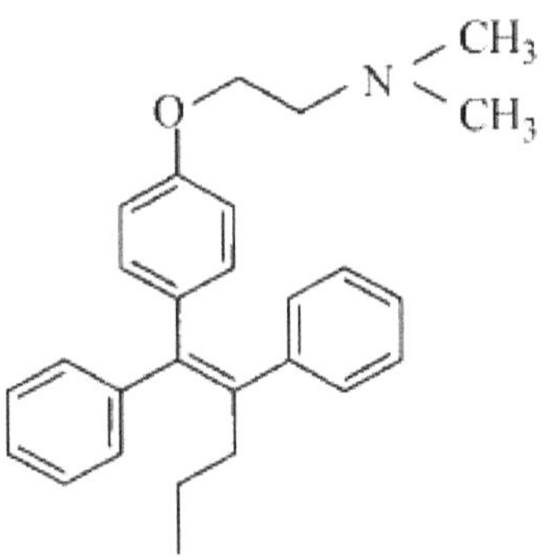

Fig. 2. Chemical Structure of Tamoxifen (Adapted from Christov et al., 2007)

systems offer a sustained release of the drug in its intact form (Fontana *et al.*, 2005 Using human breast cancer cell line, MCF7, some *in vitro* studies have shown that drug release from the tamoxifen-incorporated SLN has the same antitumoral activity as the free drug (Abbasalipourkabir *et al.*, 2011). Therefore the tamoxifen-loaded SLN as a carrier system has excellent potential in prolonged drug release in breast cancer therapy (Fundaro *et al.*, 2000).

2.1 Preparation of tamoxifen-loaded SLN

Drug-loaded SLNs can be prepared using the high-pressure homogenization technique (Abbasalipourkabir *et al.*, 2011). A mixture of Hydrogenated palm oil (Softisan 154 or S154) and Hydrogenated soybean lecithin (Lipoid S100-3, containing 90% phosphatidylcholine, including 12–16% palmitic acid, 83–88% stearic acid, oleic acid and isomers, and linoleic acid] at a ratio of 70:30 is grounded in a ceramic crucible. The mixture is heated to 65–70°C while being stirred with a PTFE-coated magnet until a clear-yellowish lipid matrix (LM) solution is obtained. A solution containing 1 mL oleyl alcohol, 0.005 g thimerosal, 4.75 g Sorbitol, and 89.25 mL bidistilled water (all w/w) at the same temperature is added to 5 g of LM. A pre-emulsion of SLN is obtained using the homogenizer (Ultra Turrax, Ika) at 13,000 rpm for 10 min and high-pressure homogenizer (EmulsiFlex-C50 CSA10, Avestin) at 1000 bar, 20 cycles, and 60 °C. The lipophilic drug tamoxifen (1 mg) is dissolved in oleyl alcohol and mixed with 5 mg of SLN pre-emulsion using the Ultra Turrax homogenizer at 13,000 rpm for 10 min. This mixture is then incubated overnight at 50–60 °C, stirred periodically with a PTFE-coated magnet at 500 rpm, and finally will expose to air to solidify.

2.2 *In Vitro* antitumoral activity of Tamoxifen-loaded solid lipid nanoparticle

Cell death basically can occur in two ways. The first is through the necrosis pathway, where traumatic injuries cause cells damage in particular cell enlarges, bursts and liberate its intracellular components into the surrounding environment. The second pathway is programmed cell death or apoptosis, which is a molecular signaling cascade, inducing a disturbance in the organization and package of the cell causing death (Fadok, 1999; Messmer & Pfeilschifter, 2000). Other mode of cell death has also been suggested, for example mitotic cell death, which plays an important role in cell death caused by ionizing radiation (Steel, 2001). Breast cancer is the most common malignancy (18% of all malignancies) in women worldwide and its occurrence is slowly increasing (Salami & Karami-Tehrani, 2003). Like many cancers, breast cancer appears to be a result of high genetic damage that caused uncontrolled cellular proliferation and unusual apoptosis. These phenomena activate proto-oncogenes and inactivate tumor suppressor genes. These events can be activated by exposure of living cells to environmental, physical, chemical and/or biological carcinogens (Russo & Russo, 2002). The antiestrogen molecule, Tamoxifen (TAM) or trans-2-[4-(1,2-diphenyl-1-butenyl) phenoxyl]-N, N-dimethylethylamine, has been widely applied in treatment of breast cancer and high risk patients. Tamoxifen can reduce the occurrence of contralateral breast cancers by at least 40% (Fontana *et al.*, 2005). Tamoxifen exhibits anti-estrogenic activity by binding to the intracellular estrogen receptor. The tamoxifen-estrogen receptor complex binds with DNA and can subsequent inhibit mRNA transcription and lead to cellular apoptosis (Chawla & Amiji, 2003). Recently nanoparticulate delivery systems in the form of nanospheres like poly- caprolactone nanoparticles and long–circulating PEG-coated poly (MePEGcyanoacrylate-co-hexadecylcyanoacrylate) nanoparticles in the form of free base have been used for tamoxifen encapsulation. The basis of this formulation is to obtain the necessary dose of drug at tumor location for a known period of time and reducing adverse effects on normal organs in the body (Chawla & Amiji, 2003). In recent years Delivering Tamoxifen within Solid Lipid Nanoparticles have been recommended. Animal models play an important role in cancer chemotherapy (Abbasalipourkabir *et al.*, 2010,2). However, today there is increasing acceptation for *in vitro* tests as the method for determining cytotoxicity and viability of chemotherapeutic drugs. The reason for change lies partly in the limitations of animal models, to include financial considerations, time, and differences between animal and human metabolism. Finally, there is the moral pressure to reduce animal experimentation. *In vitro* tests are more likely to be reproducible. In general, the procedure involves the exposure of cells to a range of concentrations of the chemicals under test for a defined time and then to test for cell viability. Such tests are most easily performed in microtitre plates, which allow rapid quantitation of the results using a micrometer plate reader (Adams, 1990). The responses of breast cancer cell lines are determined by cytotoxicity assay, cellular and nuclear morphology, apoptosis and cell cycle distribution.

2.3 Cytotoxicity effect of TAM-loaded SLN on human breast cancer cells

The TAM-loaded SLN has an equally efficient cytotoxic activity as free tamoxifen. Therefore TAM-loaded SLN preserves the antitumoral activity of the free drug. When TAM is incorporated into the SLN carrier system, its antitumoral activity is still maintained and formulating TAM by incorporating into SLN will potentially enhance the solubility of the drug through inclusion into the lipid phase and facilitating the entrapment of greater amounts of the drug in the SLN, suggesting that SLN is a good carrier for the drug (Fig. 3. & Fig. 4).

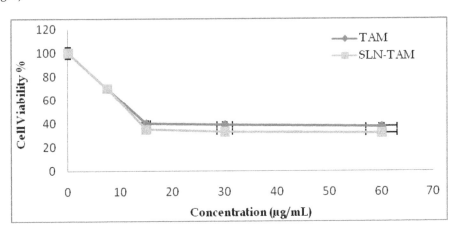

Fig. 3. MCF-7 cells viability after 72 hours incubation with TAM and TAM-Loaded SLN formulation. The percentage of cell viability is expressed as a ratio of treated cells to the untreated control cells. Each point represents the mean ± standard deviation of 5 wells. (Adapted from Abbasalipourkabir, 2010).

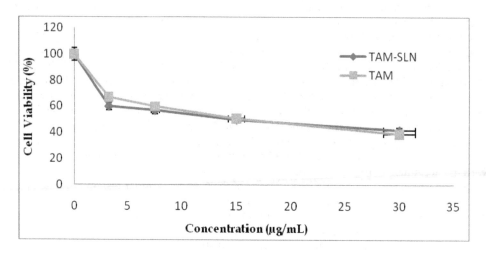

Fig. 4. MDA-MB231 cells viability after 72 hours incubation with TAM and TAM-Loaded SLN formulation. The percentage of cell viability is expressed as a ratio of treated cells to the untreated control cells. Each point represents the mean ± standard deviation of 5 wells. (Adapted from Abbasalipourkabir, 2010).

The IC_{50} of TAM and TAM-loaded SLN for MDA-MB231 cells (ER-negative or ER-independent) is higher than for MCF-7 cells (ER-positive or ER-dependent), (Tables 1-2).

Treatment	IC_{50} (µg/mL)		
	24h	48h	72h
TAM	13.45±0.46	13.00±0.98	12.50±0.91
TAM-SLN	13.18[a]±0.66	12.50[a]±1.50	11.78[b]±0.18

All value represent the means ± std. dev., (n=5) [A,b]means in each row with different superscripts are significantly different TAM = Tamoxifen; TAM-SLN = Tamoxifen-loaded solid lipid nanoparticle (Adapted from Abbasalipourkabir, 2010).

Table 1. The IC_{50} of TAM and TAM-loaded SLN on MCF-7 cells after 24, 48 and 72h.

Treatment	IC_{50} (µg/mL)		
	24h	48h	72h
TAM	17.21±1.44	16.87±1.97	15.97±0.86
TAM-SLN	16.93±0.82	16.00±0.10	15.80±0.69

All value represent the means ± std. dev., (n=5) TAM = Tamoxifen; TAM-SLN = Tamoxifen-loaded solid lipid nanoparticle (Adapted from Abbasalipourkabir, 2010).

Table 2. The IC_{50} of TAM and TAM-loaded SLN on MDA-MB231 cells after 24, 48 and 72h.

The mechanisms of ER-independent, TAM-induced apoptosis may be through the inhibition of protein kinase C. The IC_{50} value of tamoxifen for protein kinase C inhibition is 4 to 10 times the concentration for ER inhibition in ER-positive cells. Therefore, the dose of tamoxifen for treatment of patients with ER-positive breast cancer would have to be increased over the usual 20 mg per day used. High dose of tamoxifen might decrease the therapeutic index by increasing toxicity (Gelmann, 1996). It seems that improved cytotoxicity of incorporated drug is not dependent of the composition on the SLN. In fact it was reported that the IC_{50} value of drug-loaded SLN composed of different materials were lower than that of free drug solution (Yuan et al, 2008). There are at least two mechanisms that have been associated with the cytotoxicity of drug-loaded SLN. Using Doxorubicin (DOX)-loaded SLN, it was suggested that the first mechanism involves the release of DOX from DOX-SLN outside the cells, and the cytotoxicity of DOX is increased by the nanoparticles. The second mechanism suggested was, release of the drug inside the cell and thus produces greater cytotoxicity (Wong et al., 2006).

2.4 Morphological changes of TAM-loaded SLN on human breast cancer cells

Apoptotic cell death can be recognized under phase contrast and fluorescence inverted microscope after staining. This is the most practical method to identify cell morphological changes attributed to apoptotic cell death.

2.4.1 Phase contrast microscopy

TAM-loaded SLN treatments at concentrations equal to IC_{50}, causes detachment of MCF-7 and MDA-MB231 cells and loss of colony formation ability. These cells appear rounded-up and lose contact with neighboring cells (Fig. 5). The normal untreated MCF-7 and MDA-MB-231 cells, however, appear healthy and exhibiting epithelial–like features and forming a monolayer on the surface of the culture flask. In the presence of TAM and TAM-loaded SLN, the viability of the both cells diminishes and the cancer cells loss their normal morphological characteristics, detaches, aggregates, and later develops apoptotic bodies. The detachment of cells in the presence of free TAM and TAM-loaded SLN suggests that tamoxifen is cytotoxic, even when incorporated in the SLN.

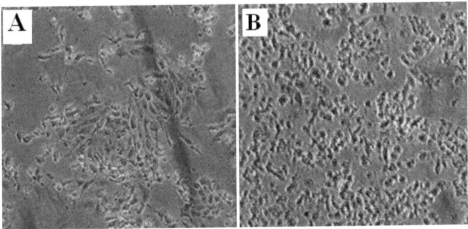

Fig. 5. Phase contrast micrographs of MCF-7 cell (A) and MDA-MB231 cell (B) treated with TAM loaded SLN (12 µg/mL) (magnification x10). (Adapted from Abbasalipourkabir, 2010).

2.4.2 Nuclear morphology

Cell death is either by physiological or pathological means. Physiological cell death is distinguished by apoptotic morphology, including chromatin condensation, membrane blebbing, internucleosomal degradation of DNA, and apoptotic body formation. Pathological cell death or necrosis is associated with cellular swelling and collapse, without severe damage to nuclei or breakdown of the DNA. In apoptosis, several cellular and molecular biological features, including cell shrinkage and DNA fragmentation are exhibited (Yu et al., 2010). To characterize the cell death, the nuclear morphology of dying cells can be examined under Hoechst dye 33258 staining. The Hoechst dye 33258 is a bis-benzimide derivatives and a fluorescent DNA-binding agent. This dye is useful for cell cycle analysis because it can be used in low concentrations, and thus minimizing the problem of toxicity. According to Latt & Stetten, (1976) the Hoechst dye binds to AT-rich regions of the DNA and when excited with an ultraviolet light produces bright fluorescence at 465 nm. TAM-loaded SLN induce death of MCF-7 and MDA-MB231 cells by apoptosis. This is evident by the typical apoptotic changes showing clear condensation of cell nuclei, nuclear fragmentation and apoptotic bodies (Fig. 6).

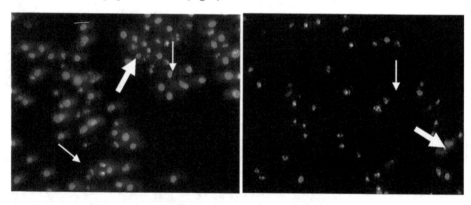

Fig. 6. Fluorescence microscopy of MCF-7 cells (A) and MDA-MB-231 cells (B) treated with TAM-loaded SLN. Cell shrinkage (thin arrow) and apoptotic cells (thick arrow) are evident (magnification x40). (Adapted from Abbasalipourkabir, 2010).

Tamoxifen-loaded SLN like free TAM display antitumoral activity against human breast cancer cells. The biological availability of drug is not affected when incorporated into SLN. Therefore SLN could be applied as a drug delivery system for cancer treatments. In conclusion, the TAM-loaded SLN, because of its small size, could not be easily phagocytosed by macrophages and therefore the nanoparticles could be potentially used in long-term circulating carrier system for breast cancer therapy.

3. Conclusion

The main challenge in cancer chemotherapy is toxic side-effects induced by chemotherapeutic drugs. Single dose or short-time application (1-2 weeks) will probably causes serious health problems, but the use of biodegradable nano-sized particles for long-term or life-time therapy may produce other serious side-effects. Increasing the encapsulation efficiency of poorly water-soluble molecules will lead to the development of improved SLN formulations. In the near future, it is expected more studies will focus on improving SLN and drug-loaded SLN formulations to increase the efficacy and reduce the side-effects of chemotherapeutic drugs for anticancer treatment. These studies should include preparation of formulations with different particle size and distributions, different matrix lipids and additional ingredients. Thus, if nanoparticulate drug delivery systems to be used effectively and routinely, the matter of toxicity of the components of nanoparticles must be addressed. Indeed, SLN requires further development before it can be used as a new drug delivery system for chemotherapy drugs in treatment of human cancers.

4. References

Abbasalipourkabir, R. Dehghan, A. Salehzadeh, A. Shams Abadi, F. and Rasedee Abdullah (2010,2) "Induction of mammary gland tumor in female Sprague-Dawley rats with LA-7 cells." *African Journal of Biotechnology,* 9(28):4491-4498

Abbasalipourkabir, R. 2010. Development, characterization, cytotoxicity and antitumor effect of tamoxifen-loaded solid lipid nanoparticles. Ph.D. thesis. Universiti Putra Malaysia. Malaysia

Abbasalipourkabir, R. Salehzadeh, A. and Rasedee Abdullah (2010,1). "Antitumor activity of tamoxifen loaded solid lipid nanoparticles on induced mammary tumour gland in Sprague-Dawley rats." *African Journal of Biotechnology,* 9(43):7337-7345

Abbasalipourkabir, R. Salehzade, A. and Rasedee A. 2011. *Pharmaceutical technology* 35(4): 74-79

Adams, R.L.P. 1990. Cell structure for biochemists. 2nd Edition. Elsevier, Amsterdam, pp. 1-22

Anderson B.O. Yip C.H. Smith R.A. Shyyan R. Sener S.F. Eniu A. Carlson R.W. Azavedo, E. and Harford, J. 2008. Guideline implementation for breast health care in low-income and middle-income countries: overview of the breast health global initiative global summit 2007. *Cancer* 113:2221-43.

Brigger, I. Chaminade, P. Marsaud, V. Appel, M. Besnard, M. Gurny, R. Renoir, M. and Couvreur, P. 2001. Tamoxifen encapsulated within polyethylene glycol-coated nanospheres. A new antiestrogen formulation. *International Journal of Pharmaceutics* 214:37-42.

Cardosa, F. Bedard, P.L. Winer, E.P. Pagani, O. Senkus-Konefka, E. Fallowfield, L.J. Kyriakides,S. Costa, A. Cufer, T. and Albain, K.S. 2009. International Guidelines for Management of Metastatic Breast Cancer: Combination vs Sequential Single-Agent Chemotherapy. *Journal of National Cancer Institute* 101: 1174-1181

Clark M.J. 2008. WITHDRAWN: Tamoxifen for early breast cancer. *Cochrane Database of Systematic Reviews* Oct 8(4):CD000486.

Cohen, I. 2004. Endometrial pathologies associated with postmenopausal tamoxifen treatment. *Gynecology & Oncology.* 94:256-266.

Coughlin, S.S. and Ekwueme, D.U. 2009. Breast cancer as a global health concern. *Cancer Epidemiology* 33:315-318.

Chawla, J.S. and Amiji, M.M. 2003. Cellular uptake and concentrations of tamoxifen upon administration in poly(-caprolactone) nanoparticles. *The American Association of Pharmaceutical Scientists Journal* 5:28-34.

Christov, K. Grubbs, C.J. Shilkaitis, A. Juliana, M.M. and Lubet. R.A. 2007. Short-Term Modulation of Cell Proliferation and Apoptosis and Preventive/Therapeutic Efficacy of Various Agents in a Mammary Cancer Model. *Clinical Cancer Research* 13:5488-5496.

Delima, G.R. Facina, G. Shida, J.Y. Chein, M.B.C. Tanaka, P. Dardes, R.S. Jordan, V.C. and Gebrim, L.H. 2003. Effects of low dose tamoxifen on normal breast tissue from premenopausal women. *European Journal of Cancer* 39:891-898.

DevCan, 2004. Probability of Developing or Dying of Cancer Software, Version 5.2. Statistical Research and Applications Branch, National Cancer Institute. http://srab.cancer.gov/devcan accessed on 11 January 2010.

Ewesuedo, RB. and Ratain, MJ. 2003. Principles of cancer therapeutics. In: Vokes EE, Golomb HM (eds) Oncologic therapies. Springer, Secaucus, NJ, pp 19–66.

Fadok, VA. 1999. Clearance: the last and often forgotten stage of apoptosis. *Journal of Mammary Gland Biology and Neoplasia* 4:203–211.

Ferlay, J. Autier, P. Boniol, M. Heanue, M. Colombet, M. and Boyle, P. 2007. Estimates of the cancer incidence and mortality in Europe in 2006. *Annals of Oncology* 18:581-592.

Fontana, G. Maniscalco, L. Schillaci, D. and Cavallaro, G. 2005. Solid Lipid Nanoparticles Containing Tamoxifen. Characterization and *in vitro* Antitumoral Activity. *Drug Delivery*, 12:385–392.

Fundaro, A. Cavalli, R. Bargoni, A. Vighetto, D. Zara, G.P. and Gasco, M.R. 2000. Non-stealth and stealth solid lipid nanospheres carrying doxorubicin; pharmacokinetics and tissue distribution after I.V. administration to rats. *Pharmacology* 42:337-343.

Gelman, E.P. 1996. Tamoxifen induction of apoptosis in estrogen receptor-negative cancers: New tricks for an old dog? *Journal of the National Cancer Institute* 88:224-226.

Gradishar, W.J. 2005. Safety considerations of adjuvant therapy in early breast cancer in postmenopausal women. *Oncology* 69:1-9.

He, J. Hou, S. Lu, W. Zhu, L. and Feng, J. 2007. Preparation, pharmacokinetics and body distribution of Silymarin-loaded Solid Lipid Nanoparticles after oral administration. *Journal of Biomedical Nanotechnology* 3:195-202

Jordan, V.C. 1994. Molecular mechanisms of antiestrogen action in breast cancer. *Breast Cancer Research and Treatment* 31:41-52.

Kayser, O. lemke, A. and Hernandez-Trejo, N. 2005. The impact of nanobiotechnology on the development of new drug delivery systems. *Current Pharmaceutical biotechnology* 6:3-5.

Latt, S.A. and Stetten, G. 1976. Spectral studies on 33258 Hoechst and related bisbenzimidazole dyes useful for fluorescent detection of deoxyribonucleic acid synthesis. *Journal of Histochemistry and Cytochemistry* 24:24-33.

Mandlekar, S. and Kong, A.N. 2001. Mechanisms of tamoxifen-induced apoptosis. *Apoptosis* 6:469-477.

Mandlekar, S. Yu, R. Tan, TH. and Kong, AN. 2000. Activation of caspase-3 and c-Jun NH2-terminal Kinase-1 signaling pathways in tamoxifen-induced apoptosis of human breast cancer cells. *Cancer Research* 60:5995-6000.

McGregor, J. and Jorda, V. 1998. Basic guide to the mechanisms of antiestrogen action. *Pharmacological Reviews* 50:151-196.

Mehnert, W. and Mäder, K. 2001. Solid lipid nanoparticles Production, characterization and applications. *Advanced Drug Delivery Reviews* 47:165-196.

Messmer, U.K. and Pfeilschifter J. 2000. New insights into the mechanism for clearance of apoptotic cells. *Bio Essays* 22:878-881.

Mirshahidi, H.R. and Abraham, J. 2004. Managing early breast cancer: prognostic features guide choice of therapy. *Postgraduate Medicine* 116:23-34.

Mourits, M.J.E. De Vries, E.G.E. Willemse, P.H.B. Ten Hoor, K.A. Hollema, H. and Van Der Zee, A.G.J. 2001. Tamoxifen treatment and gynecologic side effects: a review. *Obstetrics and Gynecology* 97:855-866.

Müller, R.H. Mäder, K. and Gohla, S. 2000. Solid lipid nanoparticles (SLN) for controlled drug delivery - a review of the state of the art. *European Journal of Biopharmaceutics* 50:161-177

Müller, R.H. Mehnert, W. Lucks, J.S. Schwarz, C. Mühlen, A. Z. Weyhers, H. Freitas, C. and Rühl, D. 1995. Solid lipid nanoparticles (SLN) - an alternative colloidal carrier system for controlled drug delivery. *European Journal of Biopharmaceutics* 41:62-69.

Paci, E. Cariddi, A. Barchielli, A. Bianchi, S. Cardona, G. Distante, V. Giorgi, D. Pacini, P. Zappa, M. and Del Turco, MR. 1996. Long-term sequelae of breast cancer surgery. *Tumori* 82:321-4.

Peters-Engle, C. Frank, W. Danmayr, E. Friedl, H.P. Leodolter, S. and Medl. M. 1999. Association between endometrial cancer and tamoxifen treatment of breast cancer. *Breast Cancer Treatment* 54:255-260.

Porter P. 2008. "Westernizing" women's risks? Breast cancer in lower-income countries. *The New England Journal of Medicine* 358:213-6.

Russo, J. and Russo, IH. 2002. Mechanisms involved in carcinogenesis of the breast. In *Breast cancer*, ed. J.R. Pasqualini, Marcel Dekker. New York. pp. 1-2.

Salami, S. and Karami-Tehrani, F. 2003. biochemical studies of apoptosis induced by tamoxifen in estrogen receptor positive and negative breast cancer cell lines. *Clinical Biochemistry* 36:247-253.

Siekmann, B. and Westesen. K. 1992. Submicron-sized parenteral carrier systems based on solid lipids. *Pharmaceutical and Pharmacological Letters* 1:123-126.

Sipos, E.P. Tyler, B. Piantadosi, S. Burger, P.C. and Brem, H. 1997. Optimizing interstitial delivery of BCNU from controlled release polymers for the treatment of brain tumors, *Cancer Chemotherapy and Pharmacology* 39:383-389.

Wong, H.L. Rauth, A.M. Bendayan, R. Manias, J.L. Ramaswamy, M. Liu, Z. Erhan, S.Z. and Wu, X.Y. 2006. A New Polymer-Lipid Hybrid Nanoparticle system Increases Cytotoxicity of Doxorubicin Against Multidrug-Resistaant Human Breast Cancer Cells. *Pharmaceutical Research* 23:1574-1584.

Wyld, D.K. Chester, J.D. and Perren, T.J. 1998. Endocrine aspect of the clinical management of the breast cancer-current issue. *Endocrine-Related Cancer* 58:97-110.

Yu, T, Lee, J. Lee, Y.G. Byeon, S.E. Kim, M.H. Sohn, E,H. Lee, Y.J. Lee, S.G. and Youl, j. 2010. *In vitro* and *in vivo* anti-inflammatory effects of ethanol extract from Acer tegmentosum. *Journal of Ethnopharmacology* 128(1):139-47).

Yuan, H. Miao, J. Du, YZ. You, J. Hu, FQ. and Su Z. 2008. Cellular uptake of solid lipid nanoparticles and cytotoxicity of encapsulated paclitaxel in A549 cancer cells. *International Journal of Pharmaceutics* 348:137-145.

4

In-Situ Versus Post-Synthetic Stabilization of Metal Oxide Nanoparticles

Georg Garnweitner

Technische Universität Braunschweig
Germany

1. Introduction

Within the last two decades, the synthesis and application of nanoparticles has evolved as one of the most active fields of research and development, promising solutions to some of humanity's most pressing needs in cases as diverse as cancer therapy, water purification, or energy storage (Duget et al., 2006; Bazito & Torresi, 2006; Centi & Perathoner, 2009; Feldmann & Goesmann, 2010; Goyal et al., 2011; Kim & van der Bruggen, 2010; Li et al., 2008; Moghimi et al., 2005; Prandeep & Anshup, 2009; Sanvicens & Pilar, 2008; Sides et al., 2002; Theron et al., 2007). Today, nanoparticles are predominantly being applied as particulate materials with enhanced properties as compared to the standard bulk material. One however envisions nanotechnology already in the near future to strongly increase in complexity, leading to the development of hierarchical nanosystems being capable of computing and robotic or even self-replicating tasks (Mallouk & Sen, 2009; Rasmussen et al., 2003; Requicha, 2003; Sánchez & Pumera, 2009). Most future applications are expected to be eventually realised via the so-called bottom-up approach, emanating from nanoparticles as tiny building blocks that self-assemble in a controlled and organised manner to a higher level of hierarchy (Shenhar & Rotello, 2003; Dong et al., 2007). The primary prerequisite for the realisation of this strategy is the availability of high-quality nanoparticles that can act as such building blocks, possessing a predefined and uniform size and shape. Consequently, research on the controlled preparation of nanoparticles has greatly intensified within the last two decades, mainly focusing on carbon-based nanostructures (Mostofizadeh et al., 2011), metals (Cushing et al., 2004; Guo & Wang, 2011), metal oxides (Cushing et al., 2004; Chen & Mao, 2007; Pinna & Niederberger, 2008), semiconductor nanostructures (Trindade et al., 2001) and organic or hybrid nanostructures (Biswas & Ray, 2001; Ballauff, 2003). Remarkable progress has been made in the synthesis of nanoparticles with defined and even complex shape for certain materials (Jun et al., 2006), although a general understanding of the formation processes of nanostructured materials appears to remain a long way off.

An aspect that however is often not treated accordingly is the second step in the fabrication chain: the control of the surface interactions of the nanoparticles to determine their secondary structure. Even though by now a large number of examples have been reported on the preparation of highly ordered superstructures for diverse nanoparticle systems (Gao & Tang, 2011), the truly rational and controlled self-assembly, which would be vital for the large-scale feasibility of many of the envisioned applications, appears still to be long out of reach. Even the first level, the control of the interaction forces between the individual nanosized building blocks to avoid uncontrolled agglomeration, is neither realised nor understood in most commercial nanotechnology products today, and one is only beginning to develop general concepts aimed to describe the interactions between ultra-small particles in the low nanometer regime, even though the models of particle interaction for larger particles have been established in colloid science almost a century ago. One concept that is particularly promising for the control of nanoparticle interaction is the use of small organic ligands – in contrast to the classical models of purely electrostatic stabilisation on the one hand and polymer-based steric stabilisation on the other hand – which in principle allows long-term stabilisation with minimum stabiliser content. This Chapter is aimed to introduce the advantages and challenges of the small molecule-based stabilisation, and will present a number of select examples for the successful application of this strategy. A special focus is set on comparing the different options of introducing the stabiliser into the system: on the one hand, the so-called in-situ stabilisation strategy, where ligands are already introduced into the system prior to or during particle formation, and on the other hand the post-synthetic stabilisation, where the ligands are added in a separate step following the synthesis, in order to give the reader significant insight into this field to allow individual judgement on the benefits and disadvantages of each strategy. The Chapter will be concluded with an outlook to current and possible future developments in the field of the small-molecule stabilisation of nanoparticles.

2. The stabilisation of nanoparticles

The systematic study of the interactions between small particles began in the first decades of the 20th century within the rapidly growing field of colloids science, following the discovery that colloidal systems consist of small units below 500 nm in size dispersed in a liquid medium. In the pioneering works of Ostwald, the inherent thermodynamic instability of such colloidal systems was recognised (Ostwald, 1915), which causes a general tendency of such systems to agglomerate and aggregate, in order to minimise the particle-fluid interface area and thus the interface free energy. A number of theories were proposed in the following decades to explain the stability of colloidal systems (Everett, 1988), providing different possibilities to change and enhance the stability, which shall be briefly summarised in the following.

2.1 Colloidal theories of particle stabilisation and their application to nanoparticles

A few decades after Graham and Ostwald, first theories to explain the stability of aqueous colloidal systems in the presence of charged species were formulated, leading to the well-known DLVO theory (Derjaguin & Landau, 1941; Verwey & Overbeek, 1948). This theory is based on previous works by Hamaker on the attractive interactions of colloidal particles, which he explained as the combined van-der-Waals attractions of the individual atoms (Hamaker, 1937). These attractive forces are balanced by the electrostatic repulsion of likely charged particles, with charges not only arising from the surface potential of the particles, but also the electrochemical double layer surrounding them. Also the Born repulsion, effective only at small distances when the electron shells of atoms start to penetrate each other, is taken into account. The DLVO theory shows that the stabilisation of particles can be easily achieved by electrostatic repulsive forces, which however are highly labile and depend on conditions such as the pH and ionic strength in the dispersion medium. Moreover, this mechanism cannot be applied to hydrophobic systems that often are preferred for the handling and manipulation of nanostructured materials, e.g. due to their lower surface tension and lower adsorption tendency to the nanoparticle surface.

Another possibility is the stabilisation by coverage of the particle surface with organic polymers. This steric stabilisation concept was investigated theoretically already briefly after the DLVO theory (Mackor, 1951). The polymer chains adsorb to the particle surface to create an organic shell around the particle. As two particles approach each other, their organic shells touch and eventually interpenetrate. At that point, the solvation shell of the polymer is disrupted, as the polymer density increases upon penetration, which creates a counteracting osmotic pressure of the solvent that forces the particles apart (Napper, 1983). Good solvent compatibility of the polymer is therefore required for an effective steric stabilisation. The electrosteric stabilisation concept involves the use of polyelectrolytes, i.e. polymers with charged groups, to attach to the particle surface, thereby combining the electrostatic and steric stabilisation concepts (Hunter, 2001).

These classic colloidal theories have been proven successful for the description of colloidal systems in numerous cases and hence, also their application to nanoparticle dispersions seems obvious. The DLVO theory has thus been applied in a number of cases to describe the behaviour of nanoparticulate systems (Schwarz & Safran, 2000; Tadmor & Klein, 2001; Reindl & Peukert, 2008; Marczak et al., 2009; Segets et al., 2011). However, some special aspects need to be taken into account when attempting to describe dispersions of nanoparticles below 20 nm in size by means of DLVO theory: On the one hand, the distance dependence of the attractive and repulsive forces is strongly dependent on particle size already in the classical DLVO concept (Verwey & Overbeek, 1948). The electrostatic interaction is simplified in colloidal models as the interaction between two planar electrochemical double layers, because the interparticle distance becomes much smaller than the particle size as the particles approach each other. For small nanoparticles, this model is no longer valid, and thus spherical double layers must be included in the calculation (Mulvaney, 1998). The interaction of small nanoparticles was studied by Wiese and Healy in 1969. They reported two significant effects when reducing the particle size to the low nanometre regime: on the one hand, the repulsive forces become very small, but on the other hand also the energy minimum becomes very shallow (Wiese & Healy, 1969). The application of the DLVO concept to such small particles generally leads to interaction potentials of only a few $k \cdot T$, which results in only small changes between a stable and an unstable system, making the prediction of the stability of a system a challenging task (Marczak et al., 2009; Segets et al., 2011). Moreover, the minimum is also shifted in position towards smaller interparticle distances and thus, the description of the particles as solid, spherical systems with defined interface becomes inappropriate. A molecular layer of organics adsorbed to the particle surface that may be neglected for large particles can become important for the stability of nanoparticle systems, as it may be sufficient to prevent the particles from reaching the critical distance. Additionally, inhomogeneities within the particles, such as an oxide shell of metal nanoparticles, can suddenly play an important role.

This is nicely exemplified in the work of Reindl and Peukert, who investigated the stabilisation of silicon nanoparticles in organic solvents, taking into account an oxidised silica shell around the particles as well as the adsorption of solvent molecules, by applying a core-shell-adsorbate model based on the DLVO theory (Reindl & Peukert, 2008). Figure 1 shows the used models as well as results of the calculation of the total energy barrier. Whilst at higher concentrations, the system would be unstable for pure Si nanoparticles, the oxide shell provides high stability against agglomeration. Moreover, it is clearly visible that the adsorbate layer formed by solvent adsorption also has a noticeable effect on the energy barrier despite the relatively large particle size of 100 nm, further increasing the stability of the system. Such core-shell models have been applied also to other systems in the meantime (Marczak et al., 2009; Segets et al., 2011).

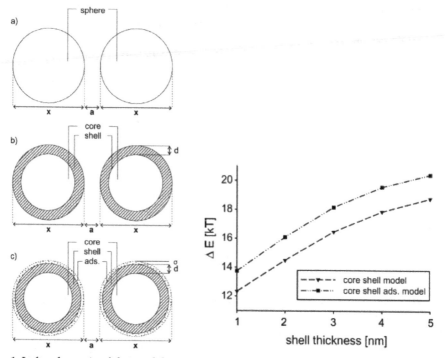

Fig. 1. Left: schematic of the model systems used for DLVO calculations on Si nanoparticles: (a) solid sphere model, (b) core-shell model, (c) core-shell-adsorbate model; right: calculated energy barrier ΔE of Si nanoparticles for the model systems in dependence of shell thickness; reproduced from (Reindl & Peukert, 2008) with kind permission of Elsevier.

On the other hand, an extension of the DLVO theory has been proposed, leading to the so-called XDLVO concept, in the works by van Oss et al. (van Oss et al., 1987, 2003). Here, Lewis acid-base interactions are additionally taken into account, and this model was shown to allow a more accurate prediction of the stability of dispersions of hectorite and chrysotile particles in various media (Wu et al., 1999). Also this concept has been applied to nanoparticle systems, being able to explain differences in the stability of iron oxide nanoparticle dispersions when adding fatty acids as stabilisers (Gyergyek et al., 2010).

The precise modelling of the steric stabilisation of nanoparticles, on the other hand, remains complex, which prevented widespread attempts to calculate particle stability by adsorption of polymers. To allow a practical application, several simplified models were presented, such as the theory of Vincent and Edwards (Vincent et al., 1986). Here, a uniform polymer segment density is assumed within the polymer shell, thus requiring a smaller number of parameters. This model has also been applied to the steric stabilisation of nanoparticles. For example, K. Lu investigated the interaction energies of Al_2O_3 nanoparticles of 20-45 nm in size stabilised with polymeric shells of different thickness (Lu, 2008). Figure 2 shows the results of these calculations, clearly revealing that with increasing layer thickness, the stability of the system clearly increases, and that a critical layer thickness is required to prevent agglomeration (in this case, amounting to 2.5 nm), according to the model.

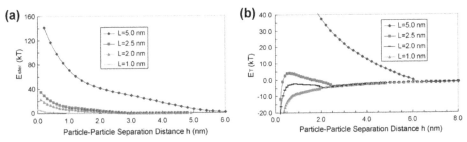

Fig. 2. Calculation of the distance-dependent steric interaction energy E_{ster} (a), and the total interaction energy E_T (b) for the stabilisation of Al_2O_3 nanoparticles protected by polymer shells of different thickness (Reproduced from (Lu, 2008) with kind permission of Elsevier).

2.2 The small-molecule stabilisation concept

Whilst the classical colloidal models of electrostatic and steric stabilisation have thus been known for decades, the treatment of nanoparticles to obtain stable dispersions in practice has always followed an empirical approach, with the well-known gold nanoparticle sols serving as typical examples that small organic molecules are sufficient to act as long-term effective stabilisers to prevent agglomeration (Turkevich et al., 1953). In many cases, the precise mechanism of stabilisation is unclear, and it is generally still being disputed whether organic ligands bound to the surface in most cases bear charges that result in electrostatic stabilisation, or whether steric repulsion effects (Peyre et al., 1997; Shah et al., 2001; Arita et al., 2009) or in fact some combined effects (Marczak et al.; 2009, Segets et al., 2011) determine the stability of the system.

Nonetheless, it has become clear that this concept of adsorption of small molecules, resulting in very thin organic layers (termed adlayers) around nanoparticles, has significant advantages over a conventional steric stabilisation with polymers: The packing of small particles is strongly influenced by the adlayer thickness Δ, basically equivalent to the chain length of the surface-bound ligands, as shown in Figure 3. For long-chain ligands, such as a polymer coating, the large adlayer thickness results in a much lower packing density of the particles as compared to short-chain ligands.

This can also be expressed as the effective volume fraction Φ_{eff} which includes the adlayer in the particle volume and is related to the true volume fraction Φ of the particles by

$$\Phi_{\mathit{eff}} = \Phi \left(1 + \frac{\Delta}{r}\right)^3 \tag{1}$$

r referring to the particle radius (Bell et al., 2005). Hence, the particles can be packed less densely, which limits the maximum achievable solids volume fraction in concentrated dispersions, leads to higher viscosity for a given volume fraction, and results in a "dilution" of core particle properties in applications such as nanocomposites.

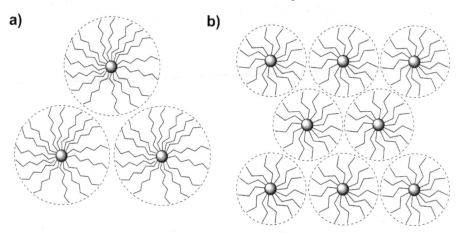

Fig. 3. Schematic (not to scale) of the packing of spherical particles with a stabilising adlayer comprised of long-chain ligands (a) and short-chain ligands (b).

3. In-situ stabilisation of nanoparticles

The stabilisation of nanoparticles can be realised experimentally most easily via the in-situ stabilisation concept. This involves the addition of stabilisers or ligands to the reaction system already during or even prior to nanoparticle formation. Hence, it is straightforward, and in principle can lead to optimum stabilisation, as it prevents particle agglomeration *a priori*, but on the other hand bears the imminent risk of an alteration of the nanoparticle formation process – an aspect that in some cases may even be desired, e.g. to obtain smaller and more uniform particles, but in many other cases is unwanted, as it makes the formation mechanism even more complex, or results in a degradation of intrinsic properties of the nanoparticles. Conceptually, the in-situ stabilisation of nanoparticles can be performed with optimised stability in the reaction medium when the processing or application of the nanoparticles in that medium is desired or feasible. In other cases, an intermediate precipitation step is performed to facilitate a change in dispersion medium, for example when a more volatile or more hydrophilic/hydrophobic medium is required than was used for the synthesis, and thus, the stabiliser is targeted to result in optimised stability in the final medium rather than the reaction solvent.

The in-situ stabilisation strategy has found widespread application, as the addition of surfactants or surface-modifying agents to a reaction mixture is commonly employed in the synthesis of metal nanoparticles (Turkevich et al., 1953; Murphy et al., 2005) as well as in the so-called hot injection method, where the precursors are injected into a hot solvent/ surfactant mixture, resulting in highly uniform and well-dispersible nanocrystals (Murray et al., 1993). This method has also been extended to transition metal oxides (Rockenberger et al., 1999; Sun et al., 2004; Epifani et al., 2008; Gilstrap Jr. et al., 2008). Despite the widespread use of in-situ stabilisation, the precise effect of the ligands on the properties of the resulting products has only been investigated in a small number of studies. Only a few studies were directed towards a systematic investigation of the effect of ligands on the particle properties, whilst in most reports the influence of the chemistry of the ligand has not been studied specifically. In the following section, some recent investigations of ligand effects on the particle properties are presented for different metal oxide nanoparticle systems.

3.1 Effects on particle crystallinity

Most ligands possess Lewis-basic groups that act as electron donors to metal centres on the nanoparticle surface, thereby enabling a firm binding of the ligands. Naturally, these groups also coordinate to molecular complexes of the same metal and hence, also to the precursor species. If the ligands are coordinated more strongly than the organics in the precursor complex, the reaction rates during the particle formation are decreased drastically, a fact that is utilised in aqueous sol-gel chemistry to deliberately decrease the rates of condensation in many cases. For strongly coordinating ligands, this may even lead to an incomplete formation, resulting in substantial organics content or a decrease in crystallinity of the resultant nanoparticles, a loss of yield, or may even impede the formation of nanoparticles altogether.

The effect of ligands on the crystallinity of in-situ stabilised titania nanoparticles has been investigated in a number of studies, because the crystallinity and phase of the particles is crucial to their performance in photocatalytic applications (Chen & Mao, 2007). Thereby, the high suitability of catechol ligands for the in-situ surface modification of TiO_2 nanoparticles, due to a very strong, covalent binding, has been reported (Niederberger et al., 2004). If a certain, critical concentration of these ligands however is exceeded and the molar Ti:ligand ratio becomes smaller than 10, a substantial decrease in crystallinity is observed (Niederberger et al., 2004). Therefore, the dopamine ligands cannot be used in high amounts, which results in the additional presence of the reaction medium benzyl alcohol on the particle surface after the synthesis (Niederberger et al., 2004). In this example, the in-situ stabilisation process did not lead to stability in benzyl alcohol but in various common organic solvents after centrifugation, washing and redispersion steps. Furthermore, different substituted catechols were investigated. Depending on the precise chemistry of the catechol side group, the particles could be made more hydrophilic or hydrophobic, even allowing stability in water when using dopamine hydrochloride as ligand (Niederberger et al., 2004).

In a more recent study, we have taken a closer look at the precise influence of the chemistry of the ligands on the crystallinity (Garnweitner et al., 2010). Different ligands with amine, alcohol and/or carboxylic acid functional groups were added prior to the synthesis of TiO_2 nanoparticles, again in benzyl alcohol as reaction medium. Interestingly, strong differences were found with respect to stability of the resulting nanoparticles, as well as their crystallinity; the differences in these properties however could not be directly correlated. Figure 4 (left) shows a series of X-ray diffraction (XRD) patterns measured from products obtained in presence of different ligands. Under the chosen experimental conditions, in the ligand-free system small anatase nanoparticles with good crystallinity and a size of about 3 nm are obtained. Whilst for glycine and malonic acid, no negative effects on crystallinity could be observed, for malic acid and glycerol a strongly decreased crystallinity was found. This could be correlated to the binding of the ligands to the particle surface by means of thermogravimetric analysis (TGA), showing that malic acid and especially glycerol bind to the nanoparticles in substantially higher quantities than the other ligands, equivalent to a higher mass loss (Figure 4, right).

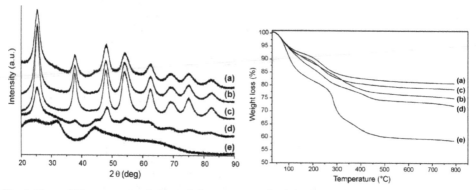

Fig. 4. X-ray diffractograms (left) and thermograms (right) of samples obtained without any present ligands (a), and in the presence of glycine (b), malonic acid (c), malic acid (d) and glycerol (e). Images reproduced from (Garnweitner et al., 2010) with kind permission of Elsevier.

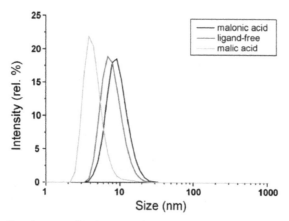

Fig. 5. Particle size distribution of TiO_2 nanoparticles modified in-situ by addition of various ligands, as determined by dynamic light scattering. Reproduced from (Garnweitner et al., 2010) with kind permission of Elsevier.

Surprisingly, the stability of dispersions of the nanoparticles could not be directly related to these results. The stability was investigated after precipitation of the particles in diethyl ether and redispersion by simple addition of de-ionised water to the precipitate. Best stability – in terms of highest solid content for minimal water addition – was observed for the glycine and malic acid systems (Garnweitner et al., 2010). However, good stability was even achieved for the ligand-free system, as in this case, the reaction of $TiCl_4$ with benzyl alcohol causes the formation of HCl as byproduct, resulting in a highly positive zeta potential. The secondary particle size distribution, as determined by dynamic light scattering (DLS), did not reveal substantial differences between the samples (Figure 5), which – in combination with the TGA results – proves that the reaction medium coordinates to the particle surface in addition to any ligand, and that for most systems, the ligand merely replaces the reaction medium rather than forming an additional shell.

3.2 Effects on particle size and morphology

Although the effect of in-situ stabilisers on the morphology of metal oxide nanoparticles has been studied in a number of publications, as these stabilisers were often purposely used to modify particle size, consistent knowledge on the precise influence of the stabilisers has not been reached, probably due to the complex interaction and the necessity of a thorough analysis of the underlying chemistry to be able to understand the effects. Gaynor et al. investigated the effect of soluble hydroxypropylcellulose on the particle size of TiO_2 nanoparticles, reporting a rather complex effect (Gaynor et al., 1997). The in-situ stabilisation of magnetite nanoparticles with L-arginine was investigated by Theerdhala et al., who showed that the particle size decreased steadily with decreasing precursor:stabiliser ratio, with however rather broad size distributions and additional changes in aspect ratio of the particles (Theerdhala et al., 2010).

In other cases, the addition of ligands even in small amounts was shown to result in dramatic changes in morphology. Again, TiO_2 nanoparticles are among the best studied systems: The influence of amino acids used as in-situ ligands in a "gel-sol" synthesis of anatase nanoparticles was investigated by Kanie and Sugimoto, showing that for certain amino acids, the initial particle-type morphology changes to rod-like due to strong adsorption of the amino acids to specific crystal faces (Kanie & Sugimoto, 2004). The synthesis of TiO_2 nanorods by hydrolysis of titanium tetraisopropoxide in presence of oleic acid was investigated by Cozzoli et al., who even proposed a model to explain different morphologies obtained for a "fast" and "slow"-type hydrolysis, again involving selective adsorption of the ligand (Cozzoli et al., 2003).

As another example, Polleux et al. investigated the synthesis of tungsten oxide nanostructures in a nonaqueous system, using benzyl alcohol as reaction medium (Polleux et al., 2006). In the pristine system, comprising only the precursor and solvent, nanoplatelets of about 5-10 nm in thickness and 30-100 nm lateral size were obtained (Figure 6, a), but after addition even of small amounts of 4-*tert*-butylcatechol, anisotropic rod-like structures with diameters of 35-40 nm and lengths of 50-800 nm were observed (Figure 6, b-d), which in fact are comprised of a highly ordered lamellar organic-inorganic hybrid nanostructure (Polleux et al., 2006). The ligand was hence inferred to act in manifold ways, not only controlling the crystal growth through selective binding to high surface energy edges but also the interaction of the individual platelets to enable their assembly (Polleux et al., 2006). Although in this example, the dispersibility of the obtained nanostructures in various organic solvents was not specifically investigated, the tremendous effect of organic ligands on the morphology of inorganic nanostructures is nicely illustrated. A broader review of these effects can be found in our earlier publication (Garnweitner & Niederberger, 2008).

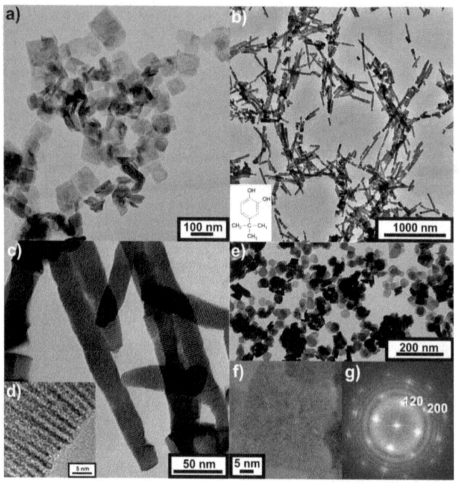

Fig. 6. TEM images of tungstite nanoparticles obtained without 4-*tert*-butylcatechol (a), and of the hybrid nanostructures obtained in the presence of 4-*tert*-butylcatechol, as-prepared (b-d) and after redispersion in water (e-g). The chemical structure of the ligand is shown in (b, Inset) (Polleux et al., 2006). Reproduced by permission of the Royal Society of Chemistry.

4. Post-synthetic stabilisation

The biggest advantage of the post-synthesis stabilisation approach is that the original synthesis is preserved, with particle properties such as size and crystallinity being unaffected by the successive stabilisation treatment that is performed at much more lenient conditions than the synthesis. Under these conditions, a possible unwanted reaction or decomposition of the stabiliser can also be excluded, thus allowing higher flexibility in terms of ligand structure and chemistry. On the other hand, after the synthesis already rather strong agglomerates may be present, and organic species from the reaction may be strongly adsorbed to the particle surface, thus limiting or preventing the stabilisation.

The post-synthetic modification of metal oxide nanoparticles has been investigated in a large number of cases, and with very different goals, not only for an enhancement of colloidal stability of the nanoparticles, but also to adjust many other properties, for example the optical and photocatalytic performance (Rehm et al., 1996; Rajh et al., 1999, 2002; Stowdam & van Veggel, 2004). Different classes of surface modifiers that have shown to be especially suitable for metal oxides include alkoxysilanes and chlorosilanes (Sanchez et al., 2001), carboxylic acids (Bourlinos et al., 2002; Arita et al., 2010), or phosphonic/phosphoric acid derivatives (Rill et al., 2007; Lomoschitz et al., 2011). Interestingly, whereas in some systems, the binding of ligands can be perceived as an exchange process of previously surface-coordinated organics (Shenhar & Rotello, 2003), in other systems a grafting of ligands to a formerly "bare" surface was observed (Arita et al., 2010). Many other groups use the term "surface modification" without further specification of the chemical surface processes. Apparently, differences between the individual systems do exist, and in contrast to the well-explored metal and semiconductor nanoparticle systems, further investigations will be required to fully understand and explain the processes during post-synthetic stabilisation of metal oxide nanostructures. Within the scope of this Chapter, only some select examples can be shown to illustrate the potentials of post-synthetic stabilisation strategies.

4.1 Exchange chemistry of metal oxide nanoparticles

The exchange chemistry of γ-Fe_2O_3 nanoparticles has been investigated in detail by Boal et al. (Boal et al., 2002). Here, alkylamine-protected maghemite nanoparticles were subjected to different tailor-made long-chain ligands (Figure 7). The amines were exchanged at least partially by all alcohol ligands used; however substantial differences in the stability of the modified particles were observed: whilst dodecanol (1)-modified nanoparticles agglomerated within a few days both when being in dispersion and in a dried state, modification of the nanoparticles with pentacosane-13-ol (2) and 13-dodecylpentacosane-13-ol (3), bearing two and three C_{12}-chains, respectively, resulted in high stability of the nanoparticles in the solid state after initial precipitation. Long-term stability in dispersion, however, was only obtained when using ligands with bidentate diol anchoring groups. Again, ligands with only a single C_{12} chain (4) showed low stability after drying, whilst double-chain bidentate ligands (5) proved most effective, leading to long-term stability both in dispersion and in the dried state (Boal et al., 2002). The authors concluded that according to their findings, a certain density of the monolayer as well as "kinetic stabilisation", in the form of multivalent ligand binding, are required in order to achieve stability, with the effect of kinetic stabilisation being especially pronounced in liquid dispersions.

In other studies, the surface modification was targeted specifically in order to adjust the polarity of the nanoparticles, allowing their redispersion in a desired organic solvent or water. For example, TiO_2 aqueous sols were modified successfully with hydrophobic DTMS and hydrophilic APTMS to make the nanoparticles redispersible in solvents of different polarity, depending on the ratio of the used silanes (Iijima et al., 2009).

4.2 Instant stabilisation through surface modification

In some cases, the addition of ligands to an agglomerated particle system can prove sufficient to result in instant stabilisation. As an example, our earlier studies on the

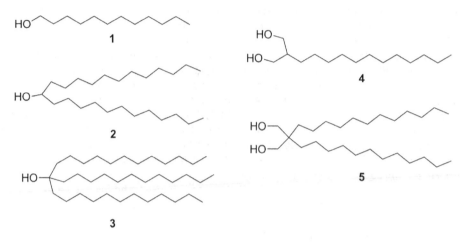

Fig. 7. Chemical structures of different ligands utilised for the modification of γ-Fe$_2$O$_3$ nanoparticles (after Boal et al., 2002).

stabilisation of ZrO$_2$ nanoparticles are presented (Garnweitner et al., 2007; Tsedev & Garnweitner, 2008). The nanoparticles were prepared in benzyl alcohol, which binds to the particle surface during the synthesis but is not capable of a full stabilisation. Hence, large agglomerates are obtained. After retrieval of the nanoparticles from the solvent by centrifugation and washing, solutions of fatty acid stabilisers were added, which resulted in instant stabilisation, visible as an immediate optical change of the mixture from turbid to transparent (Figure 8, left). TEM investigations of the nanoparticles after addition of the stabiliser show that the nanoparticles are fully stabilised, with no agglomeration occurring even during the drying process on the sample grid (Figure 8, right).

Fig. 8. Post-synthetic stabilisation of ZrO$_2$ nanoparticles by addition of fatty acids.
Left: digital images of two samples before (A) and after (B) the stabilisation treatment; right: TEM image of oleic acid-stabilised nanoparticles.

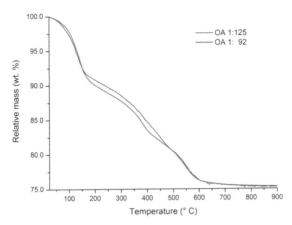

Fig. 9. Thermogravimetric analysis (TGA) of fatty acid-stabilised zirconia nanoparticles, with the stabiliser oleic acid (OA) added in different molar ratios.

In order to elucidate the binding of the fatty acid stabilisers in further detail, we performed thermogravimetric analysis of the nanoparticles before and after the stabilisation treatment. Even the as-prepared samples show a substantial amount of adsorbed organics, which can be attributed to benzyl alcohol coordinating to the particle surface during the synthesis (Niederberger et al., 2004). Figure 9 shows a comparison of two samples obtained when adding different concentrations of oleic acid (OA) as stabiliser, in a molar OA: ZrO_2 ratio of 1:125 (shown in blue), and 1:92 (shown in red). Both samples were fully stable against agglomeration. The thermograms show the same total mass loss of 24.7 wt. % at temperatures exceeding 800 °C for both samples, with however some differences in the decomposition behaviour at lower temperatures. Therefore, the total amount of organic and volatile species bound to the surface is equal, and does not rise even when increasing the amount of added stabiliser. By comparing the decomposition profile with samples measured before stabilisation and performing NMR spectroscopy (not shown), the observed differences could be attributed to a different ratio of adsorbed stabiliser to benzyl alcohol. Hence, when added above a certain concentration, the stabiliser does not increase the overall organics content of the nanoparticles, but replaces previously bound benzyl alcohol groups (Garnweitner et al., 2007).

In other investigations, however, the ligands were utilised rather to bind to a previously "bare" nanoparticle surface. This strategy is especially employed in cases of metal oxide nanoparticles synthesised by aqueous precipitation, where the resulting dispersions are charged-stabilised and highly sensitive to pH changes (Sehgal et al., 2005). Here, the binding of organic ligands can be utilised to considerably extend the range of pH stability, or achieve stability under completely different conditions. For example, the binding of poly(acrylic acid) (PAA) to CeO_2 nanoparticles was utilised to render the nanoparticles stable against agglomeration under alkaline conditions, which was attributed to an electrosteric mechanism (Sehgal et al., 2005). Figure 10 (a) shows a photograph of dispersions at different pH, proving the high stability in the alkaline regime. Additionally, DLS measurements performed by the group nicely illustrate the stabilisation process (Figure 10 (b)), with a measurable steady decrease in particle size being the result of the stabilisation process upon pH increase (Sehgal et al., 2005). In a subsequent report, Qi et al. investigated the binding of phosphonated polyethylene glycol (PPEG) oligomers on ceria nanoparticles in a comprehensive binding study (Qi et al., 2008).

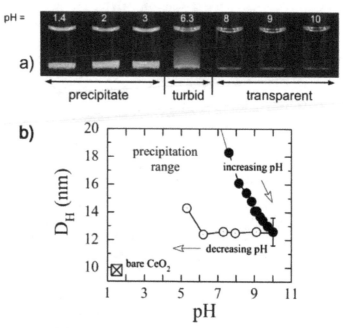

Fig. 10. Photographs of CeO$_2$-PAA solutions prepared at different pH, adjusted by NH$_4$OH (a); hydrodynamic particle diameters measured by DLS at different pH (b). Reprinted with permission from (Sehgal et al., 2005). Copyright 2005 American Chemical Society.

5. Conclusion

In this Chapter, a short review of the possibilities for the stabilisation of metal oxide nanoparticles using molecular organic stabilisers has been provided, showing select examples of the potentials of this strategy. In particular, the in-situ stabilisation technique has been compared to the post-synthetic stabilisation strategy. On the one hand, it was shown that the in-situ stabilisation strategy bears several drawbacks, as the particle properties, especially their crystallinity and morphology, may be adversely affected by the added ligands, that these problems however can be overcome by choice of the "right" ligand. The ideal in-situ stabiliser, therefore, needs to bind to the particle surface but not form too strong bonds, as this could interfere with the synthesis mechanism. For some metal oxides, such as TiO$_2$, first insights into the precise effect of different functional groups on the synthesis and stabilisation have been gained, and it can be expected that in the future, a toolbox will be available for different metal oxide systems, allowing the precise choice of an in-situ ligand for optimum effect. Then, the benefits of in-situ stabilisation can fully be exploited, such as the experimental simplicity and best stability.

Only in rare cases, a direct comparison of in-situ versus post-synthetic stabilisation has been performed, but these results clearly point to a higher efficiency of in-situ stabilisation approaches. For example, in the two-step stabilisation of magnetite nanoparticles, Theerdhala et al. achieved long-time stable dispersions only for the in-situ approach (Theerdhala et al., 2010). On the other hand, there are many cases where the initial reaction medium is not the perfect choice for further handling and storage of the nanoparticles, e.g. due to its toxicity or a too high or too low boiling point. In these cases, the post-synthetic stabilisation of nanoparticles will always be an attractive strategy, as the lack of stability of the nanoparticles allows for their simple and fast separation and recovery, and the presented examples clearly show that at least in some cases, a very simple and fast post-synthetic step is sufficient to achieve high stability of the nanoparticles, without any agglomeration even upon drying. Also in this case, it can be expected that within the next decade research on the chemical interaction of stabilisers and different metal oxide nanoparticle systems will lead to a much broader understanding than today, with general strategies available to enable the *a priori* choice of ligands at least for the most common metal oxides and reaction systems to achieve best stability with minimum addition of organics. Additionally, the increasing complexity of self-assembly tasks will also demand for structurally more complex ligands, which most likely will have to be introduced in post-synthetic steps due to their sensitivity to temperature and harsh chemical conditions. Hence, it appears clear that both the in-situ and post-synthetic approaches to the surface modification and stabilisation of nanoparticles with be applied in the future, will a more detailed knowledge being a fundamental prerequisite for the precise and rational application of experimental strategies to achieve optimum efficiency.

6. Acknowledgment

The author gratefully acknowledges funding by the Deutsche Forschungsgemeinschaft (DFG), grant GA 1492/4-2, within the framework of the priority programme SPP 1273 "Kolloidverfahrenstechnik". Furthermore, Tarik A. Cheema and Rona Pitschke, Max Planck Institute of Colloids and Interfaces, Potsdam/Germany, are acknowledged for Fig. 8, and Mario Raab is acknowledged for help in formatting the manuscript.

7. References

Arita, T.; Yoo, J.; Ueda, Y. & Adschiri, T. (2009). Size and size distribution balance the dispersion of colloidal CeO_2 nanoparticles in organic solvents. *Nanoscale*, Vol.2, No.5, (February 2010), pp. 689-693, ISSN 2040 3364

Ballauff, M. (2003). Nanoscopic Polymer Particles with a Well-Defined Surface: Synthesis, Characterization, and Properties. *Macromolecular Chemistry and Physics*, Vol.204, No. 2, (February 2003), pp. 220-234, ISSN 1022-1352

Bazito, F.F.C. & Torresi, R.M. (2006). Cathodes for Lithium Ion Batteries: The Benefits of Using Nanostructured Materials. *Journal of the Brazilian Chemical Society*, Vol.17, No.4 , (May 2006), pp. 627-642, ISSN 0103-5053

Bell, N.S.; Schendel, M.E. & Piech, M. (2005). Rheological Properties of Nanopowder Alumina coated with adsorbed Fatty Acids. *Journal of Colloid and Interface Science*, Vol.287, No.1, (July 2005), pp. 94-106, ISSN 0021-9797

Biswas, M. & Ray, S.S. (2001). Recent Progress in Synthesis and Evaluation of Polymer-Montmorillonite Nanocomposites. *Advances in Polymer Science*, Vol.155, (2001), pp. 167-221, ISSN 0065-3195

Boal, A.K.; Das, K.; Gray, M. & Rotello, V.M. (2002). Monolayer Exchange Chemistry of γ-Fe_2O_3 Nanoparticles. *Chemistry of Materials*, Vol.14, No.6, (May 2002), pp. 2628-2636, ISSN 0897-4756

Bourlinos, A.B.; Bakandritsos, A.; Georgakilas, V. & Petridis, D. (2002). Surface Modification of Ultrafine magnetic Iron Oxide Particles. *Chemistry of Materials*, Vol.14, No.8, (August 2002), pp. 3226-3228, ISSN 0897-4756

Centi, G. & Perathoner, S. (2009). The Role of Nanostructure in Improving the Performance of Electrodes for Energy Storage and Conversion. *European Journal of Inorganic Chemistry*, Vol.2009, No.26, (September 2009), pp. 3851-3878, ISSN 1099-0682

Chen, X. & Mao, S.S., (2007). Titanium Dioxide Nanomaterials: Synthesis, Properties, Modifications and Applications. *Chemical Reviews*, Vol.107, No.7, (July 2007), pp. 2891-2959, ISSN 0009-2665

Cozzoli, P.D.; Kornowski, A. & Weller, H. (2003). Low-Temperature Synthesis of Soluble and Processable Organic-Capped Anatase TiO_2 Nanorods. *Journal of the American Chemical Society*, Vol.125, No.47, (November 2003), pp. 14539-14548, ISSN 0002-7863

Cushing, B.L.; Kolesnichenko, V.L. & O'Connor, C.J. (2004). Recent Advances in the Liquid-Phase Syntheses of Inorganic Nanoparticles. *Chemical Reviews*, Vol.104, No.9, (September 2004), pp. 3893-3946, ISSN 0009-2665

Derjaguin, B. V.& Landau, L. D. (1941). Theory of the Stability of Strongly charged Lyophobic Sols and the Adhesion of strongly charged Particles in Solutions of Electrolytes. *Progress in Surface Science*, Vol.43, No.1-4, (May 1993), pp. 30-59, ISSN 0079-6816

Dong, L.; Subramanian, A.; Nelson, B.J. (2007). Carbon Nanotubes for Nanorobotics. *Nano Today*, Vol.2, No.6, (December 2007), pp. 12-21, ISSN 1748-0132

Duget, E.; Vasseur, S.; Mornet, S. & Devoisselle, J.M. (2006), Magnetic nanoparticles and their applications in medicine. *Nanomedicine*, Vol.1, No.2, (August 2006), pp.157-168, ISSN 1748-6963

Epifani, M.; Arbiol, J.; Andreu, T. & Morante, J.R. (2008). Synthesis of Soluble and Size-Controlled SnO_2 and CeO_2 Nanocrystals: Application of a General Concept for the Low-Temperature, Hydrolytic Synthesis of Organically Capped Oxide Nanoparticles. *European Journal of Inorganic Chemistry*, No.6, (February 2008), pp. 859-862, ISSN 1434-1948

Everett, D. H. (1988). *Basic Principles of Colloid Science*. Royal Society of Chemistry, ISBN 0-85186-443-0, London, England

Feldmann C. & Goesmann H. (2010). Nanoparticulate Functional Materials. *Angewandte Chemie International Edition*, Vol.49, No. 8, (February 2010), pp. 1362-1395, ISSN 1521-3773

Gao, Y. & Tang, Z. (2011). Design and Application of Inorganic Nanoparticle Superstructures: Current Status and Future Challenges. *Small*, Vol.7, No.15, (August 2011), pp. 2133-2146, ISSN 1613-6810

Garnweitner, G.; Goldenberg, L.M.; Sakhno, O.V.; Antonietti, M.; Niederberger, M. & Stumpe, J. (2007). Large-Scale Synthesis of Organophilic Zirconia Nanoparticles and their Application in Organic-Inorganic Nanocomposites for Efficient Volume Holography. *Small*, Vol.3, No.9, (September 2007), pp. 1626-1632, ISSN 1613-6810

Garnweitner, G. & Niederberger, M. (2008). Organic Chemistry in Inorganic Nanomaterials Synthesis. *Journal of Materials Chemistry*, Vol.18, No.11, (March 2008), pp. 1171-1182, ISSN 0959-9428

Garnweitner, G.; Ghareeb, H.O. & Grote C. (2010), Small-molecule in situ Stabilization of TiO_2 Nanoparticles for the facile Preparation of stable colloidal Dispersions. *Colloids and Surfaces A: Physicochemical and Engineering Aspects*, Vol.372, No.1-3, (December 2010), pp. 41-47, ISSN 0927-7757

Gaynor, A.G.; Gonzalez, R.J.; Davis, R.M. & Zallen, R. (1997). Characterization of Nanophase Titania Particles Synthesized Using In Situ Steric Stabilization. *Journal of Materials Research*, Vol.12, No. 7, (July 1997), pp. 1755-1765, ISSN 0884-2914

Gilstrap Jr., R.A.; Capozzi, C.J.; Carson, C.G.; Gerhardt, R.A. & Summers, C.J. (2008). Synthesis of a nonagglomerated Indium Tin Oxide Nanoparticle Dispersion. *Advanced Materials*, Vol.20, No.21, (November 2008), pp. 4163-4166, ISSN 0935-9648

Goyal, A.K.; Johal, E.S. & Rath, G. (2011) Nanotechnology for Water Treatment. *Current Nanoscience*, Vol.7, No. 4, (August 2011), pp. 640-654, ISSN 1573-4137

Guo, S. & Wang, E. (2011). Noble Metal Nanomaterials: Controllable Synthesis and Application in Fuel Cells and Analytical Sensors. *Nano Today*, Vol.6, No.3, (June 2011), pp. 240-264, ISSN 1748-0132

Gyergyek, S.; Makovec, D. & Drofenik, M. (2010). Colloidal Stability of Oleic- and Ricinoleic-Acid-Coated Magnetic Nanoparticles in Organic Solvents *Journal of Colloid and Interface Science*, Vol.354, No.2, (February 2011), pp. 498-505, ISBN 0021-9797

Hamaker, H. C. (1937) The London-van der Waals Attraction between Spherical Particles. *Physica*, Vol.4, No.10, (October 1937), pp. 1058-1072, ISSN 0031-8914

Heller, W. & Pugh, T.L. (1954). ``Steric Protection'' of Hydrophobic Colloidal Particles by Adsorption of Flexible Macromolecules. *Journal of Chemical Physics*, Vol.22, No.10, (1954) pp. 1778, ISSN 1089-7690

Hunter, R.J. (2001). *Foundations of Colloid Science*, Oxford University Press, ISBN 978-0-19-850502-0, Oxford, England

Iijima, M.; Kobayakawa, M. & Kamiya, H. (2009). Tuning the Stability of TiO_2 Nanoparticles in various Solvents by mixed Silane Alkoxides. *Journal of Colloid and Interface Science*, Vol.337, No.1, (September 2009), pp. 61-65, ISSN 0021-9797

Jun, Y.-W.; Choi, J.-S. ; Cheon, J. (2006). Shape Control of Semiconductor and Metal Oxide Nanocrystals through Nonhydrolytic Colloidal Routes. *Angewandte Chemie International Edition*, Vol.45, No.21, (May 2006), pp. 3414-3439, ISSN 1521-3773

Kanie, K. & Sugimoto, T. (2004). Shape Control of Anatase TiO₂ Nanoparticles by Amino Acids in a Gel-Sol System. *Chemical Communications*, Vol.2004, No.14, (July 2004), pp. 1584-1585, ISSN 1359-7345

Kim, J. & van der Bruggen, B. (2010), The Use of Nanoparticles in Polymeric and Ceramic Membrane Structures: Review of Manufacturing Procedures and Performance Improvement for Water Treatment, *Environmental Pollution*, Vol. 158, No. 7, (July 2010), pp. 2335-2349, ISSN 0269-7491

Li, Q.; Mahendra, S.; Lyon, D.Y.; Burnet, L.; Liga, M.V.; Li, D. & Alvarez, J.J. (2008). Antimicrobial nanomaterials for water disinfection and microbial control: Potential applications and implications. *Water Research*, Vol.42, No.18, (November 2008), pp 4591-4602, ISSN 0043-1354

Lomoschitz, C.J.; Feichtenschlager, B.; Moszner, N.; Puchberger, M.; Müller, K.; Abele, M. & Kickelbick, G. (2011). Directing Alkyl Chain Ordering of Functional Phosphorus Coupling Agents on ZrO₂. *Langmuir*, Vol.27, No.7, (April 2011), pp. 3534-3540, ISSN 0743-7463

Lu, K. (2008). Theoretical Analysis of Colloidal Interaction Energy in Nanoparticle Suspensions. *Ceramics International*, Vol.34, No.6, (August 2008) , pp.1353-1360, ISSN 0272-8842

Mackor, E. L. (1951) A Theoretical Approach of the Colloid-Chemical Stability of Dispersions in Hydrocarbons. *Journal of Colloid Sciences*, Vol.6, No.5, (October 1951), pp. 492-495, ISSN 0021-9797

Mallouk, T.E & Sen, A. (2009). Powering nanorobots: Catalytic Engines Enable Tiny Swimmers to Harness Fuel From Their Environment and Overcome the Weird Physics of the Microscopic World. *Scientific American*, Vol. 300, No.5 (May 2009), pp.72-77, ISSN 0036-8733

Marczak, R.; Segets, D.; Voigt, M. & Peukert W. (2009). Optimum between Purification and Colloidal Stability of ZnO Nanoparticles. *Advanced Powder Technology*, Vol.21, No.1, (January 2010), pp. 41-49, ISSN 0921-8831

Moghimi, S.M.; Hunter, A.C. & Murray J.C. (2005), Nanomedicine: Current Status and Future Prospects. *The FASEB Journal*, Vol. 19, No.3, (March 2005), pp. 311-330, ISSN 1530-6860

Mostofizadeh, A.; Li, Y.; Song, B. & Huang, Y. (2011). Synthesis, Properties, and Applications of Low-Dimensional Carbon-Related Nanomaterials. *Journal of Nanomaterials*, Vol.2011, Art. No. 685081, (2011), 21 p. , ISSN 1687-4110

Mulvaney, P. (1998) Zeta Potential and Colloid Reaction Kinetics. In: *Nanoparticles and Nanostructured Films*, J.H. Fendler, (Ed.), Wiley-VCH, ISBN 3-527-29443-0, Weinheim, Germany

Murphy, C.J.; Sau, T.K.; Gole, A.M.; Orendorff, C.J.; Gao, J.; Gou, L.; Hunyadi, S.E.; Li, T. (2005). Anisotropic Metal Nanoparticles: Synthesis, Assembly, and Optical Applications. *Journal of Physical Chemistry B*, Vol.109, No.29, (July 2005), pp. 13857-13870, ISSN 1520-6106

Murray, C.B.; Norris, D.J.; Bawendi, M.G. (1993). Synthesis and Characterization of Nearly Monodisperse CdE (E = Sulfur, Selenium, Tellurium) Semiconductor

Nanocrystallites. *Journal of the American Chemical Society*, Vol.115, No. 19, (September 1993), pp. 8706-8715, ISSN 1520-1526

Napper, D. (1983) *Polymeric Stabilization of Colloidal Dispersions*, Academic Press, ISBN 0-12513-980-2 London, England

Niederberger, M.; Garnweitner, G.; Krumreich, F.; Nesper, R.; Cölfen, H. & Antionetti, M. (2004). Tailoring the Surface and Solubility Properties of Nanocrystalline Titania by a Nonaqueous In Situ Functionalization Process. *Chemistry of Materials*, Vol.16 , No.7, (February 2004), pp. 1202-1208, ISSN 0897-4756

Ostwald, W. (1915). *Die Welt der vernachlässigten Dimensionen: Eine Einführung in die Kolloidchemie mit besonderer Berücksichtigung ihrer Anwendungen*. Steinkopff Verlag, Dresden, Germany

Peyre, D.; Spalla, O.; Belloni, L. & Nabavi, M. (1997). Stability of a Nanometric Zirconia Colloidal Dispersion under Compression: Effect of Surface Complexation by Acetylacetone. *Journal of Colloid and Interface Science*, Vol.187, No.1, (March 1997), pp. 184-200, ISSN 0021-9797.

Pinna, N. & Niederberger, M. (2008). Surfactant-free Nonaqueous Synthesis of Metal Oxide Nanostructures. *Angewandte Chemie International Edition*, Vol.47, No.29, (July 2008), pp. 5292-5304, ISSN 1433-7851

Polleux, J.; Antonietti, M. & Niederberger, M. (2006). Ligand and Solvent Effects in the Nonaqueous Synthesis of Highly Ordered Anisotropic Tungsten Oxide Nanostructures. *Journal of Materials Chemistry*, Vol.16, No.40, (July 2006), pp. 3969-3975, ISSN 0959-9428

Prandeep, T. & Anshup (2009), Noble Metal Nanoparticles for Water Purification: A Critical Review. *Thin Solid Films*, Vol.517, No.24, (October 2009), pp. 6441-6478, ISSN 1879-2731

Qi, L.; Sehgal, A.; Castaing, J.-C.; Chapel, J.-P.; Fresnais, J.; Berret, J.-F. & Cousin, F. (2008). Redispersible Hybrid Nanopowders: Cerium Oxide Nanoparticle Complexes with Phosphonated-PEG Oligomers. *ACS Nano*, Vol.2, No.5, (May 2008), pp. 879-888, ISSN 1936-0851

Rajh, T.; Nedeljkovic, J.M.; Chen, L.X.; Poluektov, O. & Thurnauer M.C. (1999). Improving Optical and Charge Separation Properties of Nanocrystalline TiO_2 by Surface Modification with Vitamin C. *Journal of Physical Chemistry B*, Vol.103, No.18, (April 1999), pp. 3515-3519, ISSN 1089-5647

Rajh, T.; Chen, L.X.; Lukas, K.; Liu, T.; Thurnauer, M.C. & Tiede, D.M. (2002). Surface Restructuring of Nanoparticles: An efficient Route for Ligand-Metal Oxide Crosstalk. *Journal of Physical Chemistry B*, Vol.106, No.41, (October 2002), pp. 10543-10552, ISSN 1089-5647

Rasmussen, S.; Chen, L.; Nilsson, M. & Abe, S. (2003). Bridging Nonliving and Living Matter. *Artificial Life*, Vol.9, No.3, (June 2003), pp. 269-316, ISSN 1064-5462

Rehm, J.M.; McLendon, G.L.; Nagasawa, Y.; Yoshihara, K.; Moser, J. & Grätzel, M. (1996). Femtosecond Electron-Transfer Dynamics at a Sensitizing Dye−Semiconductor (TiO_2) Interface. *Journal of Physical Chemistry*, Vol.100, No.23, (June 1996) pp. 9577-9578, ISSN 0022-3654

Reindl, A. & Peukert, W. (2008). Intrinsically stable Dispersions of Silicon Nanoparticles. *Journal of Colloid and Interface Science*, Vol.325, No.1, (September 2008), pp. 173-178, ISSN 0021-9797

Renger, C.; Kuschel, P.; Kristoffersson, A.; Clauss, B.; Oppermann, W. & Sigmund, W. (2006). Rheology Studies on Highly Filled Nano-Zirconia Suspensions. *Journal of the European Ceramic Society*, Vol.27, No.6, (November 2006), pp. 2361-2367, ISSN 0955-2219

Requicha, A.G.G (2003). Nanorobots, NEMS, and Nanoassembly. *Proceedings of the IEEE*, Vol.91, No.11, (November 2003), pp. 1922-1933, ISSN 0018-9219

Rill, C.; Ivanovici, S. & Kickelbick, G. (2007). Hybrid Nanoparticles Prepared by In-Situ and Post-Synthetic Surface Modification of Lanthanide-based Nanoparticles with Phosphonic Acid Derivatives, *Proceedings of Materials Research Society Symposium 1007*, pp. 207-212, ISBN: 978-155899967-1, San Francisco, USA, California, USA, April 9-13, 2007

Rockenberger, J.; Scher, E.C. & Alivisatos, A.P. (1999), A New Nonhydrolytic Single-Precursor Approach to Surfactant-Capped Nanocrystals of Transition Metal Oxides, *Journal of the American Chemical Society*, Vol.121, No.49, (December 1999), pp. 11595-11596, ISSN 0002-7863

Sanchez, C.; Soler-Illia, G.J.D.A.A.; Ribot, F.; Lalot, T.; Mayer, C.R. & Cabuil, V. (2001). Designed Hybrid Organic-Inorganic Nanocomposites from Functional Nanobuilding Blocks. *Chemistry of Materials*. Vol.13, No.10, (October 2001), pp. 3061-3083, ISSN 0897-4756

Sánchez, S. & Pumera, M. (2009). Nanorobots: The Ultimate Wireless Self-Propelled Sensing and Actuating Devices, *Chemistry-An Asian Journal*, Vol.4, No.9, (September 2009), pp. 1402-1410, ISSN 1861-471X

Sanvicens, N.& Pilar, M. (2008) Multifunctional nanoparticles – properties and prospects for their use in human medicine. *Trends in Biotechnology*, Vol.26, No.8, (August 2008), pp. 425-433, ISSN 0167-7799

Schwarz, U.S. & Safran,S.A. (2000) Phase Behavior and Material Properties of Hollow Nanoparticles, *Physical Review E*, Vol.62, No.5, (November 2000) pp. 6957–6967, ISSN 1539-3755

Segets, D.; Marczak, R.; Schäfer, S.; Paula, C.; Gnichwitz, J.F.; Hirsch, A. & Peukert, W. (2011). Experimental and Theoretical Studies of the Colloidal Stability of Nanoparticles–A General Interpretation Based on Stability Maps. *ACS Nano*, Vol.5, No.6, (May 2011), pp. 4658–4669, ISSN 1936-0851

Sehgal, A.; Lalatonne, Y.; Berret, J.-F.; Morvan, M. (2005). Precipitation-Redispersion of Cerium Oxide Nanoparticles with Poly(acrylic acid): Toward Stable Dispersions. *Langmuir*, Vol.21, No.20, (September 2005), pp. 9359-9364, ISSN: 0743-7463.

Shah, P.S.; Holmes, J.D.; Johnston, K.P. & Korgel, B.A. (2001). Size-Selective Dispersion of Dodecanethiol-coated Nanocrystals in Liquid and Supereritical Ethane by Density Tuning. *Journal of Physical Chemistry B*, Vol.106, No. 10, (March 2002), pp. 2545-2551, ISSN 1089-5647

Shenhar, R. & Rotello, V.M (2003). Nanoparticles: Scaffolds and Building Blocks. *Accounts of Chemical Research*, Vol.36, No.7, (July 2003), pp. 549-561, ISSN 0001-4842

Sides, C.R.; Li, N.; Patrissi, C.J.; Scrosati, B. & Martin, C. R. (2002). Nanoscale Materials for Lithium-Ion Batteries. *MRS Bulletin*, Vol.27, No.8, (August 2002), pp. 604-607, ISSN 0883-7694

Stowdam, J.W. & van Veggel, F.C.J.M. (2004). Improvement in the Luminescence Properties and Processability of LaF_3/Ln and $LaPO_4/Ln$ Nanoparticles by Surface Modification. *Langmuir*, Vol.20, No.26, (November 2004), pp. 11763-11771, ISSN 0743-7463.

Sun, S.; Zeng, H.; Robinson, D.B.; Raoux, S.; Rice, P.M.; Wang, S.X.; Li, G. (2004). Monodisperse MFe_2O_4 (M = Fe, Co, Mn) Nanoparticles. *Journal of the American Chemical Society*, Vol.126 No.1, pp. 273-279, ISSN 0002-7863

Tadmor, R. & Klein, J.H. (2001). Additional Attraction Between Surfactant-Coated Surfaces, *Journal of Colloid and Interface Science*, Vol.247, No.2, (March 2002), pp. 321–326, ISSN 0021-9797

Theerdhala, S.; Bahadur, D.; Vitta, S.; Perkas, N.; Zhong, Z.; Gedanken, A. (2010). Sonochemical Stabilization of Ultrafine Colloidal Biocompatible Magnetite Nanoparticles using Amino Acid, L-Arginine, for Possible Bio Applications. *Ultrasonics Sonochemistry*, Vol.17, No.4, (April 2010), pp. 730-737, ISSN 1350-4177

Theron, J.; Walker, J.A. & Cloete (2008), T.E. Nanotechnology and Water Treatment: Applications and Emerging Opportunities, *Critical Reviews in Microbiology*, Vol. 34, No.1, (February 2008), pp. 43-69, ISSN 1549-7828

Traina, C.A. & Schwartz, J. (2007). Surface Modification of Y_2O_3 Nanoparticles. *Langmuir*, Vol.23, No.18, (August 2007), pp. 9158-9161, ISSN 0743-7463

Trindade, T.; O'Brien, P. & Pickett, N.L. (2001). Nanocrystalline Semiconductors: Synthesis, Properties, and Perspectives. *Chemistry of Materials*, Vol.13, No.11, (October 2001), pp. 3843-3858, ISSN 0897-4756

Tsedev, N. & Garnweitner, G. (2008). Surface Modification of ZrO_2 Nanoparticles as Functional Component in Optical Nanocomposite Devices, *Materials Research Society Symposium Proceedings*, pp. 175-180, ISBN 978-160511046-2, San Francisco, California, USA, March 25-27, 2008

Turkevich, J.; Stevenson, P.C. & Hillier, J. (1953). The Formation of Colloidal Gold. *Journal of Physical Chemistry*, Vol.57, No.7, (July 1953), pp. 670-673, ISSN 00223654

van Oss, C.J.; Chaudhury, M.K.& Good, R.J. (1987). Monopolar Surfaces. *Advances in Colloid and Interface Science*, Vol.28, No.C, (1987)pp. 35-64, ISSN 1873-3727

van Oss, C.J. (2003). Long-Range and Short-Range Mechanisms of Hydrophobic Attraction and Hydrophilic Repulsion in Specific and Aspecific Iinteractions. *Journal of Molecular Recognition*, Vol.16, No.4, (July/August 2003), pp. 177-190, ISSN 1099-1352

Verwey, E. J. W. & Overbeek, J. T. G. (1948). *Theory of the Stability of Lyophobic Colloids*, Elsevier, ISBN 978-0-48640-929-0, New York, USA

Vincent, B.; Edwards, J.; Emmett, S. & Jones, A. (1986). Depletion Flocculation in Dispersions of sterically-stabilized Particles ("Soft Spheres"). *Colloids and Surfaces*, Vol.18, No.2-4, (June 1986), pp. 261-281, ISSN 0166-6622

Wiese, G.R. & Healy, T.W. (1969). Effect of Particle Size on Colloid Stability. *Transactions of the Faraday Society*, Vol.66, No.0, (January 1970), pp. 490-499, ISSN 0014-7672

Wu, W.; Giese, R.F. & van Oss, C.J. (1999). *Colloids and Surfaces B: Biointerfaces*, Vol.14, No.1-4, (August 1999), pp. 47-55, ISSN 0927-7765

Section 2

Drug Nanoparticles

5

Drug Nanoparticles – An Overview

Vijaykumar Nekkanti, Venkateswarlu Vabalaboina and Raviraj Pillai[*]

Dr. Reddy's Laboratories Limited, Hyderabad, India

1. Introduction

Advances in drug discovery technologies and combinatorial chemistry techniques have led to identification of a number of compounds with good therapeutic potential. However, because of their complex chemistry majority of these compounds have poor aqueous solubility resulting in reduced and variable bioavailability (Lipinski et al., 2002). The variability in systemic exposure observed often makes it difficult for dose delineation, results in fed and fast variability and in slower onset of action. These issues may lead to sub-optimal dosing and concomitantly poor therapeutic response. For compounds with poor aqueous solubility that are ionizable, preparation of salts to improve solubility/dissolution rate is a commonly used approach that had limited success. From a product development standpoint, generally a crystalline salt is preferred due to potential physical and chemical stability issues associated with the amorphous form. Identification of a crystalline salt with adequate aqueous solubility requires screening various counter-ions and solvents/crystallization conditions and at times isolation of a crystalline material is difficult. In some instances the salt formed is extremely hygroscopic posing product development and manufacturing challenges (Elaine et al., 2008).

Currently there are limited formulation approaches for compounds with poor aqueous solubility. The most commonly used approaches are micronisation and solid dispersions of the drug in water-soluble careers for filling into hard or soft gelatin capsules. Micronisation results in particles that are < 5 μm with a very small fraction that is in the sub-micron range. The decrease in particle size results in a modest increase in surface area that may not change the dissolution rate or saturation solubility to significantly impact bioavailability (Jens-Uwe et al., 2008).

Solid dispersion compositions comprise of molecular dispersion of the drug in water soluble and lipid-based surface-active carriers that can emulsify upon contact with the dissolution medium. Formation of molecular dispersions (solid solution) provides a means of reducing the particle size of the compounds to nearly molecular levels (i.e., there are no visible particles). As the carrier dissolves, the compound is exposed to the dissolution media as fine particles that are amorphous, which can dissolve rapidly and concomitantly absorbed. These formulations are filled in soft or hard gelatin capsules. There are several products using this approach in the market, e.g.,

[*] Corresponding Author

Sandimmune®/Neoral® (cyclosporin microemulsion), Norvir® (Ritnovir) and Fortovase® (Saquinavir). This approach is generally suitable for highly potent compounds and thus not applicable for moderately potent compounds where the dose requirement may be high (Merisko-Liversidge et al., 2003).

In recent years an area that is gaining popularity with formulation scientists for developing a viable dosage form for poorly soluble compounds that are moderately potent is to develop a formulation incorporating drug nanoparticles, usually less than 1 µm in diameter. For example, when the particle size of the drug is reduced from 8 µm to 200 nm there is 40-fold increase in the surface area to volume ratio. This increase in surface area can provide substantial increase in the dissolution rate if the formulation disperses into discrete particles (Liversidge et al., 1995). The nanoparticle formulation approach is proven to be very useful and invaluable in all stages of the drug product development and has opened opportunities for revitalizing marketed products with suboptimal delivery.

Nanoparticle formulation technologies have provided the pharmaceutical industry with options for addressing solubility and bioavailability issues associated with poorly soluble compounds. In new chemical entities (NCE) development, the technology has been of great value when it is used as a screening tool during preclinical efficacy and / or safety assessment studies in the early development phase. For marketed products requiring life-cycle extension opportunities, nanoparticle formulation strategies provide a means to develop a new drug-delivery platform with improved therapeutic outcome incorporating the existing drug, thus creating new avenues for addressing unmet medical needs.

2. History

Nanotechnology has a long development and application history. However, the most important scientific advancements have only taken place in the last two decades. Heterogeneous catalysts were among the first examples, developed in the early 19th century (Rogers et al., 2001). The earliest example of pharmaceutical application was Danazol that was milled using a bead mill to obtain a median particle size of 169 nm (Robertson, 1983). The Danazol nanosuspension showed enhanced oral bioavailability ($82.3 \pm 10.1\%$) as compared to the drug suspension using the "as-is" drug ($5.1 \pm 1.9\%$).

The first nanoparticle technology based product approved by FDA was Rapamune® (Sirolimus) - an immunosuppressant developed by Wyeth Pharmaceuticals (now Pfizer). The second product approved by FDA was Tricor® by Abbott Laboratories, an improved formulation of Fenofibrate (for hypercholesterolemia) incorporating drug nanoparticles that reduced the fed-fast variability resulting in no dosing restriction that allowed co-administration with other drugs used for treating lipid disorders. Another product containing Fenofibrate nanoparticles is Triglide®. The product was developed by Skye Pharma using their patented IDD-P® technology and marketed by Sciele Pharma Inc. (Atlanta, USA). Antiemetic drug, Emend® (Aprepitant) was approved by the FDA in March 2003 and launched in the United States by Merck in April 2003. Emend is a capsule containing 80 or 125 mg of Aprepitant formulated as drug nanoparticles using Elan's drug NanoCrystal® technology (Mary et al., 2005). Megace ES® (ES stands for enhanced solubility) is another product containing drug nanoparticles that was developed by Par Pharmaceutical Inc. (USA). It is an aqueous suspension of Megestrol Acetate (a synthetic progestin, anti anorexic) with a dose of

625 mg / 5 mL. The drug nanosuspension reduced the fed and fast variability similar to Tricor®. The product in nanosuspension demonstrated that aqueous nanosuspension can be produced with adequate physical stability with acceptable shelf life using this technology. A list of products developed using nanoparticle technology (Ranjita Shegokar et al., 2010; Rajesh Dubey, 2006) currently available in the market is summarized in Table 1.

Brand	Generic Name	Indication	Drug Delivery Company	Innovator	Status
Rapamune®	Rapamycin, Sirolimus	Immunosuppressant	Elan Nanosystems	Wyeth	Marketed
Emend®	Aprepitant	Anti-emetic	Elan Nanosystems	Merck & Co.	Marketed
Tricor®	Fenofibrate	Hypercholesterolemia	Abbott Laboratories	Abbott Laboratories	Marketed
Megace ES®	Megestrol	Anti-anorexic	Elan Nanosystems	Par Pharmaceuticals	Marketed
Triglide®	Fenofibrate	Hypercholesterolemia	IDD-P Skyepharma	Sciele Pharma Inc.	Marketed
Avinza®	Morphine Sulphate	Phychostimulant	Elan Nanosystems	King Pharmaceuticals	Marketed
Focalin	Dexmethyl-Phenidate HCl	Attention Deficit Hyperactivity Disorder (ADHD).	Elan Nanosystems	Novartis	Marketed
Ritalin	Methyl Phenidate HCl	CNS Stimulant	Elan Nanosystems	Novartis	Marketed
Zanaflex Capusules TM	Tizanidine HCl	Muscle Relaxant	Elan Nanosystems	Acorda	Marketed

Table 1. Overview of nanoparticle technology based products

3. Formulation theory

The basic principle of micronisation and nanonisation is based on increase in surface area leading to enhancement in dissolution rate according to Noyes-Whitney equation (Muller et al., 2000). Poor aqueous solubility correlates with slower dissolution and decreasing particle size increases the surface area with concomitant increase in the dissolution rate.

Dissolution kinetics is the primary driving force behind the improved pharmacokinetic properties of nanoparticle formulations of poorly water soluble compounds. Dissolution rate of a drug is a function of its particle size and intrinsic solubility. For drugs with poor aqueous solubility, surface area of the drug particles drives dissolution. As described by the Nernst-Brunner and Levich modification of Noyes-Whitney model the rate of drug dissolution is directly proportional to surface area;

$$dx/dt = (A \times D/\delta) \times (C - X/V) \tag{1}$$

Where X is the amount of drug in solution, t is time, A is the effective surface area, D is the diffusion coefficient of the drug, δ is the effective diffusion boundary layer, C is the saturation solubility of the drug, and V is the volume of dissolution medium.

Saturation solubility usually is a compound specific constant that depends on temperature. This understanding is true for regular particles that are above the micron range however, different for drug nanoparticles. This is because the dissolution pressure is a function of the curvature of the surface that means it is much stronger for a curved surface of nanoparticles. Below a particle size of approximately 2 µm, the dissolution pressure increases distinctly leading to an increase in the saturation solubility. In addition, the diffusional distance on the surface of drug nanoparticles is decreased, thus leading to an increased concentration gradient. The increase in surface area and concentration gradient lead to much more pronounced increase in dissolution velocity and saturation solubility compared to products containing micronized particles concomitantly resulting in improved bioavailability (Keck et al., 2006).

Increased solubility near the particle surface results in enhanced concentration gradient between the surface and the bulk solution. The high concentration gradient according to Fick's law must lead to an increased mass flux away from the particle surface (Dressman et al., 1998). As the particle diameter decreases, its surface area to volume ratio increases inversely, further leading to an increased dissolution rate. Under sink conditions in which the drug concentration in the surrounding medium approaches zero, rapid dissolution could theoretically occur.

4. Production of drug nanoparticles

There are several techniques used to produce drug nanoparticles. The existing technologies can be divided into two categories; 'bottom up' and 'top down'. The bottom-up technologies involves controlled precipitation/crystallization by adding a suitable non-solvent. The top down technologies include milling or homogenization. However, combination techniques that involves pretreatment step followed by size reduction are also being used to produce nanoparticles with the desired size distribution.

4.1 Bottom-up technologies (Precipitation methods)

Precipitation has been applied for many years for preparation of fine particles, particularly in the development of photographic film, and lately for preparation of sub-micron (nano) particles for pharmaceutical applications (Otsuka et al., 1986; Illingworth, 1972). Examples for precipitation techniques are hydrosols developed by Sucker (Sandoz, presently Novartis) and Nanomorph developed by Soliqs/Abbott (Musliner, 1974; Sjostrom et al., 1993; Gassmann et al., 1994; List et al., 1988; Sucker et al., 1994).

In this process, the drug is dissolved in a suitable solvent and the solution is subsequently added to a non-solvent. This results in high super saturation, rapid nucleation and the formation of many small nuclei. Upon solvent removal, the suspension is sterile filtered and lyophilized (Kipp et al. 2003). The mixing processes may vary considerably. Through careful control of this addition process it is possible to obtain a particle with a narrow size distribution. In the case of Nanomorph, amorphous drug nanocrystals are produced to further enhance dissolution velocity and solubility (Muller et al., 2001a).

Simple precipitation methods, however, have numerous limitations; it is very difficult to control nucleation and crystal growth to obtain a narrow size distribution. Often a metastable solid, usually amorphous, is formed which is converted to more stable crystalline forms (Violante et al., 1989; Bruno et al., 1992). Furthermore, non-aqueous solvents utilized in the precipitation process must be reduced to toxicologically acceptable levels in the end product and due to the fact that many poorly soluble drugs are sparingly soluble not only in aqueous but also in organic media. Considering these limitations, the "bottom up" techniques are not widely used for production of drug nanocrystals. Instead, "top down" technologies that include homogenization and milling techniques are more frequently used.

4.2 Top-down technologies

The two top down technology frequently used for producing drug nanoparticles include;

a. High pressure homogenization
b. Milling

a. High pressure homogenization methods

One of the disintegration method used for size reduction is high-pressure homogenization. The two-homogenization principles/homogenizer types used are;

1. Microfluidisation (Microfluidics, Inc.)
2. Piston-gap homogenizers (e.g. APV Gaulin, Avestin, etc.)

b. Microfluidisation for production of drug nanoparticles

Microfluidisation works on a jet stream principle where the suspension is accelerated and passes at a high velocity through specially designed interaction chambers. Frontal collision of fluid streams under high pressures (up to 1700 bar) inside the interaction chamber generates shear forces, particle collision, and cavitation forces necessary for particle size reduction. The Microfluidizer processor keeps a constant feed stream that gets processed by a fixed geometry which produces high shear and impact necessary to break down larger particles. This process yields smaller particles with narrow particle size distribution with repeatability and scalability.

The interaction chamber's exterior and interior is either made of stainless steel, poly-crystalline diamond (PCD) or aluminum oxide. The poly-crystalline diamond chambers typically have a lifetime 3 - 4 times longer than the aluminum oxide ceramic chambers. Single slotted interaction chambers are used for lab-scale manufacturing and multi-slotted chambers for commercial scale. Multi-slotted chambers are comprised of multiple single slots in parallel for processing larger volumes of the products. There are two types of interaction chambers: Y chamber is useful for liquid-liquid emulsions and finds application in preparing liposomes while Z-chamber is typically used for cell disruption and nanodispersion. A schematic representation of mechanism of particle size reduction in high pressure homogenizers is shown in Fig. 1. The selection of correct chamber depends upon the feed particle size, the application, and the amount of shear and impact required to carryout the operation. The Insoluble Drug Delivery – Particles (IDD-P™) technology developed by SkyePharma Canada Inc. use the Microfluidizer (Jens-Uwe et al., 2008).

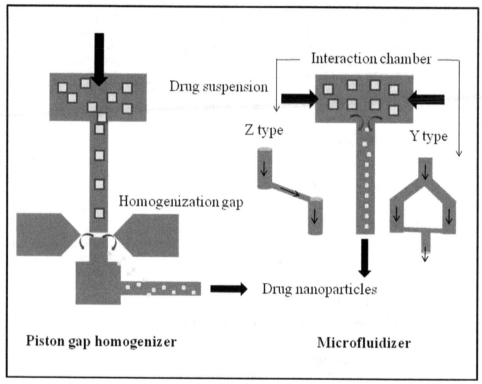

Fig. 1. Schematic representation of mechanism of particle size reduction in high pressure homogenizers

4.2.1 Process parameters affecting particle size

Studies on particle size reduction of a sparingly soluble drug (BCS class II) using the Microfluidizer (Model - Microfluidics M110-P) in our laboratory indicated that particle size reduction depends on various process parameters viz., number of homogenization cycles, homogenization pressure and, stabilizer concentration. At a constant homogenization pressure (30,000 psi) the value of mean particle size d50 decreased with increasing number of cycles from 5 to 60 (Fig. 2). Homogenization pressure has a significant effect on particle size distribution as shown in Fig. 3. At high homogenization pressure (30,000 psi) particle size reduction was significantly higher than at low homogenization pressure (10,000 psi) after 60 homogenization cycles. Surfactant concentration also plays an important role in particle size reduction through particle stabilization by forming a thin layer around the newly formed surface as evident based on the observation that at constant homogenization pressure and homogenization cycles, particle size reduced with increase in surfactant concentration from 10 mg/mL to 12 mg/mL (Fig. 4).

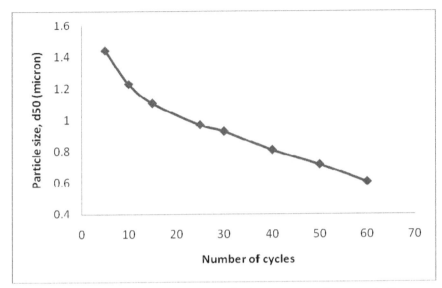

Fig. 2. Effect of number of cycles on mean particle size at constant homogenization pressure

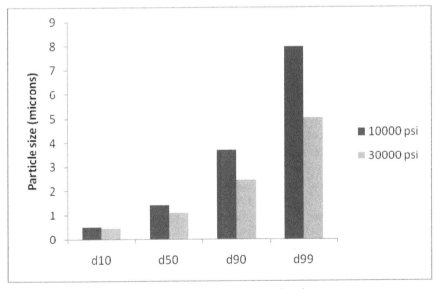

Fig. 3. Effect of homogenization pressure on particle size distribution

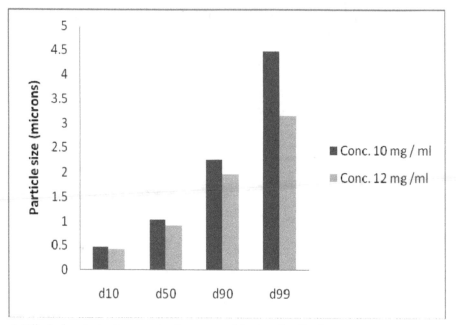

Fig. 4. Effect of surfactant concentration on particle size distribution

4.3 Piston-gap technologies

Using the microfluidisation principle, an alternative technology based on piston-gap homogenizers was developed in the middle of the 1990's for production of drug nanoparticles. Homogenization can be performed in water (DISSOCUBES®) or alternatively in non-aqueous media or water reduced media (NANOPURE®). Dissocubes® technology employs piston-gap homogenizers in which drug powder is dispersed in an aqueous surfactant solution and subsequently forced by a piston through the tiny homogenization gap (5 μm - 20 μm depending upon the viscosity of the suspension and the applied pressure) at a very high pressure (up to 4000 bar). Prior to entering the gap, the suspension is contained in a cylinder with a relatively large diameter compared to the width of the following gap. The resulting high streaming velocity of the suspension causes formation and implosion of the gas bubbles also known as cavitation which results in generation of shockwaves. The drug particle gets reduced by these high shear forces, turbulent flow and powerful shockwaves. Another approach viz., Nanopure® technology (by Pharma-Sol GmbH) is useful for particle size reduction of thermolabile drugs because it use low vapor pressure dispersion media for homogenization that helps in processing at low temperatures due to very little cavitation in the homogenization gap (Muller et al., 1999, 2001b, 2003; Muller RH & Moschwitzer JP, 2005; Jens-Uwe et al., 2008). In addition, there is also a combination process of precipitation followed by a second high-energy homogenization step (NANOEDGE®). The major limitation of this method is that nanoparticulate dispersion of low solid content (usually < 10% w/v) is produced that may be difficult for conversion to solid intermediates required for capsule filling or tableting.

4.4 Milling methods

Conventional milling and precipitation processes generally result in particles much greater than 1 µm. Milling techniques were later refined to enable milling of solid drug particles to sub-micron range. Ball mills are already known from the first half of the 20th century for the production of fine suspensions. In this method, the suspension comprising of drug and stabilizers along with milling media are charged into the grinding chamber. The reduction of particle size occurs because of the shear forces generated due to impaction of milling media. In contrast to high pressure homogenization, this is a low energy technique. Smaller or larger beads can be used as milling or attrition media. The milling media comprise of ceramics (cerium or yttrium stabilized zirconium dioxide), stainless steel or highly cross linked polystyrene resin-coated beads. Potential for erosion of the milling media during the milling process resulting in product contamination is one of the drawbacks of this technology. To overcome this issue, the milling media are often coated (Merisko-Liversidge et al., 2003). Another problem with milling process is the adherence of product to the inner surface of the mill (consisting mainly of the surface of milling media and the inner surface of milling chamber). There are two basic milling principles - either the milling medium is moved by an agitator or the complete container is moved in a complex direction leading to movement of the milling media to generate the shear forces required to fracture the drug crystals. The milling time depends on many factors such as solid content, surfactant concentration, hardness, suspension viscosity, temperature, energy input and, size of the milling media. The milling time may vary from minutes to hours or days depending on the particle size desired (Jens-Uwe et al., 2008).

In the bead milling process used for production of drug nanosuspension, the drug suspension is passed through a milling chamber containing milling media ranging from 0.2 to 3 mm. These media may be composed of glass, zirconium salts, ceramics, plastics (e.g., cross-linked polystyrene) or special polymers such as hard polystyrene derivatives. The drug concentration in the suspension may range from 5 – 40% w/v. Stabilizers such as polymers and/or surfactants are used to aid the dispersion of particles. To be effective the stabilizers must be capable of wetting the drug particles and providing steric and ionic barrier. In the absence of appropriate stabilizers, the high surface energy of the nanometer-sized particles would lead to agglomeration or aggregation of drug crystals. The concentration of polymeric stabilizers can range from 1 – 10% w/v and the concentration of surfactants is generally < 1 % w/v. If required other excipients such as buffers, salts and diluents like sugar can be added to the dispersion to enhance stability and aid further processing (Keck et al., 2006).

The milling chamber has a rotor fitted with disks that can be accelerated at the desired speed (500 – 5000 RPM). The rotation of the disk accelerates the milling media radially. The product flows axially through the milling chamber where the shear forces generated and/or forces generated during impaction of the milling media with the drug provides the energy input to fracture the drug crystals into nanometer-sized particles. The temperature inside the milling chamber is controlled by circulating coolant through the outer jacket. The process can be performed either in a batch mode or in a recirculation mode. The milled product is subsequently separated from the milling media using a separation system. A schematic of the bead milling process is shown in Fig. 5.

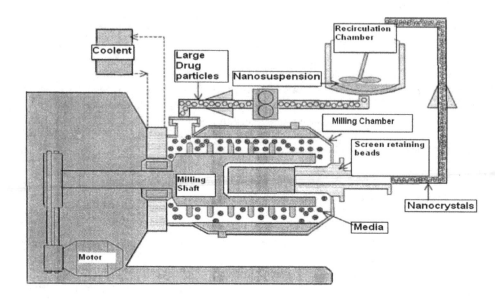

Fig. 5. Schematic of wet bead milling process used for production of drug nanoparticles

Scaling up the bead milling process is relatively easy and convenient because the process variables are scale independent. The batch size can be increased above the void volume (volume in between the hexagonal packaging of the beads) using the mill in a recirculation mode. The suspension is contained in the product container and is continuously pumped through the mill in a circular motion. This increases the batch size with concomitant increase in the milling time because the required exposure time of the drug particles per unit mass to the milling material remains unchanged.

Surfactants or stabilizers have to be added to ensure physical stability of the nanosuspensions. In the manufacturing process the drug substance is dispersed by high speed stirring or homogenizer in a surfactant/ stabilizer solution to yield a macro suspension. The choice of surfactants and stabilizers depends not only on the physical principles (electrostatic versus steric stabilization) and the route of administration. In general, steric stabilization is recommended because it is less susceptible to electrolytes in the gut or blood. Electrolytes if added can reduce the zeta potential and subsequently impair the physical stability, especially of ionic surfactants. In many cases an optimal approach is the combination of a steric stabilizer with an ionic surfactant, i.e, a combination of steric and electrostatic stabilization. There is a wide variety of bead mills available in the market, ranging from laboratory-scale to industrial-scale volumes. The ability for large-scale production is an essential prerequisite for introduction of product into market. In general, bead milling offers a convenient process for production of drug nanoparticle at high concentrations necessary for solid dosage form processing with ease of scale-up for commercial manufacturing.

5. Process optimization for the production of drug nanoparticles

Experimental design has been applied widely to formulation development, and is useful in process optimization and process validation (Fisher RA, 1926). A manufacturing process optimized using design of experiments (DOE) should result in a robust process amenable for seamless scale-up and validation (Dhananjay et al, 2010; Nekkanti et al, 2009a). The process variables in media milling can be optimized using design of experiments (DOE) to understand the effect on particle size, milling time and percentage yield (Nekkanti, et al., 2010). Though a number of statistical designs are reported, a face centered centre composite design (CCD) is often used because it provides information on direct effects, pair wise interaction effects and curvilinear variable effect (Billon et al., 2000; Vaithiyalingam & Khan, 2002; Tagne et al., 2006). For example, a design matrix prepared based on 3 variable factors at three levels (-1, 0, +1) to compute the design using statistical software program Design Expert (version No. 7.3.1) is summarized in Table 2.

S. No	Process Parameters	Level		
		Low (-1)	Center (0)	High (+1)
1	Disk Speed (RPM)	2000	2350	2750
2	Pump Speed (RPM)	40	50	60
3	Bead Volume in Milling Chamber (%)	60	70	80

Table 2. Process variables (factors) and levels

A stepwise regression can be used to generate quadratic equations for each response variable. Analysis of variance (ANOVA) and regression is used to evaluate the significant effects and model building for each response variable. Each response is then fitted to a second-order polynomial model and, the regression coefficients for each term in the model can be estimated along with R^2 and adjusted R^2 of regression model to understand how these parameters effect the critical product attributes either through non-linear, quadratic or interaction effects.

The interaction effect of pump and disc speed on milling time is shown in Fig. 6. The plot indicates that at lower disk speed the milling time (to achieve the desired particle size) increases. This may be attributed to the fact that at low disk speed the shear forces generated by accelerating beads may not be sufficient to fracture the drug crystals into smaller particles. The milling efficiency was high when the disk and pump were run at moderate speeds.

The interaction effect of pump speed and bead volume on particle size is shown in Fig. 7. The plot indicates that increase in pump speed and bead volume resulted in larger particles where as, their interaction resulted in a decrease in particle size. Both pump speed and bead volume have an effect on particle size with bead volume having a significant impact in controlling the drug particle size due to increased probability of impaction.

The interaction effect of disk speed and bead volume on yield is shown in Fig. 8. The plot indicates that the process yields obtained was significantly affected by disk speed and bead volume. At lower disk speed and higher bead volume there was a decrease in yield; this may be attributed to loss in the milling chamber due to sticking.

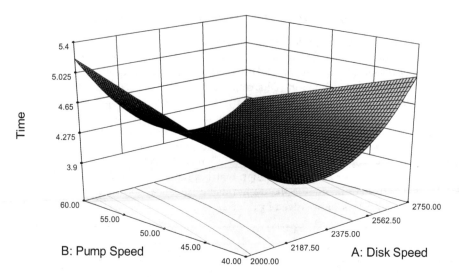

Fig. 6. Effect of disk and pump speeds on milling time

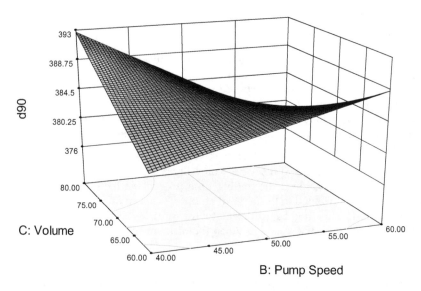

Fig. 7. Effect of pump speed and bead volume on particle size

The robustness of the model used can be validated based on confirmatory trials to ascertain difference between predicted and experimental values. The use of DOE for process optimization will result in a robust scalable manufacturing process with design space established for critical process parameters that can balance milling time, particles size and yield.

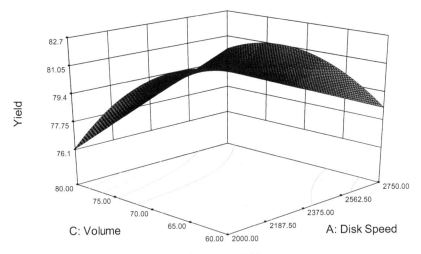

Fig. 8. Effect of disk speed and bead volume on yield

6. Conversion of nanosuspension into solid intermediate

For production of solid intermediate, the water has to be removed from the drug nanosuspension to obtain a dry powder required for tabletting or capsule filling. The objective of solid nanoparticle system is to release the drug nanoparticles in the gastrointestinal (GI) fluids as fine non-aggregated suspension and ensure physical stability upon long term storage. The solvent from nanosuspension can be removed using drying processes such as fluid bed coating / granulation, spray drying and freeze drying. Freeze drying is considered as a complex and cost-intensive process leading to a highly sensitive product. The main challenge is to preserve the re-dispersibility of the nanoparticles upon reconstitution in aqueous media and gastrointestinal fluids. The re-dispersants must be incorporated in the nanosuspensions prior or during the drying step. Commonly used re-dispersants include sugars such as lactose, sucrose and mannitol (Chu, 2000). Generally, re-dispersibility depends on the choice of re-dispersants, surfactants and polymeric stabilizers. The loading capacity of solid intermediate with drug nanoparticles can be adjusted by varying excipient concentrations.

In spray drying process the nanosuspension is atomized using a rotary or air-jet atomizer. In this process fast drying of the liquid feed happens due to the large surface area created by the atomization of the liquid feed into fine droplets and high heat transfer coefficients generated. The short drying time and consequently fast stabilisation of feed material at moderate temperatures make spray drying suitable for producing nanoparticles of drugs that are thermolabile. The spray drying process in general comprise of the following steps;

- **Atomization:** The liquid feed in the form of drug suspension is atomized into droplets by means of a nozzle or rotary atomizer. Nozzles use pressure or compressed gas to atomize the liquid feed while rotary atomizers employ an atomizer wheel rotating at high speed.

- **Drying:** Hot process gas (air or nitrogen) is brought into contact with the atomized feed using a gas disperser for drying. The balance between temperatures, feed flow rate and droplet size controls the drying process.
- **Particle formation:** As the liquid evaporates from the droplet surface the solid nanoparticle that is formed falls into the bottom of the drying chamber.
- **Recovery:** The dried nanoparticles are separated from the exhaust gas using a cyclone separator.

Depending on the spray conditions and nature of formulation, the resulting powder may be filled into hard gelatin capsules or blended with extra granular excipients and compressed in to tablets. In the case of drugs which are acid liable, the capsule or tablet can be coated with enteric polymers to protect acid labile drug from gastric fluids.

An alternative way to convert nanosuspension into solid intermediate is suspension layering onto water soluble carriers. The binders that are necessary for layering must be added before the milling process. The suspension is layered at a predetermined rate on to the water-soluble carriers using a top spray fluid bed process. Top-spraying is the most well known process for drug layering. A top-spray fluid bed processor (FBP) has three components;

- An air-handling system, which can be equipped with humidification or de-humidification and dew-point control
- A product container and expansion chamber
- An exhaust system

The nanosuspension is sprayed into the fluid bed from the top against the air flow (counter current). The granules are dried as they move upward in the fluid bed, small droplets and low viscosity of the spray medium ensures that distribution is uniform resulting in granules with a narrow size distribution (Nekkanti et al., 2008; Basa et al., 2008). The critical process variables of the top-spray layering method include the suspension spray rate, inlet air temperature, fluidization air volume, process air humidity, and the atomization air pressure (Gu et al., 2004).

7. Characterization of drug nanoparticles

There are various techniques used for characterization of drug nanoparticles. There is no single method that can be selected as the "best" for analysis. Most often the method is chosen to balance the restriction on sample size, information required, time constraints and the cost of analysis. Following methods are used commonly for characterization of drug nanoparticles.

7.1 Particle size and size distribution

The characterization of particle size of nanosuspensions is done to obtain information about its average size, size distribution and change upon storage (e.g. crystal growth and/or agglomeration). Particle size distribution of drug nanoparticles can be measured using the following techniques;

7.1.1 Spectroscopy

As nanosuspensions usually comprise of submicron particles, the appropriate method used to evaluate particle size distribution is photon correlation spectroscopy (PCS). In PCS or dynamic light scattering analyses scattered laser light from particles diffusing in a low viscosity dispersion medium (e.g. water). PCS analyze the fluctuation in velocity of the scattered light rather than the total intensity of the scattered light. The detected intensity signals (photons) are used to measure the correlation function. The diffusion coefficient D of the particles is obtained from the decay of this correlation function. Applying Stokes-Einstein equation, the mean particle size (called z-average) can be calculated. In addition, a polydispersity index (PI) is obtained as a measure for the width of the distribution. The PI value is 0 in case particles are monodisperse. Incase of narrow distribution, the PI values vary between 0.10 – 0.20, values of 0.5 and higher indicate a very broad distribution (polydispersity). From the values of z-average and PI, even small increases in size of drug nanoparticles can be evaluated. The extent of increase in particle size upon storage is a measure of instability. Therefore, PCS is considered as a sensitive instrument to detect instabilities during long-term storage (Kerker, 1969).

7.1.2 Laser Diffraction

Laser Diffractometry (LD) developed around 1980 is a very fast and used routinely in many laboratories. The instrument is also used for quantifying the amount of microparticles present, which is not possible using PCS. LD analyses the Fraunhofer diffraction patterns generated by particles in a laser beam. The first instruments were based on the Fraunhofer theory which is applicable for particle sizes 10 times larger than the wavelength of the light used for generating the diffraction pattern. For particle less than 6.3 μm (in case of using a helium neon laser, wavelength 632.8 nm) in size, the Mie theory is used to obtain the correct particle size distribution. The Mie theory requires knowledge of the actual refractive index of particles and their imaginary refractive index (absorbance of the light by the particles). Unfortunately, for most of pharmaceutical solids the refractive index is unknown. However, laser diffractometry is frequently used as a preferred characterization method for nanosuspensions because of its "simplicity" (Zhang et al., 1992; Calvo et al., 1996).

7.2 Microscopy

Microscopy based techniques can be used to study a wide range of materials with a broad distribution of particle sizes, ranging from nanometer to millimeter scale. Instruments used for microscopy based techniques include optical light microscopes, scanning electron microscopes (SEM) transmission electron microscopes (TEM) and atomic force microscopes (AFM). The choice of instrument for evaluation is determined by the size range of the particles being studied, magnification, and resolution. However, the cost of analysis is also observed to increase as the size of the particles decreases due to requirements for higher magnification, improved resolution, greater reliability and, reproducibility. The cost of size analysis also depends upon the system being studied, as it dictates the technique used for specimen preparation and image analysis. Optical microscopes tend to be more affordable and comparatively easier to operate and maintain than electron microscopes but have limited magnification and resolution (Molpeceres et al., 2000; Cavalli et al., 1997).

The surface morphology of 'as-is' drug and spray dried nanoparticles for a sparingly soluble drug, Candesartan cilexetil, examined using scanning electron microscope (Hitachi S-520 SEM, Tokyo, Japan) is shown in Fig. 9. The scanning electron micrographs of "as-is" drug and drug nanoparticles as shown in these Figures illustrate the recrystallization of water-soluble carrier around the drug creating a highly hydrophilic environment preventing particle interaction and aggregation.

Fig. 9. SEM micrographs of "as-is" drug (left); spray-dried drug nanoparticles (right).

7.3 Solid-state properties

7.3.1 Differential Scanning Calorimetry (DSC)

Differential scanning calorimetry (DSC) is used to determine the crystallinity of drug nanoparticles by measuring its glass transition temperature, melting point and their associated enthalpies. This method along with X-ray powder diffraction (XRPD) described below is used to determine the extent to which multiple phases exist in the interior and their interaction following the milling process.

7.3.2 X-ray powder diffraction (XRPD)

X-ray powder diffraction (XRD) is a rapid analytical technique primarily used for phase identification of a crystalline material and can provide information on unit cell dimensions. X-ray diffraction is based on constructive interference of monochromatic X-rays and a crystalline sample. These X-rays generated by a cathode ray tube are filtered to produce monochromatic radiation, collimated to concentrate, and directed toward the sample. The interference obtained is evaluated using Bragg's Law to determine various characteristics of the crystal or polycrystalline material (Hunter et al., 1981).

7.4 Saturation solubility

Saturation solubility evaluations to ascertain drug nanoparticles are usually carried out in buffer media at different pH conditions using a shake flask method. In this method excess amount (100 mg/mL) of drug ("as-is" and dried suspension containing microparticles or nanoparticles) is added to 25 mL of buffer medium maintained at 37°C and shaken for a period up to 24 hours. The samples are filtered using 0.10 μm pore size Millex-VV PDVF filters (Millipore Corporation, USA) prior to analysis and concentrations determined using an HPLC method. The results from saturation solubility for "as-is", micronized and spray dried nanoparticles of Candesartan cilexetil used as a model drug is summarized in Table 3 to demonstrate the impact of particle size on saturation solubility.

Solvents	Solubility (mg/mL)		
	"as-is" drug*	Micronized drug*	Spray dried drug nanoparticles
0.1 N HCl	0.011	0.016	0.134
Acetate buffer pH 4.5	0.001	0.014	0.106
Phosphate buffer pH 6.8	0.001	0.012	0.105
Water	0.000	0.001	0.073

*Solubility was tested in respective solvents containing surfactant and Stabilizer.

Table 3. Saturation solubility of "as-is", micronized and nanoparticles of Candesartan cilexetil

The saturation solubility of Candesartan cilexetil nanoparticles is significantly higher than jet-milled particles and "as-is" drug at all pH conditions. These results clearly demonstrate that reduction in particle size to sub-micron or nanometer range affects saturation solubility resulting in enhancement of dissolution rate.

The effect of particle size of Candesartan cilexetil following oral administration in male Wistar is shown in Fig. 10. As seen there is a significant enhancement in the rate and extent of drug absorption for nanosuspension. The rate and extent of drug absorption showed a 2.5-fold increase in the area under the plasma concentration - time curve (AUC_{0-t}) and a 1.7-fold increase in the maximum plasma concentration (C_{max}) and, significant reduction in the time required (1.81 hours as compared to 1.06 hours) to reach maximum plasma concentration (T_{max}) when compared to the micronized suspension (Nekkanti et al., 2009b).

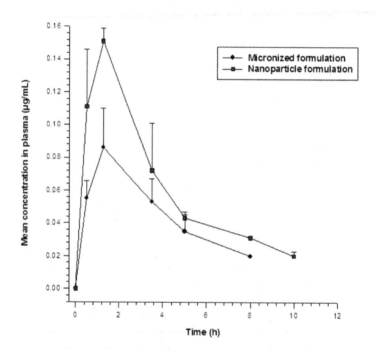

Fig. 10. Plasma concentration–time profiles following oral administration of micronized suspension and drug nanosuspension to male Wister rats

8. Conclusion

Enhancing solubility and dissolution rate of poorly soluble compounds correlates with improved pharmacokinetic (PK) profile. The approach herein can be extended to other BCS class II compounds where absorption is either solubility and/or dissolution limited. The manufacturing process used is relatively simple and scalable indicating general applicability of the approach to develop oral dosage forms of poorly soluble drugs. The enhanced bioavailability should translate into reduced dose, mitigate food effects, offer better dose delineation and result in faster onset of action that may translate into improved therapeutic outcome.

9. References

Basa, S.; Karatgi, P.; Muniyappan, T.; Raghavendra, P. & Pillai, R. (2008). Production and in vitro characterization of solid dosage form incorporating drug nanoparticles. *Drug Development and Industrial Pharmacy*, (11): 24–28.

Billon, A.; Bataille, B.; Cassanas, G. & Jacob, M. (2000). Development of spray-dried acetaminophen microparticles using experimental designs. *International Journal of Pharmaceutics*, 203, 159-168.

Bruno, JA.; Doty, BD. & Gustow, E. (1992). Method of grinding pharmaceutical substances. *US Patent* 5518187.

Calvo, P.; Vila-Jato, JL. & Alonso, MJ. (1996). Comparative in vitro evaluation of several colloidal systems, nanoparticles, nanocapsules, and nanoemulsions, as ocular drug carriers. *Journal of Pharmaceutical Sciences*, 85, 530–536.

Cavalli, R.; Caputo, O.; Carlotti, E.; Trotta, M.; Scarnecchia, C. & Gasco, MR. (1997). Sterilization and freeze-drying of drug-free and drug-loaded solid lipid nanoparticles. *International Journal of Pharmaceutics*, 148:47–54.

Chu, B. & Liu, T. (2000). Characterization of nanoparticles by scattering techniques. *Journal of Nanoparticle Research*, 2, 29–41.

Dhananjay, S.; Seshasai, M.; Gowtamrajan, K.; Giriraj, T.; Rajesh, V. & Srinivasarao, P. (2010). Optimization of formulation and process variable of nanosuspension: An industrial prospective. *International Journal of Pharmaceutics*, 402, 213–220.

Dressman, JB.; Amidon, GL.; Reppas, C. & Shah, VP. (1998). Dissolution testing as a prognostic tool for oral drug absorption: immediate release dosage forms. *Pharmaceutical Research*, 15 (1), 11-22.

Elaine, M.; Merisko-Liversidge. & Gary, L. (2008). Drug nanoparticles: Formulating poorly water-soluble compounds. *Toxicologic Pathology*, 36, 43-48.

Fisher RA. (1926). The Design of Experiments, *Oliver & Boyd*, London.

Gassmann, P.; List, M. Schweitzer, A. & Sucker, H. (1994). Hydrosols - alternatives for the parenteral application of poorly water-soluble drugs. *European Journal of Pharmaceutics and Biopharmaceutics*, 40, 64–72.

Gu, L.; Liew, CV. & Heng, PW. (2004). Wet spheronization by rotary processing: a multistage single-pot process for producing spheroids. *Drug Development and Industrial Pharmacy*, 30, 111-123.

Hunter RJ, Ed. (1981). Colloid Science: Zeta Potential in Colloid Science: Principles and Applications. London: *Academic Press*.

Illingworth, BD. (1972). Preparation of silver halide grains, *US Patent 3,655,394*.

Jens-Uwe, AH.; Junghanns. & Rainer, H. Muller. (2008). Nanocrystal technology, drug delivery and clinical applications. *International journal of Nanomedicine*, 3(3), 295–309.

Keck, CM. & Müller, RH. (2006). "Drug nanocrystals of poorly soluble drugs produced by high pressure homogenisation." *European Journal of Pharmaceutics and Biopharmaceutics*, 62(1), 3-16

Kerker, M. (1969). The scattering of light and other electromagnetic radiation. New York: *Academic Press.*

Kipp, JE.; Wong, JCT.; Doty, MJ. & Rebbeck, CL. (2003). Micro precipitation Method for preparing sub micron suspensions. *United States Patent 6,607,784*, Baxter International Inc. (Deerfield, IL), USA.

Lipinski, C. (2002). Poor aqueous solubility: an industry wide problem in drug discovery. *American Pharmaceutical Review*, 5, 82-5.

Lipinski, CA. (2004). Avoiding investment in doomed drugs, is poor solubility an industry wide problem?. *Current Drug Discovery*, 17-19.

List, M. & Sucker, H. (1988). Pharmaceutical colloidal hydrosols for injection. *GB Patent 2200048*, Sandoz Ltd.

Liversidge, GG. & Cundy, KC. (1995). Particle size reduction for improvement of oral bioavailability of hydrophobic drugs. *International Journal of Pharmaceutics*, 125, 91-97.

Mary C.; Till, Michele M.; Simkin. & Stephen Maebius. Nanotech Meets the FDA (2005): A success Story about the First Nanoparticulate drugs Approved by the FDA, *NANOTECHNOLOGY LAW & BUSINESS*, Volume 2.2

Merisko-Liversidge, E.; Liversidge, GG. & Copper, ER. (2003). Nanosizing: a formulation approach for poorly-water-soluble compounds. *European Journal of Pharmaceutical Sciences*, 18, 113-20.

Molpeceres, J.; Aberturas, MR. & Guzman, M. (2000). Biodegradable nanoparticles as a delivery system for cyclosporin: preparation and characterization. *Journal of Microencapsulation*, 17, 599-614.

Muller, RH. & Bohm, BHL. (2001a). *Dispersion techniques for laboratory and industrial scale processing, Wissenschaftliche Verlagsgesellschaft, Stuttgart.*

Müller, RH.; Becker, R. & Kruss, B. (1999). Pharmaceutical nanosuspensions for medicament administration as systems with increased saturation solubility and rate of solution. *US Patent 5858410.* USA.

Muller, RH.; Dingler, A.; Schneppe, T. & Gohla, S. (2000). Large-scale production of solid lipid nanoparticles (SLN) and nanosuspensions (DISSOCUBES). In:Wise, D.L. (Ed.), *Handbook of Pharmaceutical Controlled Release Technology. Marcel Dekker*, 359-376.

Muller, RH.; Jacobs, C. & Kayser, O. (2001b). Nanosuspensions as particulate drug formulations in therapy: Rationale for development and what we can expect for the future. *Advanced Drug Delivery Reviews*, 471, 3-19.

Müller, RH.; Jacobs, C. & Kayser, O. (2003). DissoCubes – a novel formulation for poorly soluble and poorly bioavailable drugs. In Rathbone MJ, Hadgraft J, Roberts MS (Eds). Modified-release drug delivery systems. New York: *Marcel Dekker*, 135-49.

Müller, RH. & Moschwitzer, JP. (2005). Method and apparatus for the production of ultrafine particles and coating of such particles. DE 10 2005 053 862.2 Application, Germany.

Musliner, WJ. (1974). Precipitation of metal salts. *US Patent* 3,790,387.

Nekkanti, V.; Karatgi, P.; Joshi, M. & Pillai, R. (2008) Developing nanoparticle formulations of poorly soluble drugs. *Pharmaceutical Technology Europe.* (11), 24–28.

Nekkanti, V.; Marella S.; Rudhramaraju, R. & Pillai, R. (2010). Media milling process optimization for manufacture of drug nanoparticles using design of experiments (DOE). *AAPS*, USA, Nov-10.

Nekkanti, V.; Muniyappan, T.; Marella, S.; Karatgi, P. & Pillai, R. (2009a). Spray-drying process optimization for manufacture of drug-cyclodextrin complex powder using design of experiments. *Drug Development and Industrial Pharmacy*, 5(10), 1219-29.

Nekkanti, V.; Pillai, R.; Venkateswarlu, V. & Harisudhan, T. (2009b). Development and characterization of solid oral dosage form incorporating candesartan nanoparticles. *Pharmaceutical Development Technology*, 14 (3), 290–298.

Otsuka, M. & Kaneniwa, N. (1986). Effect of seed crystals on solid state transformation of polymorphs of chloramphenicol palmitate during grinding. *Journal of Pharmaceutical Sciences*, 75, 506–511.

Rajesh Dubey. (2006). Pure Drug Nanosuspensions: Impact of nanosuspension technology on drug discovery & development. *Drug Delivery Technology*, Vol. 6, No. 5; 65-71.

Ranjita, S. & Rainer, H. Muller. (2010). Nanocrystals: Industrially feasible multifunctional formulation technology for poorly soluble actives, *International Journal of Pharmaceutics*, 399, 129-139.

Robertson, AJB. (1983). The development of ideas on heterogeneous catalysis. Progress from Davy to Langmuir. *Platinum Metals Review*, 27, 31–39.

Rogers, TL.; Johnston, KP. & Williams, III R.O. (2001). A comprehensive review: solution-based particle formation of pharmaceutical powders by supercritical or compressed fluid CO2 and cryogenic spray-freezing technologies. *Drug Development and Industrial Pharmacy*, 27, 1003–1015.

Sjostrom, B.; Kronberg, B. & Carlfors, J. (1993). A method for the preparation of submicron particles of sparingly water-soluble drugs by precipitation in oil-in-water emulsions. *Journal of Pharmaceutical Sciences*, 82, 579–583.

Sucker, H. & Gassmann, P. *(1994).* Improvements in pharmaceutical compositions, *GB-Patent 2269536A, Sandoz LTD. CH, GB.*

Tagne, PT.; Briancon, S. & Fessi, H. (2006). Spray-dried microparticles containing polymeric nanocapsules: formulation aspects, liquid phase interactions and particle characteristics. *International Journal of Pharmaceutics*, 325, 63-74.

Vaithiyalingam, S. & Khan, MA. (2002). Optimization and characterization of controlled release multi-particulate beads formulated with a customized cellulose acetate butyrate dispersion. *International Journal of Pharmaceutics*, 234, 179-193.

Violante, M.R. & Fischer, H.W. (1989). Method for making uniformly sized particles from water-insoluble organic compounds. *US Patent 4,826,689.*

Zhang, H. & Xu, G. (1992). The effect of particle refractive index on size measurement. *Powder Technology*, 70, 189–192.

6

The Development of Magnetic Drug Delivery and Disposition

Michał Piotr Marszałł
Department of Medicinal Chemistry, Collegium Medicum in Bydgoszcz,
Nicolaus Copernicus University
Poland

1. Introduction

The process of drug delivery and disposition in the modern scientific aspect is very complex. Advances in many fields are converging to make the commercialisation of advanced drug delivery concepts possible. It integrates many disciplines, including biotechnology, medicine and pharmacology. Innovative devices should protect labile active ingredients, precisely control drug release kinetics and minimise the release of the drug to non-target sites. Rapid advances in these areas led to the revolutionary change in discovery of new methods of drug delivery and disposition.

In the last two decades, interest in nanometre-size particles such dendrimers, liposomes, micelles and polymeric particles has increased significantly. As a new drug delivery system, nanoparticles resolved many drug pharmacokinetics problems related with precisely control drug release, poor stability and toxicity of active ingredient. Crucially, the nano-sized particles have very important advantage in intravenous application (Ilium et al., 1982). Generally, their circulation time is limited by the smallest blood capillaries in human vascular system (Table 1) (Arruebo et al., 2007; Yoo et al., 2011). Therefore, the particles size smaller than 1000 nm are usually objects of interest as a new potential carriers for efficient drug delivery (Figure 1). Also, the other parameters including shape and surface activity influence on their distribution and elimination from vascular system.

In the last years, a new drug targeting employing magnetic particles as a drug carriers is a promising tool for the more effective and safe chemotherapy. Initially, the micro- and nano-sized magnetic particles (MMPs; MNPs) provided an original modern technology for bioseparations, especially for ligand "fishing", protein, enzyme, DNA, RNA and cell isolation or purification (Corchero & Villaverde, 2009). Because of the superparamagnetic properties, magnetic particles can be used to simply isolate any target and can be linked with diverse manual and automated applications (Marszałł et al., 2008, 2011a).

The stability of captured protein/enzyme/drug depends on the morphology of the magnetic microspheres. There are many materials available with magnetic properties. However, the iron oxide (Fe_3O_4) or meghemite ($\gamma\text{-}Fe_2O_3$) are most commonly employed for in vivo applications. The materials, such as cobalt, chromium, nickel and other iron-based metal oxides, such as $CoFe_2O_4$, $NiFe_2O_4$, $MnFe_2O_4$ should be limited for biomedical applications due to their high toxicity and long-term changes in enzyme kinetics. The ferrite nanoparticles have large surface area to volume ratio and can adsorb plasma proteins and other biomolecules as well as agglomerate in vivo (Douziech-Eyrolles et al., 2007). Additionally, the magnetic nanoparticles without coating are rapidly cleared by macrophages in the reticulo-endothelial system. Because of the high physicochemical activity of the surface of magnetic particles and risk of corrosion process they are often encapsulated or coated with protective material: carbohydrates, lipids, gold, proteins and synthetic polymers.

Classes of blood capillaries	Permeability (size)	Location
Tight-junction capillaries	< 1nm	Blood-brain barrier (BBB)
Continuous capillaries	~ 6 nm	Muscle, lung, skin
Fenestrated capillaries	~ 5-60 nm	Kidney, intestine and some endocrine and exocrine glands
Sinusoid capillaries	~ 100-1000 nm	Liver, spleen, bone narrow

Table 1. Different classes of blood capillaries in human vascular system (Arruebo et al., 2007).

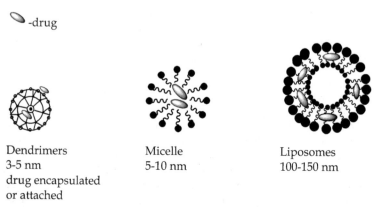

-drug

Dendrimers
3-5 nm
drug encapsulated
or attached

Micelle
5-10 nm

Liposomes
100-150 nm

Fig. 1. The most popular nano-sized drug delivery systems for in vitro and in vivo applications.

With the development of improved syntheses and techniques for determining the particle-size distribution the new name of MNPs was defined as a superparamagnetic iron oxide naoparticles (SPIONs). The main principles of SPIONs compared to ferromagnetic particles is lack of permanent magnetic dipole. The promising perspectives for biomedical application of SPIONs is that the moment of each particles fluctuates between different directions at the rate given by the temperature and the applied magnetic field (Woodward et al., 2007). This phenomenon allows for the accumulation of local chemotherapeutic agent in vivo with the use of external magnetic field which can be controlled by magnetisation expressed as:

$$M = c\mu L(\xi) = c\mu[coth(\xi) - (1 / \xi)]$$

where c is particle concentration, μ is the magnetic moment of the particle and $L(\xi)$ is the Langevin function in which $\xi = \mu H/kT$, where H is the applied magnetic field, k is the Boltzman constant and T is the temperature in Kelvin. μ is the magnetic moment of the particle which is related to the particle volume (v) by $\mu = vM_s$ where M_s is the saturation magnetization.

SPIONs have interesting properties such as high field irreversibility, high saturation field and they do not show magnetic interaction after the external magnetic field is removed (Mahmoudi et al., 2011). The advantage of SPIONs compared to ferromagnetic particles is that the standard particles tend to agglomerate due to their permanent magnetic dipoles. Thus, the suspension of SPIONs produce very stable ferrofluids which in vivo application are still investigated.

The use of solid polymer matrix or core-shell type silica-based superpara- and paramagnetic particles as a carriers for binding proteins, enzymes and drugs received increasing attention in drug delivery studies. The external multifunctional layer of magnetic beads provides reactive groups that are able to covalently conjugate aldehyde, amine, carboxy, hydroxyl, ketone, sulfhydryl and organic linkers for formation of stable linkage (Figure 2).

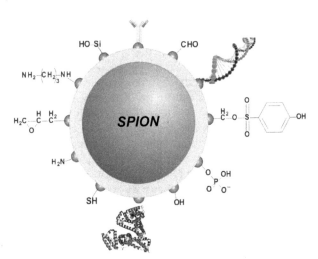

Fig. 2. The core-shell type silica-based superparamagnetic iron oxide nanoparticle (SPION) with different functional groups on the surface.

The pioneering "medical" application in the treatment of lymphatic nodes and metastases based on injecting "metallic particles" preheated in a magnetic field was first published in 1957 (Gilchrist et al., 1957). However, the potential medical application of new finding was limited by the poor performance of particles. In recent years, the new nanotechnologies allowed for the precise manufacturing of nanoparticles with specific size, surface area and biocompatibility with the view of in vivo application. Moreover, the magnetic nanoparticles offer many exciting possibilities for development not only a new potential drug carriers but also are being investigated as a magnetic contrast agents in magnetic resonance imaging (MRI) and hyperthermia agents in hyperthermia cancer treatment.

2. Magnetic drug delivery system in cancer therapy

2.1 Ionic binding of the drug to the magnetic nanoparticles

In 1978, magnetic particles were first applied as a new class of drug target (Widder et al., 1978). The superparamagnetic properties of the magnetic micro- and nanoparticles have opened promising new perspectives for in vivo application. The magnetic nanoparticles as drug carriers provide huge opportunities in cancer treatment. The use of such carriers in targeted therapy considerably reduces the side effects of conventional chemotherapy. A novel carrier system allows for intravenous drug delivery and the local accumulation of chemotherapeutic agent which is comparable to that achieved by the administration of a 100-fold higher dose of the drug. The magnetic drug targeting enables the fast and precise location of drug in the body with the use of external magnetic field. After the vascular injection, the particles can be transported and concentrated at desired location with the help of magnet (Figure 3). For drug delivery application the optimal size of nanoparticles should be in the range from 10 to 200 nm. The micrometer size below 200 allows for systematic administration in circulation system and also into targeted tissue as well as enhance the ability of nanoparticles to evade the biological particulate filters, such as the reticulo-endothelial system (Tran & Webster, 2010). To date, magnetic drug targeting has been studied mostly in pre-clinical models for cancer therapy with intravascular administration of chemotherapeutics.

The Phase I clinical trial demonstrated that the epirubicin attached to magnetic fluids (ferrofluids) could be concentrated in locally advanced tumor by magnetic field. The drug was bound to phosphate groups of starch polymers which covered the iron oxide core with particle size of 100 nm. The "magnetic epirubicin" with doses in the range of 5-100mg/m^2 was infused i.v. over 15 min into a vein located contra laterally to the tumor in patients with advanced and unsuccessfully pretreated metastatic breast cancer, chondrosarcoma and squamous cell carcinoma. The neodymium magnets were large 8x4x2cm or 3x3x1cm and were kept at a distance of 0.5 cm to tumor surface. The clinical trials provided the important information about the conditions of drug release, distribution and mechanism of action. The researchers concluded that the size of magnetic particles should be close to 1μm to increase of their accumulation in tumor site and, consequently, the concentration of the drug (Lübbe et al., 1996). The better optimization of ionic binding epirubicin on to the surface of magnetic nanoparticles also can improve the drug release in physiological parameters. Hence, a new group of novel nano- and micro magnetic particles consisting of various synthetic and natural matrices were investigated (Bergemann et al., 1999).

The idea for ionic binding of the drug to magnetic particles was continued because of the ligand/drug can be easily released from the matrix by simply changing either ionic strength or the pH. The surface of polyvinylalcohol, polyacrylates and polyacharide matrices consisted of cationic exchange groups (i.e. sulphate, phosphate) is able to ionically bind pharmaceutical active compound or biological substances. Furthermore, the modified starch nanoprticles with diethylaminoethyl (DEAE) groups can bind to negatively charged cell membrane. This tool was successful used for separation cells and gene transfer by magnetic transfection of cytokine-induced killer cells (CIK-cells) with plasmid DNA. The representative cationic and anionic nanopartcles are presented in Figure 4 (Bergemann et al., 1999).

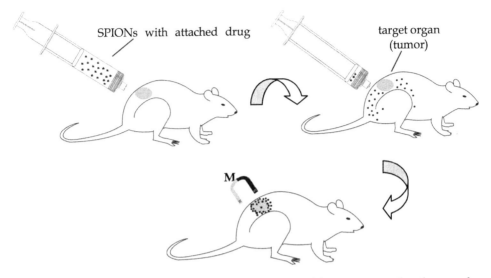

Fig. 3. The scheme of in vivo application of magnetic drug delivery system after the vascular injection; superparamagnetic iron oxide nanoparticles (SPIONs), external magnet (M).

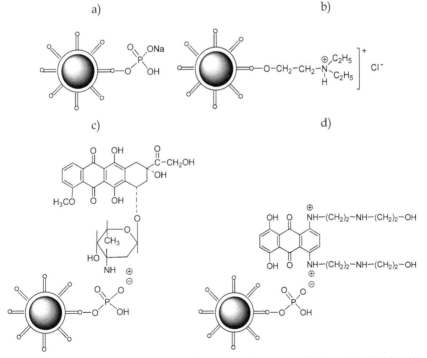

Fig. 4. The representative ion-exchange-active magnetic nanoparticles: cationic (a), anionic (b) as well as loaded with cytostatic drugs: epirubicin (c) and mitoxantrone (d).

The magnetic dug targeting approach offers a new opportunity to treat malignant tumors locoregionally. The treatment of squamous cell carcinoma in rabbits with nanoparticles covered with modified starch to which the mitoxantrone was ionically bound, caused complete and permanent remission of the cancer compared with control group (Alexiou et al., 2000). The advantage of ionically bound mitoxantrone is that the anticancer agent is able to desorb from the magnetic caries after the 30 min (half-time). Determination of time of desorption is very important because the ferrofluids have to be transferred to the tumor region by the magnetic field. Next, the drug must dissociate to act within the tumor. Generally, the total release of drug from the magnetic carriers is recommended at less than 1 h. The 100 nm particles size and strong magnetic field (1.7 Tesla) are optimal for efficient treatment of smaller animals such as mouse or rat. However, the appropriate magnetic field strength and particle size for treatment of deep body cavities and human cancer has to be optimised. The extensive biodistribution study with Iod[123] – labeled ferrofluids demonstrated that magnetic flux density is an important factor in magnetic drug targeting (Alexiou et al., 2005).

2.2 Covalent binding of the drug to the magnetic nanoparticles

On the contrary to ionic binding, the covalent binding of the drug to magnetic particles prevents unwanted drug release in a physiological environment. Hence, the new strategy is based on the covalent coupling of antibodies, nucleic acids, proteins and active compounds on to the surface of functional magnetic particles coated with polymer or silica. The most popular surface groups such as amine, carboxy and aldehyde allow for a covalent conjugate almost any custom ligand (drug) (Figure 5). The coupling efficiency depends on the functional group density and type of linker. The magnetic particles with primary amine or aldehyde functional groups on the surface are used to covalently conjugate primary amine - containing ligands. The particles coated with carboxyl functional groups on the surface can be covalently attached to primary amine- containing ligands via a stable amide bond.

This strategy was proposed for methotrexate-modified superparamagnetic nanoparticles as a drug carrier in controlled drug delivery, targeted at cancer diagnostics and therapeutics (Kohler et al., 2005). The nanoparticles were modified with (3-aminopropyl)-trimethoxysilane (APS) to form aminopropyl layer and subsequently conjugated with methotrexate through amidation using the 1-ethyl-3-[3-dimethylaminopropyl]carbodiimide hydrochloride (EDC) as a crosslinking agent.

The in vitro studies in both human breast cancer (MCF-7) and human cervical cancer (HeLa) cells demonstrated the successful internalization of used methotrexate conjugates into lysosomes as a promising tool for anticancer activity. Methotrexate is an analogue of folic acid. Folic acid is generally recognized as an effective targeting agent which receptors are overexpressed on the cell membranes of many cancer cells. Hence, the methotrexate is used as a therapeutic agent for the treatment of several forms of cancer (carcinomas, lymphomas, breast, head, leukemia, neck cancer) (Messmann & Allegra, 2001). The proposed methotrexate-modified superparamagnetic nanoparticles enable real-time monitoring of drug carriers by MRI after intravenous drug delivery. Moreover, the methotrexate is not released from magnetic carriers due to the high stability binding of drug-particles in physiological parameters. Curiously, the covalent-amide bond is broken under conditions

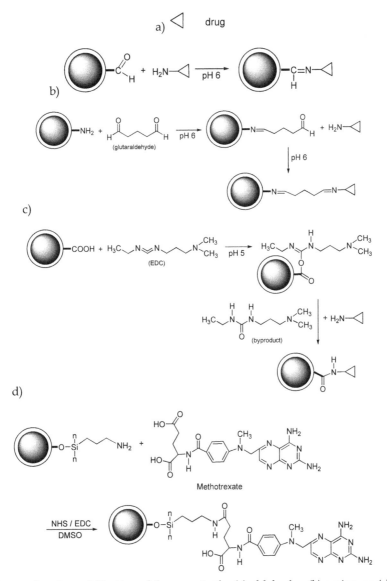

Fig. 5. The covalent immobilization of drugs on to the (a) aldehyde-, (b) amine- or (c) carboxy-terminated magnetic nanoparticles surface and metotrexate on to the surface modified with (3-aminopropyl)-trimethoxysilane (APS) (d).

present in the lysosomal compartment. The intracellular uptake of the metohrexate-magnetic nanoparticle conjugates is assumed to depend on a receptor-mediated endocytosis (Figure 6) (Kohler et al., 2005). Next, they are transported as a endosomes and fused with lyposomes containing proteases and low pH. In this condition, the peptide bond between the methotrexate and magnetic carriers is broken and anticancer drug is released inside the target cell.

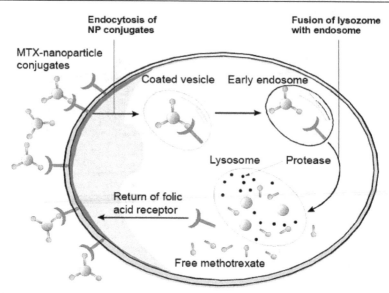

Fig. 6. The intracellular model of the uptake of methotrexate (MTX) – modified
nanoparticles into breast cancer cells (Kohler et al., 2005).
(Adapted with permission from @ 2005 American Chemical Society).

2.3 Magnetic nanoparticles for tumor imaging and therapy

Magnetic resonance imaging (MRI) is widely used as a screening non-invasive method of the
human body. It is also used to monitor cell migration to targets tissue in cell-based therapy.
The fact that the magnetic nanoparticles with labeled cells or as a drug targeting can be
visualized using MRI, they are a new alternatives and noninvasive imaging techniques for
monitoring of cell or drug migration to target tissue. For in vitro application, their advanced
development in cell manipulation/therapy, biomolecule separation, selection and purification
was found (Gijs, 2004). The ability to produce a distortion in magnetic field monitored by MRI,
allowed for the increasing application of magnetic beads in vivo. The different strategies of the
use of anionic magnetic nanoparticles (AMNPs), ultrasmall paramagnetic iron oxides
(USPIOs) and superparamagnetic iron oxide nanoparticles (SPIONs) have been demonstrated
as a contrast agents to identify magnetically labelled cells during MRI monitoring of cellular
therapies (Wilhelm & Gazeau, 2008; Modo et al., 2005). The demonstrated studies confirm the
efficacy of labelling with magnetic particles for a wide variety of mammalian cells, including
non- and phagocytic cells, different species, cell size, types and culture properties. The MRI
contrast is a result of different signal intensities of tissue, produced in response to applied
radio frequency pulses (Gijs, 2004). Labelled-specific magnetic particles provide a suitable
source of contrast and convenient tool for the non-invasive study of biological processes, such
a tumor imaging and therapy with the use of MRI.

The use of magnetic particles can significantly improve hyperthermia cancer treatment
(Marszałł, 2011b). This therapy involves raising the temperature of the target tissue to
43-46°C. In this conditions its sensitivity to chemo- and radiotherapy increases and may
additionally stimulate activities of the host immune system (Ang et al., 2007). The problem

with hypethermia therapy is the heating the large area of tissue or body in general, not only the tumor region. The healthy tissues adsorb microwave, laser and ultrasound energy which can cause burns and blisters (Phillips & Johnson, 2005). Magnetic hyperthermia is one of the anti-cancer approches based on the introduction of ferro- or superparamagnetic particles into the tumor tissue. The main advantage of magnetic particle hyperthermia is that the particles can heat previously localized target issue by external magnetic field. Next, under an applied magnetic field, energy is converted to thermal energy in tumor region which destroys cancer tissues (Figure 7) (Cole et al., 2011).

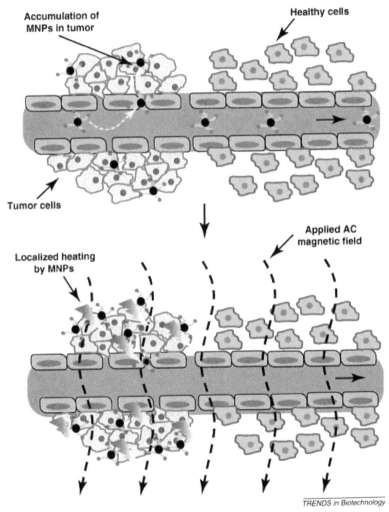

Fig. 7. Scheme of magnetic hyperthermia treatment of affected tissue; (a) accumulation of magnetic nanoparticles (MNPs) by the magnet at the tumor site; (b) exposition of tumor cells to an altering current (AC) magnetic field (Cole et al., 2011). (Adapted with permission from @ 2011 Elsevier).

Cancer-specific binding agents such as antibodies, hormones and other exo- and endogenous substances attached to magnetic nanoparticles has promising implications for magnetic fluid hyperthermia treatment (MFH) of breast, prostate and thyroid cancer or in vivo magnetic resonance imaging of acute brain inflammation (Phillips & Johnson, 2005; McAteer et al., 2007). MFH cancer treatment is based on the injecting a fluid containing magnetic nanoparticles directly into cancer region and the use of altering magnetic filed to generate the heat and destroy the tumor. Alternatively, the fluid can be injected into an artery that supplies the tumor with blood. However the location of the magnetic fluid should be precisely to minimize the effect of MFH. Attaching other cancer specific agents onto the surface of magnetic particles such as monoclonal antibodies or viruses has also promising applications. The agents can be modified to selectively bind cancerous cells. For instance, the genetic material in the viruses can be replaced with anticancer drug and than can be precisely released in cancer area at elevated temperature. The promising benefits of hyperthermia cancer treatment open a new perspectives for application of magnetic particles in cancer treatment.

3. Nanomagnetosols

Aerosol drug delivery system allows for the pulmonary drug administration delivery of therapeutic agents. The non-invasive drug delivery is mainly used for treatment of lung disorders such as asthma, chronic obstructive pulmonary disease and lung cancer. Pulmonary delivered drugs are rapidly absorbed alike to other mucosal surfaces. The appropriate size of aerosols droplets determines the passive targeting by their deposition in different lung regions. However the exact targeting to specific lung regions other than airways or the lung periphery has not been achieved to date (Dames et al., 2007). The high and effective drug concentration at disease site in standard chemotherapy in lung cancer with cytotoxic drugs is particularly difficult. The problem is that the cytotoxic potency of chemoterapeutics is not limited to cancer region.

The innovative study reports for the first time target aerosol delivery to the lung achieved with aerosol droplets comprising superparamagnetic iron oxide nanoparticles in combination with a target-directed magnetic gradient field. The high efficiency of aerosol droplets comprising SPIONs in combination with external magnetic gradient field was confirmed by computer aided simulation and demonstrated experimentally in mice. In contrast to intravenous magnetic drug targeting, the pulmonary drug administration has less limitations of the drug binding capacity of the nanoparticles. The new drug carriers, also called as a nanomagnetosols, offer high a flexibility of the system (Dames et al., 2007). The main advantage is that drug dose can be easily adjusted by the changing the drug concentration in the magnetic particles solution. Additionally, different drugs can be attached in different manner to the magnetic particles and also they can be co-delivered with other nanocarriers such as liposomes.

The successful magnetic aerosol targeting in vivo with plasmid DNA (pDNA) was achieved in intact mouse. The electromagnet with the iron circuit and the tip comprising an iron-cobalt alloy allows for accurate positioning of the SPIONs in selected site of the lung. The amount of deposited pDNA did not differ between the left and right lung in the absence of the external magnetic field. A twofold higher amount of pDNA was evaluated in the magnetized right lung than in unmagnetized left lung (Dames et al., 2007). Moreover, the authors conclude that the nanomagnetosol droplets with SPIONs are responsible for their transport, deposition the accumulation of the attached ligand/drug in affected tissue and not the single SPION as was supposed previously.

The anatomy of human lung also allows for the implementation of magnetosols in targeted therapy. The concept of non-invasive targeted delivery of magnetic aerosol droplets to the lung is presented in Figure 8. In last decade, the use of new drug delivery approach in the small animal were often demonstrated and well elaborated. The results confirm the higher deposition (~12%) of superparamagnetic nanoparticles (of ~50 nm diameter) on the magnetized airway surface and only ~4% on the unmagnetized airway area (Dames et al.,

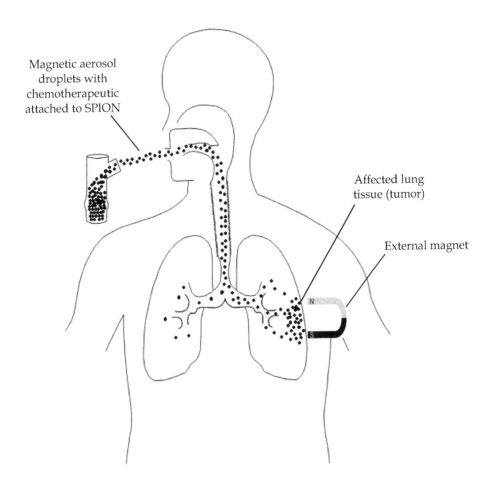

Fig. 8. The concept of non-invasive delivery of magnetic aerosol droplets to the human lungs and concentrated to the target site (tumor) with the help of magnet.

2007). Crucially, the non-deposited aerosol droplets can be transferred to further lung zones or exhaled. Regardless of the advantages for aerosol droplets with magnetic particles, the objective risk assessment studies have to be performed. Many of studies confirmed that the metal oxide particles are very toxic, especially in the pulmonary system (Machado et al., 2010). Nowadays, the in vitro studies are focused on the characterization and cytotoxic assessment of aerosol particulates, especially the nanoparticles based on heavy alloys W-Fe-Ni and W-Fe-Co. The exposure to Co, Ni, W and Fe can cause pulmonary fibrosis asthma, oedema and pneumonia among other effects.

The cytotoxic assessment studies of aerosol particulates are usually performed with the use of model for lung tissue – human epithelial cells (A549). The use of different filter exposure cytotoxicity assays allow to assess the inflammatory and related respiaratory health effects. Generally, the inhaled ~1μm particulates are moved by cilia of the bronchial epithelial cells towards the upper respiratory tract. The nanoparticles at size of <100nm can accumulate, aggregate and even cause oxidative stress and inflammation. So far it is proved that the most important parameters that can affect cytotoxicity are chemical composition, size and time of deposition of magnetic nanoparticles. Because of a lack of detailed limitations regarding the in vivo use of nanomagnetosols, it is necessary to continue the risk assessment studies and establish the well-defined regulations.

4. Magnetic drug delivery system in antimicrobial and antiviral therapy

The magnetic nanoparticles are also promising carries for antibacterial agents such copper and silver and can provide an alternative treatment for bacterial infections. As known, silver is distinguished by extraordinary inhibitory and bactericidal properties for a broad spectrum of bacterial strains. Antibacterial coatings based on hydrogen bonded multilayer containing in situ synthesized silver magnetic nanoparticles can be delivered to a specific region to localize a high concentration of antibacterial silver while maintaining a low concentration in general (Lee et al., 2005). The silver multilayer assembled on spherical support such as magnetic micro- or nanospheres were localized and focused by external magnetic field and showed excellent antibacterial properties against the Gram-positive strain *Staphylococcus epidermidis* and the Gram-negative strain *Escherichia coli*. It is expected that the modification of silver particles multilayer by the increase of surface-volume ratios can improve the antibacterial treatment (Tran & Webster, 2010).

The other studies demonstrates that the magnetic formulation of 3'-azido-3'-deoxythymidine-5'-triphosphate (AZTTP) is a new potential drug carries for targeted delivery across the blood brain barrier (Saiyed et al., 2009, 2010). Nucleoside and nucleotide analog reverse transcriptase inhibitors (NRTIs) are important part of the antiretroviral therapy (ART). Their inefficient cellular phosphorylation cause a limitation for treatment of human immunodeficiency virus (HIV). Main assumption of the in vitro study is that active NRTIs can directly bind to magnetic nanoparticles by ionic interaction and can inhibit HIV-1 replication. The development of magnetic AZTTP liposomal nanoformulation (150 nm) allows to cross blood brain barrier model by direct transport or via monocyte-mediated transport by external magnetic field. Hence, the magnetic carriers give a new opportunity to for treatment of neurological disorders.

5. Conclusion

The micro- and nanoparticle technology is highly novel and offers many possibilities for future development of new drug delivery systems. The innovative methods for drug targeting and delivery based on the micro- and nano-sized magnetic support provide a numerous advantages compared to conventional drug delivery systems. Presently, only molecular targeting ligands coupled to magnetic particles are successfully used and commercial available as contrast agents in MRI. Although the magnetic particles possess unique and useful properties for biomedical and pharmaceutical applications, they also carry potential health risks. The small size and high surface area to volume ratio of magnetic nanoparticles may have important implications due to higher biological activity per given mass compared to larger particulate forms (Helland et al., 2008). That may cause their remarkable activity and toxicity effect. Therefore, the material used for surface coating of the magnetic particles for in vivo application must not only be nontoxic but also biocompatible. So far there are no criteria and tests for evaluation of toxicity level, parameters of release as well as uptake magnetically targeted drugs. Regarding the in vivo application, it will have to be proved, that the new drug carriers possess not only useful properties but also are quit safe for environment and human health. Hence, the Food and Drug Administration issued a report to consider developing guidance for regulation of nanotechnology products and their adaptation from science to biomedical application.

6. Acknowledgment

Thanks are due to Mr. Tomasz Siódmiak, Mr. Wiktor Sroka and Ms. Beata Kochanek for technical assistance in preparing this manuscript.

7. References

Alexiou, Ch.; Arnold, W.; Klein, R.J. et al. (2000). Locoregional Cancer Treatment with Magnetic Drug Targeting. *Cancer Research*, Vol.60, (December 2000), pp. 6641-6648, ISSN 1538-7445.

Alexiou, Ch.; Jurgons, R.; Schmid, R.; Hilpert, A.; Bergemann, Ch.; Parak, F.; Iro, H. (2005). In vitro and in vivo investigations of targeted chemotherapy with magnetic nanoparticles. *Journal of Magnetism and Magnetic Materials*, Vol.293, (March 2005), pp. 389-393, ISSN 0304-8853

Ang, K.L; Venkatraman, S; Ramanujan, R.V. (2007). Magnetic PNIPA Hydrogels for Hyperthermia Applications in Cancer Therapy. *Materials Science and Engineering C.* Vol.27, No.3 (April 2007), pp. 347–351, ISSN 0928-4931

Arruebo, M.; Fernández-Pacheco, R.; Ibarra, M. R. & Santamaría. J. (2007). Magnetic nanoparticles for drug delivery. *Nanotoday*, Vol.2, No.3, (June 2007), pp. 22-32, ISSN 1748-0132

Bergemann, C.; Müller-Schulte, D.; Oster, J.; Brassard, L.; Lübbe, A.S. (1999). Magnetic ion-exchange nano- and microparticles for medical, biochemical and molecular biological applications. *Journal of Magnetism and Magnetic Materials*, Vol.194, (April 1999), pp.45-52, ISSN 0304-8853

Cole, A.J., Yang, V.C., David, A.E. (2011). Cancer theranostics: the rise of targeted magnetic nanoparticles. *Trends in Biotechnology*, Vol.29, No.7, (July 2011) pp. 323-332, ISSN 0167-7799

Corchero, J.L. & Villaverde, A. (2009). Biomedical application of distally controlled magnetic nanoparticles. *Trends in Biotechnology*, Vol.27, No.8, (June 2009) pp. 468-476, ISSN 0167-7799

Dames, P.; Gleich, B.; Flemmer, A. et al. (2007). Targeted Delivery of Magnetic Aerosol Droplets to the Lung. *Nature Nanotechnology*. Vol.2, No.8 (July 2007), pp. 495–499, ISSN 1748-3395

Douziech-Eyrolles, L.; Marchais, H.; Hervé, K.; Munnier, E. & Soucé, M. (2007). Nanovectors for anticancer agents based on superparamagnetic iron oxide nanoparticles. *International Journal of Nanomedicine*, Vol.2, No.4, pp.541-550, ISSN 1178-2013

Gijs, M.A.M. (2004).Magnetic Beads Handling on-chip: New Opportunities for Analytical Applications. *Macrofluid Nanofluid* Vol.1, No.1 (November 2004), pp. 22-40, ISNN 1613-4982

Gilchrist, R.D.; Medal, R.; Shorey, W.D.; Hanselman, R.C.; Parrott, J.C. & Taylor, C.B. (1957). Selective inductive heating of lymph nodes. *Annals of Surgery*. Vol.146, No.4, (October 1957), pp. 596-606, ISSN 1528-1140

Helland, A.; Scheringer, M.; Siegrist, M.; Kastenholz, H.G.; Wiek, A.; Scholz, R.W. (2008). Risk Assessment of Engineered Nanomaterials: A Survey of Industrial Approaches. *Environmental Science & Technology*, Vol.42, No.2 (January 2008), pp. 640-646, ISSN 1520-5851

Ilium, L.; Davis, S.; Wilson, C.; Thomas, N.; Frier, M. & Hardy, J. (1982). Blood clearance and organ deposition of intravenously administered colloidal particles. The effects of particle size, nature and shape. *International Journal of Pharmaceutics*, Vol.12, No 2–3, (October 1992) pp. 135–146, ISSN 0378-5173

Kohler, N.; Sun, C.; Wang, J.; Zhang, M. (2005). Methotrexate-modified superparamagnetic nanoparticles and their intracellular uptake into human cancer cells. *Langmuir*, Vol.21, No.19 (June 2005), pp. 8858–8864, ISSN 1520-5827

Lee, D.; Cohen, R.E.; Rubner, M.F. (2005). Antibacterial Properties of Ag Nanoparticle Loaded Multilayers and Formation of Magnetically Directed Antibacterial Microparticles, *Langmuir* Vol.21, No.21 (October 2005), pp. 9651-9659, ISSN 1520-5827

Lübbe, A.S.; Bergemann, Ch.; Riess, H. et al. (1996). Preclinical experiences with magnetic drug targeting: tolerance and efficacy. *Cancer Research*, Vol.56, (October 1996), pp. 4686-4693, ISNN 1538-7445

Mahmoudi, M.; Sant, S.; Wang, B.; Laurent, S. & Sen, T. (2011). Superparamagnetic iron oxide nanoparticles (SPIONs): Development, surfach modification and applications in chemotherapy. *Advanced Drug Delivery Reviews*, Vol.63, pp. 24–46, ISSN 0169-409X

Machado, B.I.; Murr, L.E.; Suro, R.M.; Gaytan, S.M.; Ramirez, D.A.; Garza, K.M.; Schuster, B.E. (2010). Characterization and Cytotoxic Assessment of Ballistic Aerosol Particulates for Tungsten Alloy Penetrators into Steel Target Plates. *International*

Journal of Environmental Research and Public Health. Vol.7, No.9 (September 2010), pp. 3313-3331, ISSN 1660-4601

Marszałł, M.P.; Moaddel, R.; Kole, S.; Gandhari, M.; Bernier, M. & Wainer, I.W. (2008). Ligand and protein fishing with heat shock protein 90 coated magnetic beads. *Analytical Chemistry*,Vol.80, No.19, (October 2008), pp. 7571-7575, ISSN 0003-2700

Marszałł, M.P.; Buciński, A.; Kruszewski, S.; Ziomkowska B. (2011a). A New Approach to Determine Camptothecin and Its Analogues Affinity to Human Serum Albumin. *Journal of Pharmaceutical Science,*, Vol.100, No.3, (March 2011) pp. 1142-1146, ISSN 1520-6017

Marszałł, M.P. (2011b). Application of Magnetic Particles in Pharmaceutical Sciences. *Pharmaceutical Research*, Vol.28, No.3 (March 2005), pp. 480–483, ISSN 1573-904X

McAteer, M.A.; Sibson, N.R.; von zur Muhlen, C.; Schneider, J.E.; Lowe, A.S.; Warrick, N.; Channon, K.M.; Anthony, D.C.; Choudhury R.P. (2007). *In vivo* magnetic resonance imaging of acute brain inflammation using microparticles of iron oxide. *Nature Medicine*.Vol.13, No.10 (October 2007), pp. 1253–1258, ISSN 1078-8956

Messmann, R.& Allegra, C. (2001) Antifolates. In: *Cancer Chemotherapy& Biotherapy*, Chabner, B., Longo, D. (3 ed.), ISSN 0036-8075, 139-184, Eds.; Lippincott Williams & Wilkins, Philadelphia

Modo, M; Hoehn, M; Bulte, J.W. (2005). Cellular MR Imaging. *Molecular Imaging*. Vol.4, No.3 (August 2005), pp. 143–164, ISSN 1535-3508Phillips, J. & Johnson, D.T. (2005). *Magnetic Fluid Hyperthermia*: A Topical Review, *The Journal of Science and Health at The University of Alabama*.Vol.3, (August 2005), pp. 14–18

Saiyed, Z.M.; Gandhi, N.H.; Nair, M.P.N. (2009). AZT 5′-triphosphate nanoformulation suppresses HIV-1 replication in peripheral blood mononuclearcells. *Journal of Neurovirology*. Vol.15, No.4 (July 2009), pp. 343-347, ISSN 1355-0284

Saiyed, Z.M.; Gandhi, N.H.; Nair, M.P.N. (2010). Magnetic nanoformulation of azidothymidine 5′-triphosphate for targeted delivery across the blood–brain barrier. *International Journal of Nanomedicine* Vol.5 (March 2010), pp. 157-166, ISSN 1178-2013

Tran, N. & Webster, T.J. (2010). Magnetic nanoparticles: biomedical applications and challenges. Journal of Materials Chemistry, Vol.20, No.40 2010, pp. 8760–8767, ISNN 0959-9428

Widder, K.J; Senyei, A.E. & Scarpelli, D.G. (1978). Magnetic microspheres: a model system of site specific drug delivery in vivo. *Proceedings of the Society for Experimental Biology and Medicine*, Vol.158, No.2 (June 1978) pp. 141–146, ISSN 0037-9727

Wilhelm, C.; Gazeau, F. (2008) Universal Cell Labelling with Anionic Magnetic Nanoparticles. *Biomaterials*, Vol.29, No.22 (August 2008), pp. 3161–3174, ISSN 0142-9612

Woodward, R.C.; Heeris, J.; Pierre, T.G.St.; Saunders, M.; Gilbert, E.P.; Rutnakornipituk, M.; Zhang. Q.; Riffle J.S. (2007). A comparison of methods for the measurement of the particle-size distribution of magnetic nanoparticles. *Journal of Applied Crystallography*, Vol.40, (January 2007) ISSN 0021-8898

Yoo, J-W.; Doshi, N.& Mitragotri, S. (2011) Adaptive micro and nanoparticles: Temporal control over carrier properties to facilitate drug delivery, *Advanced Drug Delivery Reviews*, in press, doi:10.1016/j.addr.2011.05.004, ISNN 0169-409X

7

Nanoparticles Based on Modified Polysaccharides

Hassan Namazi[1,2,*], Farzaneh Fathi[2] and Abolfazl Heydari[2]
*[1]Research Center for Pharmaceutical Nanonotechnology,
Tabriz University of Medical Science, Tabriz,*
[2]Research Laboratory of Dendrimers and Nanopolymers, University of Tabriz, Tabriz
Iran

1. Introduction

Nanoparticles may be comprised of several kind materials being classified as non-degradable and biodegradable. Biodegradable systems have an advantage over non-degradable systems in that they are non-toxic, biotolerabl, biocompatible, biodegradable, and water-soluble. Among these systems, the role of natural polysaccharides in developing prepared nanoparticles has significantly increased (Zhang *et al.*, 2011; Yang *et al.*, 2008a; Aumelas *et al.*, 2007; Leonard *et al.*, 2003).

On the other hand, polysaccharides are the most abundant macromolecules in the biosphere. The complex carbohydrates constituted of monosaccharides joined together by glycosidic bonds are often one of the main structural elements of plants and animals exoskeleton (cellulose, carrageenan, chitosan, chitin, etc.) or have a key role in the plant energy storage (starch, paramylon, etc.) (Aminabhavi *et al.*, 1990). Polysaccharides have a large number of reactive groups, a wide range of molecular weight, varying chemical composition, which contribute to their diversity in structure and in property. The amphiphilic nature imparted upon polysaccharides after modification gives them a wide and interesting application spectrum, for instance as rheology modifiers, emulsion stabilizers, surface modifiers for liposomes and nanoparticles and as drug delivery vehicles (Sinha and Kumria, 2001; Gurruchaga *et al.*, 2009; Chen *et al.*, 2003a; Durand *et al.*, 2002; Gref *et al.*, 2003). Recently, the hydrophobically modification of polysaccharides has been received increasing attention because they can form self-assembled nanoparticles for biomedical uses. In the aqueous phase, the hydrophobic cores of polymeric nanoparticles are surrounded by hydrophilic outer shells. Thus, the inner core can serve as a nano-container for hydrophobic drugs. Starch, chitosan, dextran, cyclodextrin, cellulose and pullulan are polysaccharides that have been modified with various reactants and after the modification step the nanoparticles based on modified polysaccharides were prepared with using various methods (Onyuksel *et al.*, 2003; Aumelas *et al.*, 2007; Ragauskas *et al.*, 2007; Kwon, 2003; Namazi and Dadkhah, 2010; Namazi. and Mosadegh, 2011).

Nanoparticles are defined as particulate dispersions or solid particles with a size in the range of 10-1000nm (P., 1988; Hamidi *et al.*, 2008). Depending upon the method of preparation, nanoparticles, nanospheres or nanocapsules can be obtained. These nano-sized objects, e.g., "nanoparticles", take on novel properties and functions such small size, modified surface, improved solubility and multi-functionality. The drug is dissolved, entrapped, encapsulated or attached to a nanoparticle matrix. Nanoparticles based on modified polysaccharides have been prepared most frequency by these methods: solvent evaporation method, spontaneous emulsification or solvent diffusion method, self-assembly of hydrophobically modified and dialysis method (Kim *et al.*, 2001; Aumelas *et al.*, 2007; Sun *et al.*, 2006; Couvreur, 1998). Modified polysaccharide could be used as stabilizers to produce stable hydrophilic nanoparticles by the o/w emulsion/evaporation technique. Modified polysaccharides were shown to exhibit surface active properties and to act as efficient emulsion stabilizers. Surface modified colloidal carriers such as nanoparticles are able to modulate the biodistribution of the loaded drug when given intravenously, but also to control the absorption of drugs administered by other routes (Durand *et al.*, 2004).

This review presents the several mechanisms to prepare polysaccharides-based nanoparticles after discusses about modification of polysaccharides with various agents. Also characterization of nanoparticles such as size particles, surface coverage, colloidal stability and enzyme degradability have been described and also provided are examples of use of the polysaccharide nanoparticles and their derivatives as medical applications.

2. Polysaccharides

Polysaccharides with polymeric carbohydrate structures, formed from repeating units joined together with glycosidic bonds. Their structures are often linear, but may contain various degrees of branching. In nature, polysaccharides have various resources from algal origin, plant origin, microbial origin and animal origin .Polysaccharides have a general formula of $C_x(H_2O)_y$ where x is usually a large number between 200 and 2500. Considering that the repeating units in the polymer backbone are often six-carbon monosaccharides, the general formula can also be represented as $(C_6H_{10}O_5)_n$ where $40 \leq n \leq 3000$.(Aminabhavi *et al.*, 1990)

2.1 Starch

Starch is made up of two types of polymers: amylose and amylopectin. Amylose is a linear homopolymer of α-1,4-linked glucose. Amylose may have a low level of branching with a α-1,6-linkage (Fig 1). Amylose makes up ~35% of starch. In solution amylose forms hydrogen bound with other amylase molecules to yield rigid gels. Amylopectin is highly branched form of "amylose". The linear α-1,4-linked glucose backbone is branched at every ~20 residues by an α-1,6-linkage which is extended by α-1,4-linked linkages (Namazi and Dadkhah, 2008; Della Valle *et al.*, 1998; Namazi *et al.*, 2009; Namazi and Dadkhah, 2010)

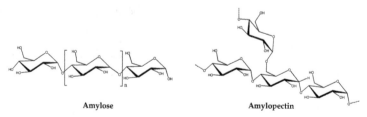

Amylose   Amylopectin

Fig. 1. Chemical structure of the starch

2.2 Chitosan and chitin

Chitosan is a linear polysaccharide composed of randomly distributed β-(1-4)-linked D-glucosamine (deacetylated unit) and N-acetyl-D-glucosamine (acetylated unit) (Fig 2). It has a number of commercial and possible biomedical uses. Chitosan is produced commercially by deacetylation of chitin, which is the structural element in the exoskeleton of crustaceans (such as crabs and shrimp) and cell walls of fungi (Thanou et al., 2005; Tharanathan and Ramesh, 2003; Yuan and Zhuangdong, 2007). Chitin $(C_8H_{13}O_5N)_n$ is a long-chain polymer of a N-acetylglucosamine, a derivative of glucose (Fig 2), and is found in many places throughout the natural world. It is the main component of the cell walls of fungi, the exoskeletons of arthropods such as crustaceans (e.g., crabs, lobsters and shrimps) and insects, the radulas of mollusks, and the beaks of cephalopods, including squid and octopuses. In terms of structure, chitin may be compared to the polysaccharide cellulose and, in terms of function, to the protein keratin. Chitin has also proven useful for several medical and industrial purposes (Kumar, 2000; Kurita, 2001).

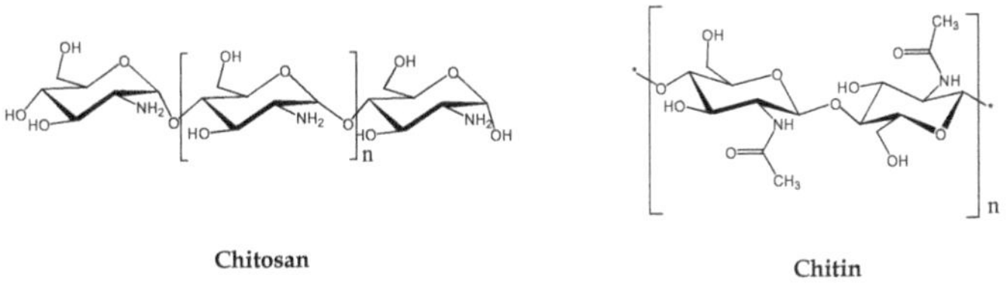

Chitosan   Chitin

Fig. 2. Chemical structure of the chitosan and chitin

2.3 Dextran

Dextran is a polysaccharide consisting of glucose molecules coupled into long branched chains, mainly through a 1,6- and some through a 1,3-glucosidic linkages as shown in Fig 3. Dextrans are colloidal, hydrophilic and water-soluble substances, inert in biological systems. It is used medicinally as an antithrombotic (anti-platelet), to reduce blood viscosity, and as a volume expander in anemia (Bertholon et al., 2006; Durand et al., 2004).

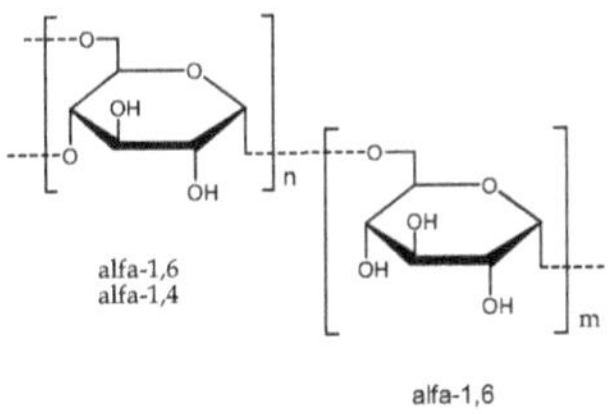

alfa-1,6
alfa-1,4

alfa-1,6

Fig. 3. Chemical structure of the dextran

2.4 Pullulan

Pullulan is a polysaccharide polymer consisting of maltotriose units, also known as α-1,4- ; α-1,6-glucan (Fig 4). Three glucose units in maltotriose are connected through an α-1,4-glycosidic bond, whereas consecutive maltotriose units are connected to each other by an α-1,6 glycosidic bond. Pullulan is produced from starch by the fungus Aureobasidium pullulans (Bataille et al., 1997; Glinel et al., 1999).

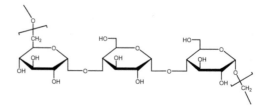

Fig. 4. Chemical structure of the pullulan

2.5 Cyclodextrins

Cyclodextrins (CDs), also using the name cycloamyloses, cyclomaltoses, or Schardinger dextrins, are natural macrocycles connected through α-(1-4)-linked glucose units in a rigid $^4C_1$ chair conformation. CDs can be produced through the enzymatic degradation of starch derived from potatoes, corn, rice or other sources. The number of glucose units per CD ring varies from 6-13, (Saenger et al., 1998; Larsen, 2002; Ueda, 2002; Hennink et al., 2009; Namazi and Kanani, 2009) as the enzyme produces a range of oligosaccharides. Because of steric factors, cyclodextrins constructed from less than six glucose units such as the five-membered cyclic oligomer, cyclomaltopentaose, has been obtained by chemical synthesis in small quantities.(T. Nakagawa et al., 1994) A chemical synthesis for other CDs has been reported, but it is too tedious for commercial production of cyclodextrins.(Ogata and Takahashi, 1995) The most common CDs contain 6, 7, and 8 D-glucose units and are known as αCD, βCD, and γCD, respectively,(Saenger, 1980) (Figure 5), while greater cyclodextrins have been reported as well.(Larsen et al., 1998; French et al., 1965; Fujiwara et al., 1990; Miyazawa et al., 1995)

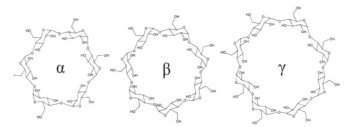

Fig. 5. Chemical structure of the cyclodextrins

2.6 Cellulose

Cellulose is an organic compound with the formula $(C_6H_{10}O_5)_n$, a polysaccharide consisting of a linear chain of several hundred to over ten thousand β(1→4) linked D-glucose units (fig. 6). Cellulose is the structural component of the primary cell wall of green plants, many forms of algae and the oomycetes. Some species of bacteria secrete it to form biofilms. Cellulose is the most common organic compound on Earth (Hinrichsen et al., 2000; Riedel and Nickel, 1999; Gassan and Bledzki, 1999).

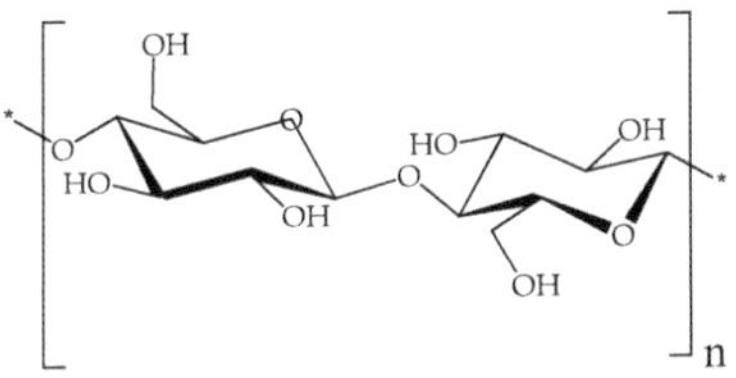

Fig. 6. Chemical structure of the cellulose

3. Modified polysaccharides (MP) for preparation of their nanoparticles

Amphiphillic polysaccharides consisting of hydrophilic and hydrophobic fragments have been modified because they can form self-assembled nanoparticles and they show unique physicochemical characteristics such as a nanoparticle structure and thermodynamic stability. Natural biopolymers have various advantages, such as availability from replenish able agricultural or marine food resources, biocompatibility, and biodegradability, therefore leading to ecological safety and the possibility of preparing a variety of chemically or enzymatically modified derivatives for specific end uses. Recently, there has been considerable interest in developing modified derivatives of polysaccharides for biodegradable nanoparticles. These nanoparticles have shown the following advantages for biomedical applications such as drug protection and ability to control the drug release. Polysaccharides have a number of positive characteristics such biotolerability, biodegradability, protein rejecting ability, receptor interaction through specific sugar moieties, and abundance of functional groups for modification or functionalization (Couvreur et al., 2004). The amphiphilic character imparted upon polysaccharides after hydrophobic modification gives them a wide and interesting use spectrum, for instance as rheology modifiers, emulsion stabilizers (Chen et al., 2003a; Durand et al., 2002), surface modifiers for liposomes and nanoparticles (Vyas and Sihorkar, 2001) and as drug delivery vehicles (Rodrigues et al., 2003; Leonard et al., 2003).

3.1 Modified starch

Starch is one of the polysaccharide that it has been modified with various reactants for preparation of nanoparticles. The use of starch nanoparticles is receiving a significant amount of notice because of the plentiful availability of natural polymer, inexpensive, renewability, biocompatibility, biodegradability and nontoxicity. Chemical modification of starch has been widely studied for producing modified starch by way of chemical reaction with hydroxyl groups in the starch molecule. Starch esters are a kind of modified starches which are synthesized with various reactants such as acid anhydrides octenyl succinic anhydride (OSA), dodecenyl succinic anhydride (DDSA) fatty acids and fatty acid chlorides (Tukomane and Varavinit, 2008; Wang et al., 2007a; Borredon et al., 1999; Fowler et al., 2002). Hydroxyethyl starch was esterified with the long chain fatty acids under mild reaction conditions using DCC and DMAP (Mader et al., 2007). The synthesis of modified hydrophobic starch using fatty acids was done by means of potassium persullphate as catalyst in DMSO (Abraham and Simi, 2007). Several substituted starches were prepared by acylation of starch with fatty acid chlorides in organic solvents, such as pyridine or dimethylacetamide (Kapusniak and Siemion, 2007; Wang et al., 2008). Hydrophilic

Starch	Grafting agent	References
Amylopectin (from waxy corn)	Lactic acid	(Hong-Wei Lua and Li-Ming Zhanga, 2011)
Modification: Amylopectin and aqueous lactic acid (LA) were added to a three-necked flask equipped. After the stirring at 75 °C for 30 min, the temperature of the reaction system was thermostated to be 100 °C. Then a required amount of Sn(Oct)₂ was added to the flask. Then the product was further purified by Soxhlet extraction to remove completely the unreacted LA monomer as well as PLA homopolymer that may be formed during the reaction.		
Amylopectin-rich waxy maize starch	Stearic acid	(Dufresne et al., 2004; Dufresne et al., 2006)
Modification: Chemical modification of the nanoparticles was performed in a round-bottomed reaction flask under a nitrogen atmosphere while constantly stirring with amagnetic stir bar. The stearate modification was performed by the reaction of dry starch nanocrystals with stearic acid chloride in methyl ethyl ketone.		
Amylomaize starch	n-Butanol	(Lim and Kim, 2009)
Modification: The amylomaize starch (0.5%, w/v) was dispersed DMSO solution with heating and stirring in a boiling water bath, and then magnetic-stirred at room temperature for 24 h. An aliquot of the starch solution was allowed to gravimetrically pass through a membrane filter into the bottom compartment filled with n-butanol. The precipitate in the butanol layer was collected by centrifugation, and then washed three timesin the n-butanol.		
Cassava starch	Monochloroacetic acid (MAC) $ClCH_2COONa$	(Wu et al., 2011)
Modification: Cassava starch in anhydrous ethanol was placed in a glass reactor. An aqueous solution of sodium hydroxide was added drop wise to the starch–solvent mixture under stirring until the whole amount of sodium hydroxide were added. Then, the solution of MAC was added drop wise to the starch–solvent–sodium hydroxide mixture under ultrasonic irradiation.		
Waxy corn starch		(Fowler et al., 2004; Namazi and Dadkhah, 2010)
Modification: Starch esterification was carried out in two steps. In the first step, starch nanocrytstals dispersed in the reaction medium were alkali treated at room temperature with mechanical stirring under an atmosphere of N₂ for 10 min and in the second step, 0.5 mol equivalents of the required acid chloride was added drop wise and the reaction mixture was stirred for 20 min.		
Cassava starch	$CH_3(CH_2)_7CH=CH(CH_2)_7COOH$	(Abraham and Simi, 2007)
Modification: For the graft copolymerization, about 1g starch was dissolved in 10 ml DMSO and was taken in a round bottom flask. Oleic acid, weighing was added and potassium per sulphate was the catalyst.		

Hydroxyethyl starch (HES)	Fatty acid	(Mader *et al.*, 2007)
Modification: HES was dried for 2 h before dissolving in 20 mL of dry DMSO. To the solution were added the fatty acid, DCC, and DMAP, and they dissolved for 24 h. The formed precipitate (dicyclohexyl urea, DCU) was removed by filtration, and the filtrate was added to 200 mL of precipitating solvent mixture.		
Potato starch		(Namazi and Dadkhah, 2008; Dufresne *et al.*, 1996)
Modification: A mixture of starch nanoparticle (1 g) and CL (2 mL) was first added to flask. A determined amount of Sn(Oct)$_2$ of total amount of reagents was then introduced via a conditioned syringe. Polymerization was stopped by fast cooling to room temperature.		

Table 1. Functional molecules for modification of starch

amylopectin was modified by grafting hydrophobic poly (lactic acid) chains (Hong-Wei Lua and Li-Ming Zhanga, 2011). Since 1950, considerable effort has gone into hydrophobically modified derivatives of hydrophilic polysaccharides(Namazi *et al.*, 2011).

Recent studies have been carried out to investigate the synthesis and the application of polysaccharide-based nanoparticles. In Table 1 functional molecule that used for modification of starch are listed which have been used for preparation of their nanoparticles.

3.2 Modified chitosan and chitin

Biopolymer chitosan with a lot of primary amino groups is a polysaccharide derived from deacetylation of chitin. Due to the excellent film-forming ability, biocompatibility, nontoxicity, high mechanical strength, cheapness of chitosan, it is used for synthesis and the application of polysaccharide-based nanoparticles (Payne *et al.*, 2005; Kwon *et al.*, 2003). Chitosan is one of the polysaccharides that modified with various groups such as 5β-cholanic acid, linoleic acid, Monomethoxy poly (ethyleneglycol) and etc. After modification process, modified chitosan are used for preparation of their nanoparticles. These groups are listed in Table 2.

Grafted chitosan has been studied by many researchers. These studies have been intensified since 1992 because chitin and chitosan show excellent biological properties such as biodegradation in the human body. Modification can marginally improve the solubility of chitosan. As a polymeric amphiphile, grafted-chitosan with monomethoxy poly (ethyleneglycol) can aggregate into core–shell nanoparticles in aqueous media because in the aqueous phase, the hydrophobic cores of chitosan nanoparticles are encircled by hydrophilic outer shells. Thus, the internal core can serve as a nano-container for hydrophobic drugs. Modified chitosan is appropriate for decreasing severe side effects such as cytotoxicity in usual tissue (Fang *et al.*, 2006; Gorochovceva *et al.*, 2005; Opanasopit *et al.*, 2006).

Chitosan	Grafting agent	References
Glycol chitosan	5β-Cholanic acid	(Kwon *et al.*, 2006; Kwon *et al.*, 2004; Kwon *et al.*, 2003)
Modification: Glycol chitosan was hydrophobically modified with cholanic acid in methanol/water. To activate the carboxylic acid groups of cholanic acid, equal amounts of 1-ethyl-3-(3-dimethylaminopropyl) - carbodiimide ydrochloride and N-hydroxysuccinimide were added.		
Chitosan of 100 mesh	Linoleic acid	(Lu *et al.*, 1994; Ichinose *et al.*, 2000)
Modification: Chitosan was dissolved in aqueous acetic acid solution and diluted of methanol. LA was added to the chitosan solution glucosamine residue of chitosan followed by a dropwise addition of 15 mL of EDC methanol solution (0.07 g/L) while stirring.		
chitosan	α-Cyclodextrin	(Sakairi *et al.*, 1998; Martel *et al.*, 2001; Aoki *et al.*, 2003)
Modification: Sakairi prepared α-CD linked chitosan using 2-O-formylmethyl-α-CD by reductive N-alkylation and confirmed the host-guest complex with p-nitrophenol.		
Chitosan	ε-Caprolactone	(Albertsson *et al.*, 1999; Yang *et al.*, 2008b)
Modification: The PCL-graft-chitosan copolymers were synthesized by coupling the hydroxyl end-groups on preformed PCL chains and the amino groups present on 6-O-triphenylmethyl chitosan and by removing the protective 6-O-triphenylmethyl groups in acidic aqueous solution		
Biomedical grade chitosan	Monomethoxy poly(ethyleneglycol)	(Zhang *et al.*, 2005; Yang *et al.*, 2008b)
Modification: Chitosan was completely dissolved in formic acid by stirring and a suitable amount of mPEG was added. After 15 min, enough formaldehyde solution was added to the above mixture and was stirred for 12 h.		

Table 2. Functional molecules for modification of chitosan

3.3 Modified dextran

The development of existing materials to prepare modified dextran is the subject of numerous researches due to their surface-active properties and potential pharmaceutical, biochemical and medical applications. Modified dextran gives a large range of properties, allowing the selection of the carrier which proves the most useful for a particular drug encapsulation and release. Dextran is one of the water-soluble polysaccharides that have been modified to obtain amphiphilic polymers capable of forming micellar structures and binding organics solutes in the hydrophobic domain. Also, it is amphiphilic block copolymers that can self- assemble in selective solvents to form micelles with a core and a shell containing insoluble and soluble blocks (Lu *et al.*, 1994; Ichinose *et al.*, 2000; Lu and Tjerneld, 1997). Core-shell type nanoparticles of a poly (DL-lactide-co-glycolide) (PLGA) grafted-dextran copolymer are prepared with varying graft ratio of PLGA. The DexLG copolymer was able to form nanoparticles in water by self-aggregating process (Song *et al.*,

2006). Dextran was chemically modified by the covalent attachment of hydrocarbon groups (aliphatic or aromatic) via the formation of ether links. According to the extent of modification, either water-soluble or water-insoluble dextran derivatives were obtained. The latter exhibited solubility in organic solvents like tetrahydrofuran or dichloromethane saturated with water (Bertholon *et al.*, 2006; Durand *et al.*, 2004; Leonard *et al.*, 2003; Aumelas *et al.*, 2007; Leonard *et al.*, 2000; Osterberg *et al.*, 1995). Biodegradable hydrogel nanoparticles were prepared from glycidyl methacrylate dextran (GMD) and dimethacrylate poly(ethylene glycol) (DMP). GMD was synthesized by coupling of glycidyl methacrylate to dextran in the presence of 4-(*N,N*-dimethylamino)pyridine (DMAP) using dimethylsulfoxide (DMSO) as an aprotic solvent (Kim *et al.*, 2000; Vandijkwolthuis *et al.*, 1995). Dextran also was modified using click-chemistry. Each reaction step was done under aqueous conditions, including the introduction of azide functionalities to the backbone of the polysaccharide. The reaction consisted of the synthesis of 1-azido-2,3-epoxypropane, which was etherified onto the backbone of the polysaccharide using base-catalysis in water/isopropanol mixture at ambient temperature (Fringuelli *et al.*, 1999; Seppala *et al.*, 2010). Modified dextran was synthesized by conjugating the various groups to dextran such as poly (lactic-co-glycolic acid, p-hexylbenzoyl chloride. These groups are listed in Table 3.

Dextran	Grafting agent	References
Dextran (average molecular weights: 77,000)	Poly(lactic-co-glycolic acid)	(Tiera *et al.*, 2003)
Modification: The DexLG graft copolymer was synthesized by conjugating the carboxylic acid end of PLGA and the hydroxyl group of dextran using DCC as a coupling agent.		
Dextran T40 ðMw < 40; 000	P-Hexylbenzoyl chloride	(Tiera *et al.*, 2003; Bertholon *et al.*, 2006)
Modification: Dextran was dissolved under stirring in 5 ml of water containing 1.8 g of triethylamine. The resulting solution was heated at 20 8C and 1.4 g of p-hexylbenzoyl chloride was added under vigorous stirring for 1h.		
Dextran	1,2- Epoxy-3-phenoxypropan	(Durand *et al.*, 2002; Sun *et al.*, 2006; Song *et al.*, 2006)
Modification: Water-soluble amphiphilic dextran, i.e. dextran with lowsubstitution ratio – here DexP15 – was obtaiobtained after reaction with 1,2- epoxy -3-phenoxypropane in 1M NaOH as previously described.		
Dextran (Mw) 30 200	Bile acid	(Melo *et al.*, 1999; Akiyoshi *et al.*, 1993)
Modification: modified dextran were obtained by reacting dextran (*Mw*) 30 200, *Mw*/*Mn*) 1.112) with a bile acid in the presence of *N,N*-dicyclohexylcarbodiimide as a coupling agent and 4-(*N,N*-dimethylamino)pyridine as a catalyst.		
Dextran methoxypolyethylene	Glycol/poly (ε-caprolactone)	(Zhang *et al.*, 2008; Cao *et al.*, 2005)
Modification: A series of amphiphilic copolymers, dextran-graft-methoxypolyethylene glycol/poly (ε-caprolactone) (Dex-g-mPEG/PCL) were synthesized by grafting both PCL and mPEG chains to dextran, and subsequently the micellar self-assembly behavior of resultant copolymers was investigated.		

Table 3. Functional molecules for modification of dextran

3.4 Modified pullulan

Due to their amphiphilic structure, modified pullulan has potential high surface and interfacial properties. They diffuse through the bulk phase and adsorb at the interface, inducing a sharp reduction in the surface or interfacial tension of a polymer solution (Muller et al., 2003). Like other polysaccharides pullulan have been used to modify with various groups for preparation of their nanoparticles (table 4). Pullulan which is partly modified by relatively higher hydrophobic groups such as cholesteryl groups, it shows unique association behavior. Cholesterol-bearing pullulans have been studied in detail by Akiyoshi and Sunamoto. It was designed as a self-aggregate to form monodisperse and stable nanogels due to the hydrophobic moieties in an aqueous solution. The nanogels formed complexes with various drugs and proteins by hydrophobic interaction and released them upon exposure to specific proteins (Akiyoshi et al., 1997; Akiyoshi et al., 1993; Cheng et al., 2008). Hydrophobically-modified pullulans of moderate molar mass and differing in hydrophobic modification ratio, charge ratio and the nature of the hydrophobic chains were prepared (Bataille et al., 1997; Glinel et al., 1999; Fischer et al., 1998). Poly ($_{DL}$-lactide-co-glycolide)-grafted pullulan can form self-assembling nanospheres and controll adriamycin release. Pullulan acetate (PA) is the other important hydrophobized pullulan, which can form self-aggregation nanoparticles as well as its modified materialsn (Zhang et al., 2009; Na et al., 2007).

Pullulan	Grafting agent	References
Pullulan	C(O)NH(CH₂)₆NHC(O)O	(Akiyoshi et al., 1998)
Modification: Cholesterol-bearing pullulan forms a spherical andmonodisperse nanoparticle which is a self-aggregate of 10–12 CHP molecules. This nanoparticlehas several hydrophobic domains of four to five associated cholesteryl moieties.		
carboxymethylpullulan	Alkyl bromide (octyl, decyl or dodecyl)	(Bataille et al., 1997; Glinel et al., 1999)
Modification: Hydrophobically-modified carboxymethylpullulans (HMCMPs) were obtained by a synthetic pathway adapted from that used by Della Valle for gellan and Fischer et al. for pectin.		
Pullulan with molecular weight of 50,000–100,000 (g/mol)	Poly($_{DL}$-lactide-co-glycolide)	(Jeong et al., 2006)
Modification: Pullulan (1 g) was dissolved in DMSO (15 ml) for 3 h. Various amounts of PLGA were dissolved in DMSO (5 ml) with a 1.3 equiv. amount of DCC and DMAP.		
Pullulan (Mw = 200,000)	Acetic anhydride	(Na et al., 2007; Zhang et al., 2009)
Modification: 2 g of pullulan, suspended in 20ml of formamide, was dissolved by vigorous stirring at 54 °C. To this solution, 6ml pyridine and 15 ml, 10ml or 7.5ml of acetic anhydride were added to change the acetylation degree.		

Table 4. Functional molecules for modification of pullulan

3.5 Modified cyclodextrins

Modifications to the cyclodextrins(Namazi *et al.*, 2005; Namazi and Kanani, 2009) lead to a wide range of photochemistry of cyclodextrin complexes, through which the improvement of guest reactivity occurs; in addition, light harvesting molecular devices and photochemical frequency switches may be constructed. A few amphiphilic β-CD derivatives such as β-CDC$_6$ modified on the secondary face with 6C aliphatic esters and 6-N-CAPRO-β-CD modified on the primary face with a 6C aliphatic amide were demonstrated to give stable nanoparticles of high drug loading capacity and reduction of burst effect during the drug release process when nanoparticles are prepared directly from preformed drug/amphiphilic CD inclusion complex (Lemos-Senna *et al.*, 1998). A new nanoparticle carrier system was obtained from amphiphilic cyclodextrin bearing fatty acids (with a chain length of either 6 or 12 carbon atoms) grafted O$_2$ and O$_3$ position of the cyclodextrin. Nanoparticles with a mean diameter of several hundred nm were prepared by dispersion. Amphiphilic cyclodextrins (CDs) are obtained by the chemical per-modification of natural CDs (β-CD or γ-CD) by the selective substitution of aliphatic chains of varying length (2C to 18C), structure (linear or branched) linked with varying bonds (ester, ether, amide, thio, fluoro) of high purity. These CD derivatives were demonstrated to yield nanospheres or nanocapsules spontaneously using the nanoprecipitation technique with or without the presence of surfactants. Carboxymethyl-β-cyclodextrin modified nanoparticles were fabricated for removal of copper ions from aqueous solution by grafting CM-β-CD onto the magnetite surface via carbodiimide method. The grafted Carboxymethyl-β-cyclodextrin on the Fe$_3$O$_4$ nanoparticles contributes to an enhancement of the adsorption capacity because of the strong abilities of the multiple hydroxyl and carboxyl groups in CM-β-CD to adsorb metal ions. Double hydrophilic copolymers with one polyethylene glycol (PEG) block and one β-cyclodextrin (β-CD) flanking block (PEG-β-PCDs) were synthesized through the post-modification of macromolecules. The self-assembly of PEG-β-PCDs in aqueous solutions was studied by a fluorescence technique(Choisnard *et al.*, 2006).

3.6 Modified celullose

Modified celullose have received wide applications for the stabilization of disperse systems, in particular suspensions and emulsions (Namazi and Rad, 2004). The most important types of associating polymers are water-soluble amphiphilic polymers, notably block or graft copolymers, with hydrophobic blocks or grafts. Cellulose is the most abundant polysaccharide available worldwide and exhibits attractive structure and single properties, which are quite attractive for both academic and industrial researchers. Recently, cellulose based polymers have been widely investigated for its positive characteristics such as safety, biodegradability, biocompatibility, and protein rejecting ability, and so on(Namazi and Jafarirad, 2008). However, there have been few reports on the utilization of self-assembled micelles based on amphiphilic cellulose derivatives as delivery carriers for poorly water-soluble pharmaceutical active ingredients (Klemm *et al.*, 2005; Cheng *et al.*, 2008; Dong *et al.*, 2008). Poly (ε-caprolactone) (PCL) and poly (L-lactic acid) (PLLA) are biodegradable polymers that are potential candidates as matrixes in biocomposites. Several studies have been conducted on the PCL and PLLA modification of soluble cellulose and its derivate (Nishio and Teramoto, 2003; Nishio *et al.*, 2002; Burt and Shi, 2003). Modified cellulose was prepared with hydrophilic groups that it can be self-assemble into polymeric vesicle or as nontoxic surfactants. Sulfate was firstly introduced as hydrophilic groups, then the hydrophobic groups for cellulose derivatives. The aqueous self-assembly of the modified cellulose was investigated using transmission electron microscopy (TEM) and dynamic laser scattering (DLS). Results showed that modified cellulose were capable of forming polymeric micelles in water with an average particle diameter ranging from 20 to 67 nm (Cheng *et al.*, 2008). Novel modified cellulose derivatives were synthesized long chain alkyl groups as hydrophobic moieties and quaternary ammonium groups as hydrophilic moieties. The results of measurements (DLS, TEM) revealed that modified cellulose can be self-assembled into cationic micelles in distilled water with the average hydrodynamic radius of 320–430 nm (Zhou *et al.*, 2011).

4. Prepration methods and characterization of polysaccharide-based nanoparticle

As for polysaccharide-based nanoparticles, Alonso et al. (Alonso *et al.*, 2001) and Prabaharan et al. (Prabaharan and Mano, 2005) have made excellent reviews in 2001 and 2005, respectively, focusing on the preparation and application of chitosan nanoparticle carriers. Many studies have demonstrated that nanoparticles have a number of advantages over microparticles (Panyam and Labhasetwar, 2003). It has been reported that micro particles are less effective drug delivers than particle having size ranging in between nanometers for e.g. Nanoparticles having size range greater than 230 nm acquire in the spleen shown by body distribution studies (Kreuter, 1991). As time goes on, more polysaccharide-based nanoparticles emerge, which greatly enriches the versatility of nanoparticle carriers in terms of category and function. In this section, several mechanisms are introduced to prepare these nanoparticles, that is, emulsification solvent evaporation method, solvent diffusion method, self-assembly of hydrophobically modified, dialysis method and other methods. The select of method depends on a number of factors, such as, particle size, particle size distribution, area of application and etc. Particle size is the greatest important characteristics of nanoparticles. Some methods for the determining particle size are (Labhasetwar *et al.*, 2003)

a.   Photon-correlation spectroscopy.
b.   Dynamic light scattering.
c.   Brownian motion and light scattering properties.
d.   Scanning or transmission electron microscopy (SEM or TEM).

They determine the in vivo distribution, biological fate, toxicity and targeting ability of these delivery systems. In addition, they can influence drug loading, drug release and stability of the nanoparticles.

4.1 Polysaccharides-based nanoparticles through emulsification solvent evaporation method

Emulsification solvent evaporation is the most widely employed technique to prepare nanoparticles of polymers in the current literature on techniques using a dispersion of preformed polymers (Vanderhoff *et al.*, 1979). In the conventional methods, two main strategies are being used for the formation of emulsions: the preparation of single-emulsions, e.g., oil-in-water (o/w) or double-emulsions, e.g., (water-in-oil)-in-water, (w/o/w).

In a single emulsification solvent evaporation process, polymer dissolved in a volatile water-immiscible organic solvent such as dichloromethane, chloroform, ethyl acetate, which is also used as the solvent for dissolving the hydrophobic surfactant. This solution is emulsified in an aqueous phase containing a surfactant or stabilizer (emulsifying agent) resulting in oil-in-water (o/w) emulsion.(ODonnell and McGinity, 1997; I. *et al.*, 2004; Lee, 2001) The coalescence of the organic droplets can be avoided by continuous stirring. Emulsification can also be enhanced by using sonication or microfluidization with a homogenizer, which reduces the droplet size of the organic dispersed phase. After the formation of stable emulsion, the organic solvent is evaporated either under stirring at room temperature or by rotary evaporation under reduced pressure to transform the nano-emulsion into a nanoparticle suspension. Formed nanoparticles are harvested from the aqueous slurry by lyophilization.

For the water-soluble surfactants, a double-emulsion (water-oil-water) variation of the process is utilized. An aqueous solution of the active agent (internal water phase, w1) is emulsified into an organic solution containing the biodegradable polymer and lipophilic surfactant (oil phase, o) for resulting primary emulsification. Then, this emulsion (w1/o) is added to the large aqueous phase with emulsifier (external water phase, w2) to create w1/o/w2 double emulsion. The emulsifier amount is much higher in the first emulsion than in the second emulsion, because the droplet size of the first emulsion needs to be much smaller than in the second outer emulsion. The organic solvent is removed by evaporation or extraction and solid nanoparticles are formed. The nanoparticles are collected by centrifugation or filtration and are subsequently lyophilized.

Wouessidjewe and coworks(Lemos-Senna *et al.*, 1998) using this method for preparing nanospheres from an amphiphilic 2,3-di-O-hexanoyl-γ -cyclodextrin (γCDC$_6$). This preparation method involves in emulsifying an organic phase having the cyclodextrin in an aqueous phase containing Pluronic F68 as surfactant. This solution was dispersed in aqueous phase by using a high speed homogenizer. Afterward, the organic solvent was evaporated by mechanical stirring at room temperature. The influence of the process parameters, i.e. surfactant concentration and initial γCDC$_6$ content, on the characteristics of nanosphere preparation, as well as on the nanosphere loading of a hydrophobic drug, progesterone, was calculated. Cyclodextrin nanospheres presenting a mean diameter varying from 50 to 200 nm were obtained, even in the presence of low surfactant concentration.

Nanoparticles of dextran (Aumelas *et al.*, 2007) could be simply prepared by the o/w emulsion solvent evaporation method, with using a low modified dextran (DexP$_{15}$) as polymeric surfactant in the water phase and a highly modified dextran in the CH$_2$Cl$_2$ phase. After emulsification and solvent evaporation, core-shell particles with a dense dextran core and a dextran surface coverage are expected. Dextran segments originating from DexP$_{15}$ chains which are not embedded in the dextran core are assumed to extend freely toward the aqueous solution and to form a hydrophilic shell. The size of DexP$_{130}$ nanoparticles prepared by o/w emulsion process decreases as the amount of DexP$_{15}$ in the water phase increases. Unpredictably, dextran nanoparticles were also obtained without any polymeric surfactant in the aqueous phase. For comparison, when poly (lactic acid) was used instead of hydrophobically modified dextran, it was not possible to obtain nanoparticles without the presence of surfactant in the aqueous phase. This specific result can be explained assuming a limited solubility of highly hydrophobized dextrans in water. This solubility can be due to the presence of a fraction of low substituted dextran molecules in the final product or to partitioning of the highly substituted sample. This water-soluble fraction could act as a stabilizer for the transient oil droplets. Generally speaking, the size of bare dextran nanoparticles, i.e. prepared in the absence of $DexP_{15}$, increases with the substitution ratio of dextran, for example from 370 nm for $DexP_{65}$ nanoparticles to 850 nm for $DexP_{210}$ nanoparticles. Other dextran particles, in the size range 150–250 nm, were obtained in the presence of $DexP_{15}$. The colloidal stability of suspensions was also examined at various NaCl concentrations. For the targeted nanoparticles, surface coverage by hydrophilic loops is essential to provide a convenient colloidal stability in physiological conditions (especially with regard to the ionic strength).

In the o/w emulsion process, we showed that the size of particles is strongly related to the concentration of surfactive polymer in the aqueous phase. In generally, parameters in the emulsification solvent evaporation process that affect particle size, zeta potential, hydrophilicity, and drug loading include:

1. Homogenization intensity and duration.
2. Type and amounts of emulsifier, polymer and drug.
3. Particle hardening (solvent removal) profile (Zambaux *et al.*, 1998).

4.2 Polysaccharides-based nanoparticles through solvent diffusion method

Spontaneous emulsification or solvent diffusion method is a modified version of solvent evaporation method. The different process variants are all based upon the use of solvents which are of limited water miscibility and capable of spontaneous emulsion formation. This method thus offers the advantage of the use of pharmaceutically acceptable solvents and does not require the use of high-pressure homogenizers for the formation of the o/w emulsion as the preliminary stage of nanoparticle formation (Allemann *et al.*, 1998; Leroux *et al.*, 1995). In this method, the water-miscible solvent along with a small amount of the water-immiscible organic solvent is used as an oil phase. Due to the spontaneous diffusion of solvents an interfacial turbulence is created between the two phases leading to the formation of small particles. In this technique, the phase separation is accompanied by vigorous stirring. On the opposite with o/w, the size of nanoparticles obtained using the solvent-diffusion method is poorly affected by the concentration of polymeric surfactant added to the aqueous phase. A reduction in particle size can be gained by increasing the concentration of water miscible solvent.

Nanoparticles of dextran (Aumelas *et al.*, 2007) could be prepared by solvent-diffusion method. Dextran nanoparticles of similar size were obtained with or without using stabilizer such as $DexP_{15}$. This process avoids the use of any high energy input step. The colloidal stability of suspensions was also examined at various NaCl concentrations. The particular colloidal stability of $DexC_{1052}$ nanoparticles up to high ionic strengths without $DexP_{15}$ can be justified by assuming that the water-soluble fraction contained in that polymer is higher than in the others. Also this method was employed to prepare pullulan acetate (PA) nanoparticles.(Zhang *et al.*, 2009) This technique had some advantages compared with other methods. It is a straightforward technique and the particle size increased from 185.7 nm to

423.0 nm with the degree of acetylation increasing from 2.71 to 3.0. Briefly, PA is readily soluble in dimethyl sulfoxide (DMSO), DMF, tetrahydrofuran (THF), dichloromethane, chloroform, acetone, and pyridine. To make nanoparticles by solvent diffusion method, only water-miscible solvents were considered because the solvents could diffuse into aqueous phase. The solvent selected to dissolve the polymer, as well as the type of polymer can influence the formation of nanoparticles, due to differences in the polymer-solvent and water–solvent interactions. It was supposed that the diffusion-stranding process might be altered, thus inducing changes in the mean size. Therefore, solvents are of primary importance in the formation of nanoparticles by the solvent diffusion method. In other study, five water-miscible solvents, i.e., DMSO, DMF, acetone, THF and pyridine were used. 0.5% poly (vinyl alcohol) [PVA] or distilled water served as aqueous phase. PA2 could form nanoparticles in anyone of the five organic solvents added to water or 0.5% PVA. However, PA1 could do only in DMSO and DMF added to 0.5% PVA. Really, PA2 led to the smallest nanoparticles (185.7 nm), and the largest was PA1 nanoparticles (423.0 nm).

4.3 Polysaccharides-based nanoparticles through self-assembly method

The literature survey showed that several studies have been carried out to investigate the synthesis and the application of polysaccharide based self- aggregate nanoparticles as drug delivery systems. When hydrophilic polymeric chains are grafted with hydrophobic segments, amphiphilic copolymers are formed. Upon contact with an aqueous environment, polymeric amphiphiles spontaneously form micelles or micelle-like aggregates via undergoing intra- or intermolecular associations between hydrophobic moieties, primarily to minimize interfacial free energy. These polymeric micelles display unique characteristics, such as small hydrodynamic radius (less than microsize) with core-shell structure, unusual rheology feature, thermodynamic stability, depending on the hydrophilic/hydrophobic constituents. In specific, polymeric micelles have been recognized as a promising drug carrier, since their hydrophobic domain, surrounded by a hydrophilic outer shell, can serve as a preservatory for various hydrophobic drugs (Letchford and Burt, 2007). Usually, these hydrophobic molecules can be divided into linear, cyclic hydrophobic molecules, hydrophobic drug, polyacrylate family, etc.

4.3.1 Linear hydrophobic molecules

Poly (ε-caprolactone) (PCL) is biodegradable industrial polyester with excellent mechanical strength, non-toxicity, and biocompatibility. It has been frequently used as implantable carriers for drug delivery systems or as surgical repair materials. It is hopeful to combine chitosan with the biodegradable polyester to create amphiphilic copolymer applicable to drug delivery systems. In 2002 and 2003, (Gref et al., 2002; Lemarchand et al., 2003) synthesized amphiphilic dextran by coupling between carboxylic function present on preformed PCL monocarboxylic acid and the hydroxyl groups on dextran. The comb-like copolymers (dextran-PCLn) consisted of a dextran back bone on to which preformed PCL blocks were grafted. Nanoparticles of less than 200 nm were successfully prepared by using the new materials (Rodrigues et al., 2003). Further, bovine serum albumin and lectin were incorporated in the nanoparticles. Lectins could also be adsorbed onto the surface of the nanoparticles. Surface-bound lectin conserved its hemagglutinating activity, suggesting the possible application of this type of surface-modified nanoparticles for targeted oral administration. Caco-2 cellular viability was higher than 70% when put in contact with the nanoparticles, even at concentrations as high as 660 mg/ml (Rodrigues et al., 2003). In addition, it was found that the modification of the surface with dextran significantly reduced the cytotoxicity towards J774 macrophages. Biodegradable amphiphilic PCL-graft-chitosan copolymer was synthesized (Jing et al., 2006). The copolymers could form spherical or elliptic nanoparticles in water.

Poly (ethylene glycol) has been employed extensively in pharmaceutical and biomedical fields because of its outstanding physicochemical and biological properties including hydrophilic property, solubility, non-toxicity, ease of chemical modification and absence of antigenicity and immunogenicity. Therefore, poly (ethylene glycol) is widely used as a pharmacological polymer with high hydrophilicity, biocompatibility and biodegradability. In recent years, derivative poly (ethylene glycol)-g-derivative chitosan to obtain nanoparticles has been studied by many researchers (Ouchi et al., 1998; Jung et al., 2006) (Park et al., 2008) (Yang et al., 2008b) (Opanasopit et al., 2007). The grafted poly (ethylene glycol) methyl ether onto N-Phthaloyl chitosan chains, aggregated to obtain sphere-like nanoparticles (an et al., 2004). When the chain length of poly (ethylene glycol) methyl ether was as high as 5×10^3 Da, the sphere size became as small as 80-100 nm. By simply adjusting the hydrophobicity/hydrophilicity of the chitosan chain, stable nanospheres could be obtained directly. Also methoxy poly (ethylene glycol)-grafted chitosan to develop polymeric micelles for the drug delivery to brain tumor was synthesized.(Jung et al., 2006) Methoxy poly (ethylene glycol)-grafted-chitosan conjugates by formaldehyde linking method was synthesized(Yang et al., 2008b). The conjugates formed monodisperse self-aggregated nanoparticles with a roughly spherical shape and a mean diameter of 261.9 nm. A poorly water-soluble anticancer drug, methotrexate was physically entrapped inside the nanoparticles. Other group synthesized amphiphilic grafted copolymers, N-phthaloyl chitosan- grafted poly (ethylene glycol) methyl ether (Opanasopit et al., 2007). These copolymers could form micelle-like nanoparticles. The CMC of these nanoparticles in water was similar (28 µg/ml). The nanoparticles exhibited a regular spherical shape with core-shell structure with sizes in the range of 100-250 nm. Camptothecin as a model drug was loaded into the inner core of the micelles.

For modifying polysaccharides have been used some long-chain fatty acids such as hexanoic acid, decanoic acid, linoleic acid, linolenic acid, palmitic acid, stearic acid, and oleic acid. Choisnard et al. (Choisnard et al., 2006) prepared decanoate β-cyclodextrin esters (DS, 2-7) and hexanoate β-cyclodextrin esters (DS, 4-8) biocatalyzed by thermolysin from native β-cyclodextrin and vinyl hexanoate or vinyl decanoate used as acyldonors. Both esters self - organized into nanoparticles by a nanoprecipitation method. Chen et al. (Chen et al., 2003a) modified chitosan by coupling with linoleic acid through the 1-ethyl-3-(3-dimethylamino-propyl)-carbodiimide-mediated reaction to increase its amphipathicity for enhanced emulsification. The micelle formation of linoleic acid-modified chitosan in the 0.1 M acetic acid solution was improved by o/w emulsification with methylene chloride, an oil phase, the self-aggregation concentration from 1.0 g/L to 2.0 g/L. The addition of 1 M sodium chloride promoted the self-aggregation of linoleic acid-chitosan molecules both with and without emulsification. The micelles formed nanosize particles ranging from 200 to 600 nm. The nanoparticles encapsulated a lipid soluble model compound, retinal acetate, with 50% efficiency. The similar group modified chitosan with linolenic acid (the DS 1.8%) using the same reaction. The self-aggregated nanoparticles of linolenic acid-chitosan were also used to immobilize trypsin using glutaraldehyde as crosslinker. Results indicated that the activity of trypsin immobilized onto the nanoparticles increased with increasing concentration of glutaraldehyde up to 0.07% (v/v) and then decreased with increasing amount of glutaraldehyde. On the other side, particle size increased (from 523 to 1372 nm) with the increasing concentration of glutaraldehyde (from 0.03 to 0.1% v/v) (Liu *et al.*, 2005).

Water-soluble N-palmitoyl chitosan was prepared by swollen chitosan coupling with palmitic anhydride in dimethyl sulfoxide, which could procedure micelles in water (Jiang *et al.*, 2006). The DS of N-palmitoyl chitosan was in the range of 1.2-14.2% and the CMC of N-palmitoyl chitosan micelles was in the range of 2.0×10^{-3} to 37.2×10^{-3} mg/ml. The loading capacity of hydrophobic model drug ibuprofen in the micelles was about 10%. Also stearic acid grafted chitosan oligosaccharide by 1-ethyl-3-(3-dimethylaminopropyl) carbodiimide-mediated coupling reaction was synthesized (Hu *et al.*, 2006). The CMC of the copolymer was approximately 0.06, 0.04, 0.01 mg/ml respectively. To increase the stability of the micelle in vivo and controlled drug release, the shells of micelles were cross-linked by glutaraldehyde. Paclitaxel was used as a model drug to incorporate into the micelles, and the surfaces of the micelles were further cross-linked by glutaraldehyde to form drug loaded and shell cross-linked nanoparticles. The higher drug entrapment efficiencies (above 94%) were observed in all cases. Zhang et al. (Zhang *et al.*, 2007) developed self-assembled nanoparticles based on oleoyl-chitosan with a mean diameter of 255.3 nm. Doxorubicin was efficiently loaded into the nanoparticles with an encapsulation efficiency of 52.6%. The drug was rapidly and completely released from the nanoparticles at pH 3.8, whereas at pH 7.4 there was a sustained release after a burst release. Amylose-conjugated linoleic acid complexes were synthesized to serve as molecular nanocapsules for the protection and the delivery of linoleic acid (Shimoni *et al.*, 2005).

Pluronic tri-block copolymers collected of poly (ethylene oxide)-poly (propylene oxide) - poly (ethylene oxide) show lesser critical solution temperature behaviors over a broad temperature range depending on the composition and MW. They self-assemble to procedure a spherical micellar structure above the lower critical solution temperature by hydrophobic interaction of the poly (propylene oxide) middle block in the structure. Pluronic/heparin composite nanocapsules, which displayed a 1000-fold volume transition (ca. 336 nm at 25 °C; ca. 32 nm at 37 °C), and a reversible swelling and de-swelling behavior when the temperature was cycled between 20 and 37 °C is prepared (Choi *et al.*, 2006). Core/shell nanoparticles with the poly (lactide-co-glycolide) core and the polymeric shell made-up of pluronics and hyaluronic acid was synthesized (Yuk *et al.*, 2005).

4.3.2 Cyclic hydrophobic molecules

Cholesterol is an essential lipid in animals, which not only participates the formation of cell membranes but also works as a raw material for the synthesis of bile acids, vitamin D and steroid hormones. Conjugating hydrophobic cholesterol to hydrophilic polysaccharides may form amphiphilic copolymer which may further form self-assembly nanoparticles in aqueous solution. cholesterol-modified chitosan conjugate with succinyl linkages was synthesized (Wang *et al.*, 2007c). The conjugates formed monodisperse self-aggregated nanoparticles with a roughly spherical shape and a mean diameter of 417.2 nm by probe sonication in aqueous media. Epirubicin, as a model anticancer drug, was physically entrapped inside the nanoparticles by the remote loading technique. Epirubicin-loaded nanoparticles were almost spherical in shape and their size increased from 338.2 to 472.9 nm with the epirubicin-loading content increasing from 7.97% to 14.0%. Also was prepared self-aggregated nanoparticles of cholesterol-modified O-carboxymethyl chitosan (Wang et al., 2007b).

Various cholesterol-bearing pullulans with different MWs of the parent pullulan and DS of the cholesteryl moiety was synthesized (Nishikawa et al., 1996; Akiyoshi et al., 1997). Irrespective of the MW of the parent pullulan and the DS, all of cholesterol-pullulans provided unimodal and mono-disperse self-aggregates in water. The size of the self-aggregate reduced with an increase in the DS of the cholesteryl moiety (hydrodynamic radius, 8.4-13.7 nm). However, the aggregation number of cholesterol-pullulans in one nanoparticle was almost independent of the DS. The polysaccharide density within the self-aggregate (0.13– 0.50 g/ml) was affected by both the MW and the DS of cholesterol-pullulans. The characteristic temperature to cause a structural change of the nanoparticles decreased with an increase in the DS and the ionic strength of the medium. Moreover, they also prepared thermo-responsive nanoparticles by self-assembly of two different hydrophobically modified polymers, namely, cholesterol-pullulan and a copolymer of N-isopropylacrylamide and N-[4-(1-pyrenyl) butyl]-N-n-octadecylacrylamide via their hydrophobic moieties (Akiyoshi et al., 2000) , as well as hexadecyl group-bearing pullulan self-assembly nanoparticles (Kuroda et al., 2002).

Bile acids such as deoxycholic acid and 5β-cholanic acid are known to form micelles in water as a result of their amphiphilicity, which plays an important role in the emulsification, solubilization, and absorption of cholesterol, fats, and liphophilic vitamins in human body. Therefore, it is expected that the introduction of deoxycholic acid or 5β-cholanic acid into chitosan would induce self-association to form self-aggregates. Covalently conjugated deoxycholic acid to chitosan via carbodiimide-mediated reaction to generate self-aggregated nanoparticles was prepared (Lee et al., 1998; Jeong et al., 1998). Adriamycin was physically entrapped inside the self- aggregates. The size of adriamycin-loaded self-aggregates increased with increasing the loading content of adriamycin (Lee et al., 2000).

Chemically modified chitosan oligosaccharides with deoxycholic acid was reported (Chae et al., 2005). Owing to the amphiphilic characters, the deoxycholic acid-chitosan formed self-aggregated nanoparticles in aqueous milieu. The particle size of the nanoparticles was in the range of 200-240 nm. Furthermore, deoxycholic acid-chitosan showed great potential for gene carrier with the high level of gene transfection efficiencies, even in the presence of serum. Deoxycholic acid-heparin amphiphilic conjugates with different degree of substitution of deoxycholic acid was synthesized (Park et al., 2004), which provided monodispersed self-aggregates in water, with mean diameters (120-200 nm) decreasing with increasing DS. Increasing DS enhanced the hydrophobicity of the self-aggregate inner core.

However, chitosan -based self-aggregates were difficult to be widely applied for drug delivery systems because chitosan aggregates are insoluble in biological solution (pH7.4) and they are readily precipitated within a few days. Recently, water-soluble chitosan derivatives have been used to increase their stability in biological solution and decrease the cytotoxicity induced by acidic solution, where chitosan is soluble. Covalently modified glycol chitosan with deoxycholic acid self-aggregates as a new drug delivery system was prepared (Kim et al., 2005) and investigated in detail the effect of deoxycholic acid attached to glycol chitosan on the formation, physicochemical characteristics, and stability of self-aggregates in aqueous media. The same group (Kwon et al., 2003; Park et al., 2007) covalently attached the 5β-cholanic acid to glycol chitosan through amide formation using carbodiimide as catalyzer. The 5β-cholanic acid-glycol chitosan formed self-aggregates (210-859 nm in diameter) in an aqueous phase by intra- or intermolecular association between hydrophobic 5β-cholanic acids attached to glycol chitosan.

FITC is a widely used hydrophobic fluorescein, the isothiocyanato of which can readily react with free amine to incorporate fluorescence labeling. Doxorubicin is an anti-tumor antibiotic, which can inhibit the synthesis of RNA and DNA and has a therapeutic effect on many tumors. FITC and doxorubicin themselves are hydrophobic cyclic molecules, which can be conjugated onto hydrophilic polysaccharides form amphiphilic copolymers. Hydrophobically modified glycol chitosans by chemical conjugation of FITC or doxorubicin to the backbone of glycol chitosan was prepared (Lee et al., 2006; Son et al., 2003). Biodistribution of self-aggregates (300 nm in diameter) was evaluated using tissues obtained from tumor-bearing mice, to which self-aggregates were systemically administered via the tail vein. Na et al. (Na et al., 2003) introduced vitamin H to pullulan acetate and prepared corresponding self-assembled nanoparticles (~100 nm) in order to improve their cancer-targeting activity and internalization. Three samples of biotinylated pullulan acetate, comprising 7, 20 and 39 vitamin H groups per 100 anhydroglucose units, were synthesized. In addition, synthesized successfully N-succinyl-chitosan, which could be self-assembly of well-dispersed and stable nanospheres in distilled water with 50-100 nm in diameter (Zhu et al., 2006). Experimental results indicated that a hydrophobic domain formed within these nanospheres. The assembly mechanisms were believed to be the intermolecular H-bonding of N-succinyl-chitosan and hydrophobic interaction among the hydrophobic moieties in N-succinyl-chitosan macromolecules. Park et al. (Park et al., 2006) described N-acetyl histidine-conjugated glycol chitosan self-assembled nanoparticles as a promising system for intracyto-plasmic delivery of drugs.

4.3.3 Polyacrylate-based nanoparticles applicable as biomaterials

Poly (methyl methacrylate) and poly (isobutyl cyanoacrylate) (PIBCA) all belong to polyacrylate family and they were widely used for biomaterials. Containing carboxylic ester groups in their structures, they are hydrophobic. The efficient uptake of injected nanoparticles by cells of the mononuclear phagocyte system limits the development of long-circulating colloidal drug carriers. The complement system plays a major role in the opsonization and recognition processes of foreign materials. Since heparin is an inhibitor of complement activation, nanoparticles bearing heparin covalently bound to poly (methyl methacrylate) and evaluated their interactions with complement was prepared (Passirani et al., 1998a). Nanoparticles bearing covalently bound dextran instead of heparin were weak activators of complement as compared with cross-linked dextran or bare poly (methyl methacrylate) nanoparticles. In addition to the specific activity of bound heparin, the protective effect of both polysaccharides is hypothesized to be due to the presence of a dense brush-like layer on the surface of the particles. Dextran nanoparticles were also eliminated very slowly over 48 h. bare poly (methyl methacrylate) nanoparticles were found to have a half-life of only 3 min. Both types of nanoparticles proved to be long-circulating. The potent capacity for opsonization of the poly (methyl methacrylate) core was hidden by the protective effect of either polysaccharide, probably due to a dense brush -like structure. In the case of heparin nanoparticles, the "stealth" effect was probably increased by its inhibiting properties against complement activation (Passirani et al., 1998b).

PIBCA-chitosan nanoparticles by emulsion polymerization of IBCA in the presence of chitosan as a polymeric stabilizer at low pH were prepared (Yang et al., 2000). Nimodipine as a model drug was successfully incorporated into the nanoparticles with mean particle diameter of 31.6 nm and a positive charge. Also PIBCA-chitosan, PIBCA-dextran and PIBCA-dextran sulfate core-shell nanoparticles by redox radical or anionic polymerization of IBCA in the presence of chitosan, dextran or dextran sulfate was prepared (Bertholon et al., 2006). Bravo-Osuna et al. (Bravo-Osuna et al., 2006; Bravo-Osuna et al., 2007a; Bravo-Osuna et al., 2007c; Bravo-Osuna et al., 2007b) developed PIBCA-thiolated chitosan nanoparticles by radical emulsion polymerization. The nanoparticles had mean hydrodynamic diameter around 200 nm and positive zeta potential values, indicating the presence of the cationic thiolated chitosan at the nanoparticle surface. Polysaccharide-coated nanoparticles by radical emulsion polymerization of IBCA in the presence of various polysaccharides (dextran, dextran sulfate, heparin, chitosan, hyaluronic acid, pectin) was synthesized (Chauvierre et al., 2003). They also measured the complement activation induced by different polysaccharide-coated nanoparticles and of the antithrombic activity of heparin. These nanoparticles maintained the heparin antithrombic properties and inhibited complement activation. This work demonstrated the hemoglobin loading on nanoparticle surface, rather than being encapsulated. With a size of 100 nm, these drug delivery systems made suitable tools in the treatment of thrombosis oxygen deprived pathologies (Chauvierre et al., 2004). In addition, they investigated for the first time the mobility of dextran chains on the PIBCA nanoparticles with electronic paramagnetic resonance. This technique opens an interesting prospect of investigating surface properties of polysaccharide-coated nanoparticles by a new physicochemical approach to further correlate the mobility of the polysaccharide chains with the fate of the nanoparticles in biological systems (Vauthier et al., 2004).

4.4 Polysaccharides-based nanoparticle through dialysis method

The preparation of nanoparticles was performed by a dialysis method without the use of any surfactant or emulsifiers. Dialysis offers a simple and effective method for the preparation of small, narrow-distributed polymer nanoparticle (Fessi et al., 1989; Jeong et al., 2001; Kostog. M et al., 2010; Jeon et al., 2000). Polymer is dissolved in an organic solvent and placed inside a dialysis tube with proper molecular weight cut-off. Dialysis is performed against a non-solvent miscible with the former miscible. The displacement of the solvent inside the membrane is followed by the progressive aggregation of polymer due to a loss of solubility and the formation of homogeneous suspensions of nanoparticles.

Paclitaxel-loaded HGC (PTX-HG C) nanoparticles were simply prepared by this method (Kwon et al., 2006). The incorporation of PTX into the HGC nanoparticles occurred simultaneously during dialysis. The loading efficiency of PTX into HGC nanoparticles was determined by varying the feed weight ratio of PTX to HGC nanoparticles. When the feed ratio was less than 0.1, the loading efficiency was above 90%. Importantly, the PTX-HG C nanoparticles were well dispersed in an aqueous medium. However, if the feed ratio was above 0.1, the loading efficiency significantly decreased to about 42% and the excess of PTX molecules precipitated during dialysis. Thus, the maximum loading content of PT X into HGC nanoparticles was determined to 10 wt%.

To make core-shell type nanoparticles, poly (DL-lactide-co-glycolide) (PLGA) grafted-dextran (DexLG) graft copolymer was dissolved in DMSO and the core-shell type nanoparticles were prepared by dialysis method against water. The morphology of core-shell type nanoparticles of DexLG copolymer was observed by SEM and the particle size was evaluated by DLS. Core-shell type nanoparticles of DexLG copolymer has spherical shapes in their morphology and particle size was around 50-200 nm.

Starch ester nanoparticles were prepared by the dialysis method. Appropriate amount of graft polymer was dissolved in DMSO, the sample was dialysed against water using a dialysis membrane of MW 12,000 g mol-1 cut off. Starch nanoparticle formed was studied by atomic force microscopy. Nanoparticles in DMSO water solution were transferred to freshly cleaved mica sheet by drop and analyzed by tapping mode. Size of the particles was found to be in the range of 65–75 nm (diameter), and 17–19 nm (height).

5. Medical applications of polysaccharide-based nanoparticles

Polysaccharide-based nanoparticles have received considerable attention in recent years as one of the most promising nanoparticulate drug delivery systems owing to their unique potentials. Nanoparticle drug delivery systems are defined as particulate dispersions or solid particles with a size in the range of 10-1000nm and with various morphologies, including nanospheres, nanocapsules, nanomicelles, nanoliposomes, and nanodrugs, etc. The drug is dissolved, entrapped, encapsulated or attached to a nanoparticle matrix (Kommareddy et al., 2005; Lee and Kim, 2005). Drug delivery systems of nanoparticles have several advantages, such as high drug encapsulation efficiency, efficient drug protection against chemical or enzymatic degradation, unique ability to create a controlled release, cell internalization as well as ability to reverse the multidrug resistance of tumor cells (Soma et al., 1999). The use of starch nanoparticles is receiving a significant amount of attention due to their good hydrophilicity, biocompatibility and biodegradability. Starch nanocrystals have also been found to be excellent reinforcements (Elvira et al., 2002). Hydrophobic grafted and cross-linked starch nanoparticles were used for drug delivery and Indomethacin was taken as the model drug (Abraham and Simi, 2007). Hydrophilic amylopectin was modified by grafting hydrophobic poly (lactic acid) chains (PLA) for the fabrication of polymeric micelles for drug delivery. When these spherical nano-aggregates were used as the drug carrier, it was found that they had a good loading capacity and in vitro release properties for hydrophobic indomethacin drug (Brecher et al., 1997; Dufresne et al., 2006).

A novel amphiphilic copolymer (dextran-g-polyethyleneglycol alkyl ether) was synthesized which resulted in polymeric micelle formation, encapsulating cyclosporine in the hydrophobic core and providing a hydrophilic corona (Na et al., 2003; Francis et al., 2003).

Nanoparticles of poly (DL-lactide-co-glycolide)-grafted dextran were synthesized for use as a nanoparticulate oral drug carrier. These nanoparticles were able to form nanoparticles in water by self-aggregating process, and their particle size was around 50 nm~300 nm. Core-shell type nanoparticles of DexLG copolymer can be used as a colonic drug carrier (Tiera *et al.*, 2003). Superparamagnetic chitosan–dextran sulfate hydrogels as drug carriers was synthesized. The 5- aminosalicylic acid was chosen as model drug molecule (Saboktakin *et al.*, 2010). Dextran sulphate–chitosan nanoparticles were prepared to overcome the pharmacokinetic problems and to obtain the full benefits of the drug (Anitha *et al.*, 2011). Self-assembled hydrogel nanoparticles composed of dextran and poly (ethylene glycol) was synthesized and prepared nanoparticles used for drug carrier with hydrophobic model drug in vitro (Kim *et al.*, 2000).

Hydrophobized pullulan has been used as drug delivery carrier, Specifically, cholesterol-pullulan and a copolymer of N-isopropylacrylamide and N-[4-(1-pyrenyl)butyl]-N-n-octadecylacrylamide via their hydrophobic moieties, as well as hexadecyl group-bearing pullulan self-assembly nanoparticles (Akiyoshi *et al.*, 1998; Akiyoshi *et al.*, 1993; Jung *et al.*, 2004). These hydrophobized pullulan self-associate to form colloidally stable nanoparticles with inner hydrophobic core. This hydrophobic core can only encapsulate hydrophobic substances like insoluble drugs and proteins (Gupta and Gupta, 2004). Amphiphilic polysaccharides composed of pullulan and poly (DL-lactide-coglycolide) (PLGA) were synthesized to give amphiphilicity and biodegradability as novel drug carriers. Due to its biodegradability, PLGA is commonly used for the controlled release of drugs (Jeong *et al.*, 2006). Hydrophobically modified glycol chitosan (HGC) nanoparticles showed potential as carriers for anticancer peptides and anticancer drugs because of their biocompatible in vivo (Kwon *et al.*, 2003; Yoo *et al.*, 2005). Modified chitosan derivatives, are emerging as novel carriers of drugs because of their solubility and biocompatibility in vivo (Sinha *et al.*, 2004; Jiang *et al.*, 2006; Chen *et al.*, 2003b). Nanoparticles of carboxymethyl chitosan (CM-chitosan) as carriers for the anticancer drug, were prepared by gelification with calcium ions and Doxorubicin (DOX) was chosen as a model drug.

6. Conclusions and future trends

The literature survey showed that in the last decades a lot of attention has been focused to the combination of polysaccharides based polymers with inorganic nanoparticles, to benefit from the advantages of both organic and inorganic composite components. As this chapter showed the use of polysaccharides-based nanoparticles is receiving a significant amount of interests because of the plentiful availability of natural polymer, inexpensive, renewability, biocompatibility, biodegradability and nontoxicity. Therefore, a number of formulations of such bionanocomposites exhibits some excellent characteristics such as magnetic, optical, antimicrobial functionalities, size particles, surface coverage, colloidal stability, enzyme degradability and interesting applications of the polysaccharide based nanoparticles and their derivatives for biotechnological and biomedical applications was explained. The preparation of this kind of materials strongly relies on earlier steps of their production and modification steps which emphasises the relevance of preparative strategies that take in consideration their final applications. With this respect, we introduced various methods for the preparation of polysaccharides-based nanoparticles such as: solvent evaporation method, spontaneous emulsification or solvent diffusion method, self-assembly of hydrophobically modified and dialysis method. On the other hand, the modified polysaccharides exhibit considerable potentials to utilize as stabilizers to produce stable hydrophilic nanoparticles through the o/w emulsion/evaporation technique. Modified polysaccharides were shown to exhibit surface active properties and to act as efficient emulsion stabilizers. Surface modified colloidal carriers such as nanoparticles are able to modulate the biodistribution of the loaded drug when given intravenously, but also to control the absorption of drugs administered by other routes. The amphiphilic character imparted upon polysaccharides after hydrophobic modification gives them a wide and interesting use spectrum, for instance as rheology modifiers, emulsion stabilizers, surface modifiers for liposomes and nanoparticles and as drug delivery vehicles. The recent attempts toward finding new methods for the earlier diagnosis of diseases and more effective therapies to synthesize the new generation of multifunctional nanostructured materials based on polysaccharides, modified polysaccharides and polysaccharide-based dendrimers is very fast emerging. As time goes on, more polysaccharide-based nanoparticles emerge, which greatly enriches the versatility of nanoparticle carriers agents in terms of category and function.

7. Acknowledgments

Authors are greatly acknowledging the Research Center for Pharmaceutical Nanotechnology and the University of Tabriz for their financial supports of this work.

8. References

Abraham, T. E. & Simi, C. K. (2007). Hydrophobic grafted and cross-linked starch nanoparticles for drug delivery. *Bioprocess and Biosystems Engineering* 30(3): 173-180.

Akiyoshi, K., Deguchi, S., Moriguchi, N., Yamaguchi, S. & Sunamoto, J. (1993). Self-Aggregates of Hydrophobized Polysaccharides in Water - Formation and Characteristics of Nanoparticles. *Macromolecules* 26(12): 3062-3068.

Akiyoshi, K., Deguchi, S., Tajima, H., Nishikawa, T. & Sunamoto, J. (1997). Microscopic structure and thermoresponsiveness of a hydrogel nanoparticle by self-assembly of a hydrophobized polysaccharide. *Macromolecules* 30(4): 857-861.

Akiyoshi, K., Kang, E. C., Kurumada, S., Sunamoto, J., Principi, T. & Winnik, F. M. (2000). Controlled as sociation of amphiphilic polymers in water: thermosensitive nanoparticles formed by self-assembly of hydrophobically modified pullulans and poly(N-isopropylacrylamides) *Macromolecules* 33: 3244-3249.

Akiyoshi, K., Kobayashi, S., Shichibe, S., Mix, D., Baudys, M., Kim, S. W. & Sunamoto, J. (1998). Self-assembled hydrogel nanoparticle of cholesterol-bearing pullulan as a carrier of protein drugs: complexation and stabilization of insulin. *Journal of Controlled Release* 54: 313-320.

Albertsson, A. C., Qu, X. & Wirsen, A. (1999). Synthesis and characterization of pH-sensitive hydrogels based on chitosan and D,L-lactic acid. *Journal of Applied Polymer Science* 74(13): 3193-3202.

Allemann, E., Quintanar-Guerrero, D., Fessi, H. & Doelker, E. (1998). Preparation techniques and mechanisms of formation of biodegradable nanoparticles from preformed polymers. *Drug Development and Industrial Pharmacy* 24(12): 1113-1128.

Alonso, M. J., Janes, K. A. & Calvo, P. (2001). Polysaccharide colloidal particles as delivery systems for macromolecules. *Adv Drug Deliv Rev* 47(1): 83-97.

Aminabhavi, T. M., Balundgi, R. H. & Cassidy, P. E. (1990). A Review on Biodegradable Plastics. *Polymer-Plastics Technology and Engineering* 29(3): 235-262.

an, R. Y., Matsusaki, M., Akashi, M. & Chirachanchai, S. (2004). Controlled hydrophobic/ hydrophilic chitosan: colloidal phenomena and nanosphere formation. *Colloid Polym. Sci.* 282: 337-342.

Anitha, A., Deepagan, V. G., Divya Rani, V. V., Deepthy Menon, Nair, S. V. & Jayakumar, R. (2011). Preparation, characterization, in vitro drug release and biological studies of curcumin loaded dextran sulphate-chitosan nanoparticles. *Carbohydrate Polymers* 84: 1158-1164.

Aoki, N., Nishikawa, M. & Hattori, K. (2003). Synthesis of chitosan derivatives bearing cyclodextrin and adsorption of p-nonylphenol and bisphenol A. *Carbohydrate Polymers* 52(3): 219-223.

Aumelas, A., Serrero, A., Durand, A., Dellacherie, E. & Leonard, M. (2007). Nanoparticles of hydrophobically modified dextrans as potential drug carrier systems. *Colloids Surf B Biointerfaces* 59(1): 74-80.

Bataille, I., Huguet, J., Muller, G., Mocanu, G. & Carpov, A. (1997). Associative behaviour of hydrophobically modified carboxymethylpullulan derivatives. *International Journal of Biological Macromolecules* 20(3): 179-191.

Bertholon, I., Vauthier, C. & Labarre, D. (2006). Complement activation by core-shell poly(isobutylcyanoacrylate)-polysaccharide nanoparticles: influences of surface morphology, length, and type of polysaccharide. *Pharmaceutical Research* 23: 1313-1323.

Borredon, E., Aburto, J. & Alric, I. (1999). Preparation of long-chain esters of starch using fatty acid chlorides in the absence of an organic solvent. *Starch-Starke* 51(4): 132-135.

Bravo-Osuna, I., Millotti, G., Vauthier, C. & Ponchel, G. (2007a). In vitro evaluation of calcium binding capacity of chitosan and thiolated chitosan poly(isobutyl cyanoacrylate) core-shell nanoparticles. *Int J Pharm* 338(1-2): 284-290.

Bravo-Osuna, I., Ponchel, G. & Vauthier, C. (2007b). Tuning of shell and core characteristics of chitosan-decorated acrylic nanoparticles. *Eur J Pharm Sci* 30(2): 143-154.

Bravo-Osuna, I., Schmitz, T., Bernkop-Schnurch, A., Vauthier, C. & Ponchel, G. (2006). Elaboration and characterization of thiolated chitosan-coated acrylic nanoparticles. *Int J Pharm* 316(1-2): 170-175.

Bravo-Osuna, I., Vauthier, C., Farabollini, A., Palmieri, G. F. & Ponchel, G. (2007c). Mucoadhe-sion mechanism of chitosan and thiolated chitosan-poly(isobutyl cyanoacrylate) core-shell nanoparticles. *Biomaterials* 28: 2233-2243.

Brecher, M. E., Owen, H. G. & Bandarenko, N. (1997). Alternatives to albumin: Starch replacement for plasma exchange. *Journal of Clinical Apheresis* 12: 146-153.

Burt, H. M. & Shi, R. W. (2003). Synthesis and characterization of amphiphilic hydroxypropylcellulose-graft-poly(epsilon-caprolactone). *Journal of Applied Polymer Science* 89(3): 718-727.

Cao, A. I., Yang, J., Yu, Y. H., Li, Q. B. & Li, Y. (2005). Chemical synthesis of biodegradable aliphatic polyesters and polycarbonates catalyzed by novel versatile aluminum metal complexes bearing salen ligands. *Journal of Polymer Science Part a-Polymer Chemistry* 43(2): 373-384.

Chae, S. Y., Son, S., Lee, M., Jang, M. K. & Nah, J. W. (2005). Deoxycholic acid-conjugated chitosan oligosaccharide nanoparticles for ef ficient gene carrier. *Journal of Controlled Release* 109: 330-344.

Chauvierre, C., Labarre, D., Couvreur, P. & Vauthier, C. (2003). Novel polysaccharide-decorated poly(isobutyl cyanoacrylate) nanoparticles. *Pharm Res* 20(11): 1786-1793.

Chauvierre, C., Marden, M. C., Vauthier, C., Labarre, D., Couvreur, P. & Leclerc, L. (2004). Heparin coated poly(alkylcyanoacrylate) nanoparticles coupled to hemoglobin: a new oxygen carrier. *Biomaterials* 25(15): 3081-3086.

Chen, X. G., Lee, C. M. & Park, H. J. (2003a). OM emulsification for the self-aggregation and nanoparticle formation of linoleic acid-modified chitosan in the aqueous system. *Journal of Agricultural and Food Chemistry* 51: 3135-3139.

Chen, X. G., Lee, C. M. & Park, H. J. (2003b). OM emulsification for the self-aggregation and nanoparticle formation of linoleic acid-modified chitosan in the aqueous system. *Journal of Agricultural and Food Chemistry* 51: 3135-3139.

Cheng, F., Wei, Y. P., Hou, G. & Sun, S. F. (2008). Amphiphilic cellulose: Surface activity and aqueous self-assembly into nano-sized polymeric micelles. *Reactive & Functional Polymers* 68(5): 981-989.

Choi, S. H., Lee, J. H., Choi, S. M. & Park, T. G. (2006). Thermally reversible pluronic/heparin nanocapsules exhibiting 1000-fold volume transition. *Langmuir* 22(4): 1758-1762.

Choisnard, L., Geze, A., Putaux, J. L., Wong, Y. S. & Wouessidjewe, D. (2006). Nanoparticles of beta-cyclodextrin esters obtained by self-assembling of biotransesterified beta-cyclodextrins. *Biomacromolecules* 7: 515-520.

Couvreur, P. (1998). Polyalkylcyanoacrylates as colloidal drug carriers. *Crit Rev Ther Drug Carr Syst* 5: 1-20.

Couvreur, P., Lemarchand, C. & Gref, R. (2004). Polysaccharide-decorated nanoparticles. *European Journal of Pharmaceutics and Biopharmaceutics* 58(2): 327-341.

Della Valle, G., Buleon, A., Carreau, P. J., Lavoie, P. A. & Vergnes, B. (1998). Relationship between structure and viscoelastic behavior of plasticized starch. *Journal of Rheology* 42(3): 507-525.

Dong, H., Xu, Q., Li, Y., Mo, S., Cai, S. & Liu, L. (2008). The synthesis of biodegradable graft copolymer cellulose-graft-poly(L-lactide) and the study of its controlled drug release. *Colloids Surf B Biointerfaces* 66(1): 26-33.

Dufresne, A., Angellier, H., Choisnard, L., Molina-Boisseau, S. & Ozil, P. (2004). Optimization of the preparation of aqueous suspensions of waxy maize starch nanocrystals using a response surface methodology. *Biomacromolecules* 5(4): 1545-1551.

Dufresne, A., Cavaille, J. Y. & Helbert, W. (1996). New nanocomposite materials: Microcrystalline starch reinforced thermoplastic. *Macromolecules* 29(23): 7624-7626.

Dufresne, A., Thielemans, W. & Belgacem, M. N. (2006). Starch nanocrystals with large chain surface modifications. *Langmuir* 22(10): 4804-4810.

Durand, A., Marie, E., Rotureau, E., Leonard, M. & Dellacherie, E. (2004). Amphiphilic polysaccharides: Useful tools for the preparation of nanoparticles with controlled surface characteristics. *Langmuir* 20(16): 6956-6963.

Durand, A., Rouzes, C., Leonard, M. & Dellacherie, E. (2002). Surface activity and emulsification properties of hydrophobically modified dextrans. *Journal of Colloid and Interface Science* 253(1): 217-223.

Elvira, C., Mano, J. F., San Román, J. & Reis, R. L. (2002). Starch-based biodegradable hydrogels with potential biomedical applications as drug delivery systems. *Biomaterials*. 23: 1955-1966.

Fang, Y., Huang, M. F., Liu, L., Zhang, G. B. & Yuan, G. B. (2006). Preparation of chitosan derivative with polyethylene glycol side chains for porous structure without specific processing technique. *International Journal of Biological Macromolecules* 38(3-5): 191-196.

Fessi, H., Puisieux, F., Devissaguet, J., Ammoury, N. & Benita, S. (1989). Nanocapsule formation by interfacial polymer deposition following solvent displacement. *Int J Pharm* 55: 1-4.

Fischer, A., Houzelle, M. C., Hubert, P., Axelos, M. A. V., Geoffroy-Chapotot, C., Carre, M. C., Viriot, M. L. & Dellacherie, E. (1998). Detection of intramolecular associations in hydrophobically modified pectin derivatives using fluorescent probes. *Langmuir* 14(16): 4482-4488.

Fowler, P. A., Fang, J. M., Sayers, C. & Williams, P. A. (2004). The chemical modification of a range of starches under aqueous reaction conditions. *Carbohydrate Polymers* 55(3): 283-289.

Fowler, P. A., Fang, J. M., Tomkinson, J. & Hill, C. A. S. (2002). The preparation and characterisation of a series of chemically modified potato starches. *Carbohydrate Polymers* 47(3): 245-252.

Francis, M., Lavoie, L., Winnik, F. & Leroux, J. C. (2003). Solubilization of cyclosporin A in dextran-g-polyethyleneglycolalkylether polymeric micelles. *Eur. J. Pharm. Sci.* 56: 337-346.

French, D., Pulley, A. O., Effenberger, J. A., Rougvie, M. A. & Abdullah, M. (1965). The schardinger dextrin: molecular size and structure of the δ-, ε-, ζ-, and η-dextrin. *Arch. Biochem. Biophys.* 111: 153-160.

Fringuelli, F., Piermatti, O., Pizzo, F. & Vaccaro, L. (1999). Ring opening of epoxides with sodium azide in water. A regioselective pH-controlled reaction. *Journal of Organic Chemistry* 64(16): 6094-6096.

Fujiwara, T., Tanaka, N. & Kobayashi, S. (1990). Structure of Delta-Cyclodextrin 13.75h2o. *Chemistry Letters* (5): 739-742.

Gassan, J. & Bledzki, A. K. (1999). Composites reinforced with cellulose based fibres. *Progress in Polymer Science* 24(2): 221-274.

Glinel, K., Huguet, J. & Muller, G. (1999). Comparison of the associating behaviour between neutral and anionic alkylperfluorinated pullulan derivatives. *Polymer* 40(25): 7071-7081.

Gorochovceva, N., Naderi, A., Dedinaite, A. & Makuska, R. (2005). Chitosan-N-poly(ethylene glycol) brush copolymers: Synthesis and adsorption on silica surface. *European Polymer Journal* 41(11): 2653-2662.

Gref, R., Rodrigues, J. & Couvreur, P. (2002). Polysaccharides grafted with polyesters: Novel amphiphilic copolymers for biomedical applications. *Macromolecules* 35(27): 9861-9867.

Gref, R., Rodrigues, J. S., Santos-Magalhaes, N. S., Coelho, L. C. B. B., Couvreur, P. & Ponchel, G. (2003). Novel core (polyester)-shell(polysaccharide) nanoparticles: protein loading and surface modification with lectins. *Journal of Controlled Release* 92(1-2): 103-112.

Gupta, M. & Gupta, A. (2004). Hydrogel pullulan nanoparticles encapsulating pBUDLacZ plasmid as an efficient gene delivery carrier. *Journal of Controlled Release* 99: 157-166.

Gurruchaga, M., Silva, I. & Goni, I. (2009). Physical blends of starch graft copolymers as matrices for colon targeting drug delivery systems. *Carbohydrate Polymers* 76(4): 593-601.

Hamidi, M., Azadi, A. & Rafiei, P. (2008). Hydrogel nanoparticles in drug delivery. *Adv Drug Deliv Rev* 60(15): 1638-1649.

Hennink, W. E., van de Manakker, F., Vermonden, T. & van Nostrum, C. F. (2009). Cyclodextrin-Based Polymeric Materials: Synthesis, Properties, and Pharmaceutical/Biomedical Applications. *Biomacromolecules* 10(12): 3157-3175.

Hinrichsen, G., Mohanty, A. K. & Misra, M. (2000). Biofibres, biodegradable polymers and biocomposites: An overview. *Macromolecular Materials and Engineering* 276(3-4): 1-24.

Hong-Wei Lua & Li-Ming Zhanga (2011). Carbohydrate Preparation and properties of new micellar drug carriers based on hydrophobically modified amylopectin *Polymers* 83: 1499-1506.

Hu, F. Q., Ren, G. F., Yuan, H., Du, Y. Z. & Zeng, S. (2006). Shell cross-linked stearic acid grafted chitosan oligosaccharide self-aggregated micelles for controlled release of paclitaxel. *Colloids Surf B Biointerfaces* 50(2): 97-103.

I., B., S., H. & R., K. M. (2004). PLGA nano particles in drug delivery: the state of the art. *Crit. Rev. Ther. Drug Carrier Syst.* 21(5): 387-422.

Ichinose, K., Tomiyama, N., Nakashima, M., Ohya, Y., Ichikawa, M., Ouchi, T. & Kanematsu, T. (2000). Antitumor activity of dextran derivatives immobilizing platinum complex (II). *Anti-Cancer Drugs* 11(1): 33-38.

Jeon, H., Jeong, Y., Jang, M., Park, Y. & Nah, J. (2000). Effect of solvent on the preparation of surfactant-free poly(DL-lactide-co-glycolide) nanoparticles and norfloxacin release characteristics. *Int J Pharm* 207: 99-108.

Jeong, S. Y., Lee, K. Y., Kwon, I. C., Kim, Y. H. & Jo, W. H. (1998). Preparation of chitosan self-aggregates as a gene delivery system. *Journal of Controlled Release* 51(2-3): 213-220.

Jeong, Y., Cho, C., Kim, S., Ko, K., Kim, S., Shim, Y. & Nah, J. (2001). Preparation of poly(DL-lactide-co-glycolide) nanoparticles without surfactant. *Journal of Applied Polymer Science* 80: 2228-2236.

Jeong, Y., Na, H. S., Oh, J. S., Choi, K. C., Song, C. & Lee, H. (2006). Adriamycin release from self-assembling nanospheres of poly(DL-lactide-co-glycolide)-grafted pullulan. *International Journal of Pharmaceutics* 322: 154-160.

Jiang, G. B., Quan, D., Liao, K. & Wang, H. (2006). Novel polymer micelles prepared from chitosan grafted hydrophobic palmitoyl groups for drug delivery. *Mol Pharm* 3(2): 152-160.

Jing, X. B., Yu, H. J., Wang, W. S., Chen, X. S. & Deng, C. (2006). Synthesis and characterization of the biodegradable polycaprolactone-graft-chitosan amphiphilic copolymers. *Biopolymers* 83(3): 233-242.

Jung, S., Jeong, Y. I., Kim, S. H., Jung, T. Y., Kim, I. Y., Kang, S. S., Jin, Y. H., Ryu, H. H., Sun, H. S., Jin, S. G., Kim, K. K. & Ahn, K. Y. (2006). Polyion complex micelles composed of all-trans retinoic acid and poly (ethylene glycol)-grafted-citosan. *J Pharm Sci* 95(11): 2348-2360.

Jung, S. W., Jeong, Y. I., Kim, Y. H. & Kim, S. W. (2004). Self- assembled nanoparticles of poly (ethylene glycol) grafted pullulan acetate as a novel drug carrier. *Arch. Pharm. Res.* 27: 562-569.

Kapusniak, J. & Siemion, P. (2007). Thermal reactions of starch with long-chain unsaturated fatty acids. Part 2. Linoleic acid. *Journal of Food Engineering* 78(1): 323-332.

Kim, J. H., Kwon, H. Y., Lee, J. Y., Choi, S. W. & Jang, Y. S. (2001). Preparation of PLGA nanoparticles containing estrogen by emulsification-diffusion method. *Colloids and Surfaces a-Physicochemical and Engineering Aspects* 182(1-3): 123-130.

Kim, K., Kwon, S., Park, J. H., Chung, H., Jeong, S. Y., Kwon, I. C. & Kim, I. S. (2005). Physicochemical characterizations of self-assembled nanoparticles of glycol chitosan-deoxycholic acid conjugates. *Biomacromolecules* 6(2): 1154-1158.

Kim, S. H., Kim, I. S. & Jeong, Y. I. (2000). Self-assembled hydrogel nanoparticles composed of dextran and poly(ethylene glycol) macromer. *International Journal of Pharmaceutics* 205(1-2): 109-116.

Klemm, D., Heublein, B., Fink, H. P. & Bohn, A. (2005). Cellulose: Fascinating biopolymer and sustainable raw material. *Angewandte Chemie-International Edition* 44(22): 3358-3393.

Kommareddy, S., Tiwari, S. & Amiji, M. (2005). Long-circulating polymeric nanovectors for tumor-selective gene delivery. *Technol Cancer Res Treat* 4: 615-625.

Kostog. M, Kohler. S, Liebert. T & Heinze. T (2010). Pure cellulose nanoparticles from trimethylsilyl cellulose. *Macromol Symp* 294(2): 96-106.

Kreuter, J. (1991). Peroral administration of nanoparticles. *Adv Drug Deliv Rev* 7(1): 71-86.

Kumar, M. N. V. R. (2000). A review of chitin and chitosan applications. *Reactive & Functional Polymers* 46(1): 1-27.

Kurita, K. (2001). Controlled functionalization of the polysaccharide chitin. *Progress in Polymer Science* 26(9): 1921-1971.

Kuroda, K., Fujimoto, K., Sunamoto, J. & Akiyoshi, K. (2002). Hierarchical self-assembly of hydrophobically modified pullulan in water: gelation by networks of nanoparticles. *Langmuir* 18: 3780-3786.

Kwon, G. S. (2003). Polymeric micelles for delivery of poorly water-soluble compounds. *Critical Reviews in Therapeutic Drug Carrier Systems* 20(5): 357-403.

Kwon, I. C., Kim, J. H., Kim, Y. S., Kim, S., Park, J. H., Kim, K., Choi, K., Chung, H., Jeong, S. Y., Park, R. W. & Kim, I. S. (2006). Hydrophobically modified glycol chitosan nanoparticles as carriers for paclitaxel (Reprinted from Journal of Controlled Release, vol 109, pg 1, 2005). *Journal of Controlled Release* 111(1-2): 228-234.

Kwon, I. C., Kwon, S., Park, J. H., Chung, H., Jeong, S. Y. & Kim, I. S. (2003). Physicochemical characteristics of self-assembled nanoparticles based on glycol chitosan bearing 5 beta-cholanic acid. *Langmuir* 19(24): 10188-10193.

Kwon, I. C., Park, J. H., Kwon, S. G., Nam, J. O., Park, R. W., Chung, H., Seo, S. B., Kim, I. S. & Jeong, S. Y. (2004). Self-assembled nanoparticles based on glycol chitosan bearing 5 beta-cholanic acid for RGD peptide delivery. *Journal of Controlled Release* 95(3): 579-588.

Labhasetwar, V., Panyam, J., Dali, M. A., Sahoo, S. K., Ma, W. X., Chakravarthi, S. S., Amidon, G. L. & Levy, R. J. (2003). Polymer degradation and in vitro release of a model protein from poly(D,L-lactide-co-glycolide) nano- and microparticles. *Journal of Controlled Release* 92(1-2): 173-187.

Larsen, K. L. (2002). Large cyclodextrins. *Journal of Inclusion Phenomena and Macrocyclic Chemistry* 43(1-2): 1-13.

Larsen, K. L., Endo, T., Ueda, H. & Zimmermann, W. (1998). Inclusion complex formation constants of alpha-, beta-, gamma-, delta-, epsilon-, zeta-, eta- and theta-cyclodextrins determined with capillary zone electrophoresis. *Carbohydrate Research* 309(2): 153-159.

Lee, K. Y., Jo, W. H., Kwon, I. C., Kim, Y. H. & Jeong, S. Y. (1998). Structural determination and interior polarity of self-aggregates prepared from deoxycholic acid-modified chitosan in water. *Macromolecules* 31: 378-383.

Lee, K. Y., Kim, J. H., Kwon, I. C. & Jeong, S. Y. (2000). Self-aggregates of deoxycholic acid modified chitosan as a novel carrier of adriamycin. *Colloid and Polymer Science* 278: 1216-1219

Lee, M., Cho, Y. W., Park, J. H., Chung, H. S., Jeong, S. Y., Choi, K. W., Moon, D. H., Kim, S. Y., Kim, I. S. & Kwon, I. C. (2006). Size control of self-assembled nanoparticles by an emulsion/solvent evaporation method. *Colloid and Polymer Science* 284(5): 506-512.

Lee, M. & Kim, S. (2005). Polyethylene glycol-conjugated copolymers for plasmid DNA delivery. *Pharmaceutical Research* 22: 1-10.

Lee, V. H. L. (2001). Encyclopedia of Controlled Drug Delivery. *Journal of Controlled Release* 71(3): 353-354.

Lemarchand, C., Couvreur, P., Besnard, M., Costantini, D. & Gref, R. (2003). Novel polyester-polysaccharide nanoparticles. *Pharm Res* 20(8): 1284-1292.

Lemos-Senna, E., Wouessidjewe, D., Lesieur, S. & Duchene, D. (1998). Preparation of amphiphilic cyclodextrin nanospheres using the emulsification solvent evaporation method. Influence of the surfactant on preparation and hydrophobic drug loading. *International Journal of Pharmaceutics* 170: 119-128.

Leonard, M., Rouzes, C., Durand, A. & Dellacherie, E. (2003). Influence of polymeric surfactants on the properties of drug-loaded PLA nanospheres. *Colloids and Surfaces B-Biointerfaces* 32(2): 125-135.

Leonard, M., Rouzes, C., Gref, R., Delgado, A. D. & Dellacherie, E. (2000). Surface modification of poly(lactic acid) nanospheres using hydrophobically modified dextrans as stabilizers in an o/w emulsion/evaporation technique. *Journal of Biomedical Materials Research* 50(4): 557-565.

Leroux, J. C., Allemann, E., Doelker, E. & Gurny, R. (1995). New Approach for the Preparation of Nanoparticles by an Emulsification-Diffusion Method. *European Journal of Pharmaceutics and Biopharmaceutics* 41(1): 14-18.

Letchford, K. & Burt, H. (2007). A review of the formation and classi fication of amphiphilic block copolymer nanoparticulate structures: micelles, nanospheres, nanocap-sules and polymersomes. *Eur. J. Pharm. Biopharm.* 65: 259-269.

Lim, S. T. & Kim, J. Y. (2009). Preparation of nano-sized starch particles by complex formation with n-butanol. *Carbohydrate Polymers* 76(1): 110-116.

Liu, C. G., Desai, K. G. H., Chen, X. G. & Park, H. J. (2005). Preparation and characterization of nanoparticles containing trypsin based on hydrophobically modified chitosan. *Journal of Agricultural and Food Chemistry* 53: 1728-1733.

Lu, M., Albertsson, P. A., Johansson, G. & Tjerneld, F. (1994). Partitioning of Proteins and Thylakoid Membrane-Vesicles in Aqueous 2-Phase Systems with Hydrophobically-Modified Dextran. *Journal of Chromatography A* 668(1): 215-228.

Lu, M. & Tjerneld, F. (1997). Interaction between tryptophan residues and hydrophobically modified dextran - Effect on partitioning of peptides and proteins in aqueous two-phase systems. *Journal of Chromatography A* 766(1-2): 99-108.

Mader, K., Besheer, A., Hause, G. & Kressler, J. (2007). Hydrophobically modified hydroxyethyl starch: Synthesis, characterization, and aqueous self-assembly into nano-sized polymeric micelles and vesicles. *Biomacromolecules* 8(2): 359-367.

Martel, B., Devassine, M., Crini, G., Weltrowski, M., Bourdonneau, M. & Morcellet, M. (2001). Preparation and sorption properties of a beta-cyclodextrin-linked chitosan derivative. *Journal of Polymer Science Part a-Polymer Chemistry* 39(1): 169-176.

Melo, E., Nichifor, M., Lopes, A. & Carpov, A. (1999). Aggregation in water of dextran hydrophobically modified with bile acids. *Macromolecules* 32(21): 7078-7085.

Miyazawa, I., Ueda, H., Nagase, H., Endo, T., Kobayashi, S. & Nagai, T. (1995). Physicochemical Properties and Inclusion Complex-Formation of Delta-Cyclodextrin. *European Journal of Pharmaceutical Sciences* 3(3): 153-162.

Muller, G., Duval-Terrie, C. & Huguet, J. (2003). Self-assembly and hydrophobic clusters of amphiphilic polysaccharides. *Colloids and Surfaces a-Physicochemical and Engineering Aspects* 220(1-3): 105-115.

Na, K., Lee, T. B., Park, K. H., Shin, E. K., Lee, Y. B. & Cho, H. K. (2003). Self-assembled nanoparticles of hydrophobically-modified polysaccharide bearing vitamin H as a targeted anti-cancer drug delivery system. *Eur. J. Pharm. Sci.* 18: 165-173.

Na, K., Park, K. H., Song, H. C., Bom, H. S., Lee, K. H., Kim, S., Kang, D. & Lee, D. H. (2007). Ionic strength-sensitive pullulan acetate nanoparticles (PAN) for intratumoral administration of radioisotope: Ionic strength-dependent aggregation behavior and (99m)Technetium retention property. *Colloids and Surfaces B-Biointerfaces* 59(1): 16-23.

Namazi, H., Bahrami, S. & Entezami, A. A. (2005). Synthesis and controlled release of biocompatible prodrugs of beta-cyclodextrin linked with PEG containing ibuprofen or indomethacin. *Iranian Polymer Journal* 14(10): 921-927.

Namazi, H. & Dadkhah, A. (2008). Surface modification of starch nanocrystals through ring-opening polymerization of epsilon-caprolactone and investigation of their microstructures. *Journal of Applied Polymer Science* 110(4): 2405-2412.

Namazi, H. & Dadkhah, A. (2010). Convenient method for preparation of hydrophobically modified starch nanocrystals with using fatty acids. *Carbohydrate Polymers* 79(3): 731-737.

Namazi, H., Fathi, F. & Dadkhah, A. (2011). Hydrophobically modified starch using long-chain fatty acids for preparation of nanosized starch particles . *Scientia Iranica, Transactions C: Chemistry and Chemical Engineering* 18: 439-445.

Namazi, H. & Jafarirad, S. (2008). Preparation of the New Derivatives of Cellulose and Oligomeric Species of Cellulose Containing Magneson II Chromophore. *Journal of Applied Polymer Science* 110(6): 4034-4039.

Namazi, H. & Kanani, A. (2009). Investigation diffusion mechanism of beta-lactam conjugated telechelic polymers of PEG and beta-cyclodextrin as the new nanosized drug carrier devices. *Carbohydrate Polymers* 76(1): 46-50.

Namazi, H. & Mosadegh, M. (2011) Bio-nanocomposites based on naturally occurring common polysaccharides chitosan, cellulose and starch with their biomedical applications. In Tiwari, A. (Eds) Recent developments in bio-nanocomposites for biomedical applications (pp. 379-397)

Namazi, H. & Mosadegh, M. (2011). Preparation and Properties of Starch/Nanosilicate Layer/Polycaprolactone Composites. J Polym Environ 19: 980-987

Namazi, H., Mosadegh, M. & Dadkhah, A. (2009). New intercalated layer silicate nanocomposites based on synthesized starch-g-PCL prepared via solution intercalation and in situ polymerization methods: As a comparative study. *Carbohydrate Polymers* 75(4): 665-669.

Namazi, H. & Rad, S. J. (2004). Synthesis of block and grafted copolymers containing spacer-linked chromophore based on cellulose and polyethylene glycol. *Journal of Applied Polymer Science* 94(3): 1175-1185.

Nishikawa, T., Akiyoshi, K. & Sunamoto, J. (1996). Macromolecular complexation between bovine serum albumin and the self-assembled hydrogel nanoparticle of hydrophobized polysaccharides. *Journal of the American Chemical Society* 118(26): 6110-6115.

Nishio, Y. & Teramoto, Y. (2003). Cellulose diacetate-graft-poly(lactic acid)s: synthesis of wide-ranging compositions and their thermal and mechanical properties. *Polymer* 44(9): 2701-2709.

Nishio, Y., Teramoto, Y., Yoshioka, M. & Shiraishi, N. (2002). Plasticization of cellulose diacetate by graft copolymerization of epsilon-caprolactone and lactic acid. *Journal of Applied Polymer Science* 84(14): 2621-2628.

ODonnell, P. B. & McGinity, J. W. (1997). Preparation of microspheres by the solvent evaporation technique. *Adv Drug Deliv Rev* 28(1): 25-42.

Ogata, T. & Takahashi, Y. (1995). *Carbohydrate Research* 138: C5.

Onyuksel, H., Krishnadas, A. & Rubinstein, I. (2003). Sterically stabilized phospholipid mixed micelles: In vitro evaluation as a novel carrier for water-insoluble drugs. *Pharmaceutical Research* 20(2): 297-302.

Opanasopit, P., Ngawhirunpat, T., Chaidedgumjorn, A., Rojanarata, T., Apirakaramwong, A., Phongying, S., Choochottiros, C. & Chirachanchai, S. (2006). Incorporation of camptothecin into N-phthaloyl chitosan-g-mPEG self-assembly micellar system. *European Journal of Pharmaceutics and Biopharmaceutics* 64(3): 269-276.

Opanasopit, P., Ngawhirunpat, T., Rojanarata, T., Choochottiros, C. & Chirachanchai, S. (2007). Camptothecin-incorporating N-phthaloylchitosan-g-mPEG self-assembly micellar system: effect of degree of deacetylation. *Colloids Surf B Biointerfaces* 60(1): 117-124.

Osterberg, E., Bergstrom, K., Holmberg, K., Schuman, T. P., Riggs, J. A., Burns, N. L., Vanalstine, J. M. & Harris, J. M. (1995). Protein-Rejecting Ability of Surface-Bound Dextran in End-on and Side-on Configurations - Comparison to Peg. *Journal of Biomedical Materials Research* 29(6): 741-747.

Ouchi, T., Nishizawa, H. & Ohya, Y. (1998). Aggregation phenomenon of PEG-grafted chitosan in aqueous solution. *Polymer* 39(21): 5171-5175.

P., C. (1988). Polyalkylcyanoacrylates as colloidal drug carriers. *Crit Rev Ther Drug Carr Syst* 5: 1-20.

Panyam, J. & Labhasetwar, V. (2003). Biodegradable nanoparticles for drug and gene delivery to cells and tissue. *Adv Drug Deliv Rev* 55(3): 329-347.

Park, J. S., Han, T. H., Lee, K. Y., Han, S. S., Hwang, J. J., Moon, D. H., Kim, S. Y. & Cho, Y. W. (2006). N-acetyl histidine-conjugated glycol chitosan self-assembled nanoparticles for intracytoplasmic delivery of drugs: endocytosis, exocytosis and drug release. *Journal of Controlled Release* 115(1): 37-45.

Park, J. S., Koh, Y. S., Bang, J. Y., Jeong, Y. I. & Lee, J. J. (2008). Antitumor effect of all-trans retinoic acid-encapsulated nanoparticles of methoxy poly(ethylene glycol)-conjugated chitosan against CT-26 colon carcinoma in vitro. *J Pharm Sci* 97(9): 4011-4019.

Park, K., Kim, J. H., Nam, Y. S., Lee, S., Nam, H. Y., Kim, K., Park, J. H., Kim, I. S., Choi, K., Kim, S. Y. & Kwon, I. C. (2007). Effect of polymer molecular weight on the tumor targeting characteristics of self-assembled glycol chitosan nanoparticles. *Journal of Controlled Release* 122(3): 305-314.

Park, K., Kim, K., Kwon, I. C., Kim, S. K., Lee, S., Lee, D. Y. & Byun, Y. (2004). Preparation and characterization of self-assembled nanoparticles of heparin-deoxycholic acid conjugates. *Langmuir* 20(26): 11726-11731.

Passirani, C., Barratt, G., Devissaguet, J. P. & Labarre, D. (1998a). Interactions of nanoparticles bearing heparin or dextran covalently bound to poly(methyl methacrylate) with the complement system. *Life Sci* 62(8): 775-785.

Passirani, C., Barratt, G., Devissaguet, J. P. & Labarre, D. (1998b). Long-circulating nanoparticles bearing heparin or dextran covalently bound to poly(methyl methacrylate),. *Pharmaceutical Research* 15: 1046-1050.

Payne, G. F., Yi, H. M., Wu, L. Q., Bentley, W. E., Ghodssi, R., Rubloff, G. W. & Culver, J. N. (2005). Biofabrication with chitosan. *Biomacromolecules* 6(6): 2881-2894.

Prabaharan, M. & Mano, J. F. (2005). Chitosan-based particles as controlled drug delivery systems. *Drug Delivery* 12(1): 41-57.

Ragauskas, A. J., Zhang, J. G., Elder, T. J. & Pu, Y. Q. (2007). Facile synthesis of spherical cellulose nanoparticles. *Carbohydrate Polymers* 69(3): 607-611.

Riedel, U. & Nickel, J. (1999). Natural fibre-reinforced biopolymers as construction materials - new discoveries. *Angewandte Makromolekulare Chemie* 272: 34-40.

Rodrigues, J. S., Santos-Magalhaes, N. S., Coelho, L. C. B. B., Couvreur, P., Ponchel, G. & Gref, R. (2003). Novel core (polyester)-shell(polysaccharide) nanoparticles: protein loading and surface modification with lectins. *Journal of Controlled Release* 92: 103-112.

Saboktakin, M., Tabatabaie, R., Maharramov, A. & Ramazanov, M. (2010). Synthesis and characterization of superparamagnetic chitosan-dextran sulfate hydrogels as nano carriers for colon-specific drug delivery. *Carbohydrate Polymers* 81: 372-376.

Saenger, W. (1980). *Angewandte Chemie-International Edition* 19: 344-362.

Saenger, W. R., Jacob, J., Gessler, K., Steiner, T., Hoffmann, D., Sanbe, H., Koizumi, K., Smith, S. M. & Takaha, T. (1998). Structures of the common cyclodextrins and their larger analogues - Beyond the doughnut. *Chemical Reviews* 98(5): 1787-1802.

Sakairi, N., Tojima, T., Katsura, H., Han, S. M., Tanida, F., Nishi, N. & Tokura, S. (1998). Preparation of an alpha-cyclodextrin-linked chitosan derivative via reductive amination strategy. *Journal of Polymer Science Part a-Polymer Chemistry* 36(11): 1965-1968.

Seppala, J., Pahimanolis, N., Vesterinen, A. H. & Rich, J. (2010). Modification of dextran using click-chemistry approach in aqueous media. *Carbohydrate Polymers* 82(1): 78-82.

Shimoni, E., Lalush, I., Bar, H., Zakaria, I. & Eichler, S. (2005). Utilization of amylose-lipid complexes as molecular nanocapsules for conjugated linoleic acid. *Biomacromolecules* 6(1): 121-130.

Sinha, V. R. & Kumria, R. (2001). Polysaccharides in colon-specific drug delivery. *International Journal of Pharmaceutics* 224(1-2): 19-38.

Sinha, V. R., Singla, A. K., Wadhawan, S., Kaushik, R., Kumria, R., Nansal, K. & Dhawan, S. (2004). Chitosan microspheres as a potential carrier for drugs. *Int J Pharm* 274: 1-33.

Soma, C. E., Dubernet, C., Barratt, G., Nemati, F., Appel, M., Benita, S. & Couvreur, P. (1999). Ability of doxorubicin-loaded nanoparticles to overcome multidrug resistance of tumour cells after their capture by macrophages. *Pharmaceutical Research* 16: 1710-1716.

Son, Y. J., Jang, J. S., Cho, Y. W., Chung, H., Park, R. W., Kwon, I. C., Kim, I. S., Park, J. Y., Seo, S. B., Park, C. R. & Jeong, S. Y. (2003). Biodistribution and anti-tumor ef fi cacy of doxorubicin loaded glycol-chitosan nanoaggregates by EPR effect. *Journal of Controlled Release* 91: 135-145.

Song, C. E., Jeong, Y. I. & Choi, K. C. (2006). Doxorubicin release from core-shell type nanoparticles of poly(DL-lactide-co-glycolide)-grafted dextran. *Archives of Pharmacal Research* 29(8): 712-719.

Sun, K., Tang, M. H. & Dou, H. J. (2006). One-step synthesis of dextran-based stable nanoparticles assisted by self-assembly. *Polymer* 47(2): 728-734.

T. Nakagawa, K., Ueno, M., Kashiw, J. & Watanabe, J. (1994). Preparation of a novel cyclodextrin homologue with d.p. five. *Tetrahedron Letters* 35: 1921-1924.

Thanou, M., Kean, T. & Roth, S. (2005). Trimethylated chitosans as non-viral gene delivery vectors: Cytotoxicity and transfection efficiency. *Journal of Controlled Release* 103(3): 643-653.

Tharanathan, R. N. & Ramesh, H. P. (2003). Carbohydrates - The renewable raw materials of high biotechnological value. *Critical Reviews in Biotechnology* 23(2): 149-173.

Tiera, M. J., Vieira, N. A. B., Moscardini, M. S. & Tiera, V. A. D. (2003). Aggregation behavior of hydrophobically modified dextran in aqueous solution: a fluorescence probe study. *Carbohydrate Polymers* 53(2): 137-143.

Tukomane, T. & Varavinit, S. (2008). Influence of octenyl succinate rice starch on rheological properties of gelatinized rice starch before and after retrogradation. *Starch-Starke* 60(6): 298-304.

Ueda, H. (2002). Physicochemical properties and complex formation abilities of large-ring cyclodextrins. *Journal of Inclusion Phenomena and Macrocyclic Chemistry* 44(1-4): 53-56.

Vanderhoff, J., El Aasser, M. & Ugelstad, J. (1979).Polymer emulsification process. In *US Patent*, Vol. 4,177,177.

Vandijkwolthuis, W. N. E., Franssen, O., Talsma, H., Vansteenbergen, M. J., Vandenbosch, J. J. K. & Hennink, W. E. (1995). Synthesis, Characterization, and Polymerization of Glycidyl Methacrylate Derivatized Dextran. *Macromolecules* 28(18): 6317-6322.

Vauthier, C., Chauvierre, C., Labarre, D. & Hommel, H. (2004). Evaluation of the surface properties of dextran-coated poly(isobutylcyanoacrylate) nanoparticles by spin-labelling coupled with electron resonance spectroscopy. *Colloid and Polymer Science* 282(9): 1016-1025.

Vyas, S. P. & Sihorkar, V. (2001). Potential of polysaccharide anchored liposomes in drug delivery, targeting and immunization. *Journal of Pharmacy and Pharmaceutical Sciences* 4(2): 138-158.

Wang, P. X., Chi, H., Xu, K., Wu, X. L., Chen, Q., Xue, D. H., Song, C. & Zhang, W. (2008). Effect of acetylation on the properties of corn starch. *Food Chemistry* 106(3): 923-928.

Wang, P. X., Chi, H., Xu, K., Xue, D. H., Song, C. L. & Zhang, W. D. (2007a). Synthesis of dodecenyl succinic anhydride (DDSA) corn starch. *Food Research International* 40(2): 232-238.

Wang, Y. S., Jiang, Q., Liu, L. R. & Zhang, Q. Q. (2007b). The interaction between bovine serum albumin and the self- aggregated nanoparticles of cholesterol -modified O-carboxymethyl chitosan. *Polymer* 48: 4135-4142.

Wang, Y. S., Liu, L. R., Jiang, Q. & Zhang, Q. Q. (2007c). Self-aggregated nanoparticles of cholesterol-modified chitosan conjugate as a novel carrier of epirubicin. *European Polymer Journal* 43: 43-51.

Wu, H. Y., Gao, W. X., Lin, X. Q., Lin, X. P., Ding, J. C. & Huang, X. B. (2011). Preparation of nano-sized flake carboxymethyl cassava starch under ultrasonic irradiation. *Carbohydrate Polymers* 84(4): 1413-1418.

Yang, L. Q., Kuang, J. L., Li, Z. Q., Zhang, B. F., Cai, X. & Zhang, L. M. (2008a). Amphiphilic cholesteryl-bearing carboxymethylcellulose derivatives: self-assembly and rheological behaviour in aqueous solution. *Cellulose* 15(5): 659-669.

Yang, S. C., Ge, H. X., Hu, Y., Jiang, X. Q. & Yang, C. Z. (2000). Formation of positively charged poly (butyl cyanoacrylate) nanoparticles stabilized with chitosan. *Colloid and Polymer Science* 278: 285-292.

Yang, X. D., Zhang, Q. Q., Wang, Y. S., Chen, H., Zhang, H. Z., Gao, F. P. & Liu, L. R. (2008b). Self-aggregated nanoparticles from methoxy poly(ethylene glycol)-modified chitosan: Synthesis; characterizat ion; aggregation and methotrexate release in vitro,. *Colloids Surf.* 61: 125-131.

Yoo, H. S., Lee, J. E., Chung, H., Kwon, I. & Jeong, S. Y. (2005). Self-assembled nanoparticles containing hydrophobically modified glycol chitosan for gene delivery. *Journal of Controlled Release* 103: 235-243.

Yuan & Zhuangdong (2007). Study on the synthesis and catalyst oxidation properties of chitosan bound nickel(II) complexes. *Journal of Agricultural and Food Chemistry* 21(5): 22-24.

Yuk, S. H., Han, S. K., Lee, J. H., Kim, D. & Cho, S. H. (2005). Hydrophilized poly(lactide-co-glycolide) nanoparticles with core/shell structure for protein delivery. *Science and Technology of Advanced Materials* 6(5): 468-474.

Zambaux, M., Zambaux, X. F., Gref, R., Maincent, P., Dellacherie, E., Alonso, M., Labrude, P. & Vigneron, C. (1998). Influence of experimental parameters on the characteristics of poly(lactic acid) nanoparticles prepared by a double emulsion method. *Journal of Controlled Release* 50: 31-40.

Zhang, H. Z., Gao, F. P., Liu, L. R., Li, M. M., Zhou, Z. M., Yang, X. D. & Zhang, Q. Q. (2009). Pullulan acetate nanoparticles prepared by solvent diffusion method for epirubicin chemotherapy. *Colloids and Surfaces B-Biointerfaces* 71(1): 19-26.

Zhang, J., Chen, X. G., Li, Y. Y. & Liu, C. S. (2007). Self-assembled nanoparticles based on hydrophobically modified chitosan as carriers for doxorubicin. *Nanomed-Nanotechnol.* 3: 258-265.

Zhang, L. M., Lu, H. W., Liu, J. Y. & Chen, R. F. (2008). Synthesis of an amphiphilic polysaccharide derivative and its micellization for drug release. *Journal of Bioactive and Compatible Polymers* 23(2): 154-170.

Zhang, L. M., Lu, H. W., Wang, C. & Chen, R. F. (2011). Preparation and properties of new micellar drug carriers based on hydrophobically modified amylopectin. *Carbohydrate Polymers* 83(4): 1499-1506.

Zhang, M., Bhattarai, N. & Matsen, F. A. (2005). PEG-grafted chitosan as an injectable thermoreversible hydrogel. *Macromolecular Bioscience* 5(2): 107-111.

Zhou, J. P., Song, Y. B., Zhang, L. Z., Gan, W. P. & Zhang, L. N. (2011). Self-assembled micelles based on hydrophobically modified quaternized cellulose for drug delivery. *Colloids and Surfaces B-Biointerfaces* 83(2): 313-320.

Zhu, A. P., Chen, T., Yuan, L. H., Wu, H. & Lu, P. (2006). Synthesis and characterization of N-succinyl-chitosan and its self-assembly of nanospheres. *Carbohydrate Polymers* 66(2): 274-279.

8

Polysaccharide-Based Nanoparticles for Controlled Release Formulations

A. Martínez[1], A. Fernández[2], E. Pérez,
M. Benito[3], J.M. Teijón[2] and M.D. Blanco[2]
*[1]Departamento de Farmacología, Facultad de Farmacia,
Universidad Complutense de Madrid,
[2]Departamento de Bioquímica y Biología Molecular, Facultad de Medicina,
Universidad Complutense de Madrid,
[3]Centro Universitario San Rafael-Nebrija. Ciencias de la Salud, Madrid
Spain*

1. Introduction

Nanoscience is the science of the phenomena peculiar to matter on the scale from 1 to several hundred nanometers (10^{-9} m). Some unique features of matter emerge when features are on the nanoscale, and the appreciation of these new properties opens new opportunities. Ignored in the past decades due to the lack of technology, these new emerging opportunities offered by nanoscience have been one of the most important areas of researching from the middle of the twentieth century to nowadays (Tibbals, 2010).

New opportunities have been realized in a wide variety of areas of technology, ranging from intelligent nanoscale materials, faster electronics or nanomotors, to medicine and biology, where first nanotechnology applications have demonstrated an enormous potential.

While medical nanotechnology was improving a wide range of medical resources and practice, the concept of nanomedicine was taking shape. Nanomedicine has recently been referred by the National Institutes of Health as the applications of nanotechnology for treatment, diagnosis, monitoring, and control of biological systems (Moghimi et al., 2005). Although this term has been defined in the literature in many ways, nanomedicine means essentially applying nanotechnology to medicine.

In contrast with other therapies, nanomedicine attempts to use sophisticated approaches to either kill specific cells or repair them one cell at a time. This approach also offers new possibilities towards the development of personalized medicine (Gurwitz & Livshits, 2006). Because nanomedicine inherits its focus on certain diseases which are currently being investigated, its primary aims have been towards non-infectious diseases, especially cancer, and on degenerative diseases in order to characterize them in the increasingly sedentary and aging populations of the wealthiest countries that lead in medical research (Tibbals, 2010).

One of the most important and hopeful tools employed in nanomedicine for medical applications are nanoparticles. Nanoparticles are solid, colloidal particles consisting of macromolecular substances that vary in size from 10 nm to 1000 nm. However, particles >200 nm are not heavily pursued and nanomedicine often refers to devices <200 nm (i.e.,width of microcapillaries). Depending on the method of preparation nanoparticles, nanospheres, or nanocapsules can be constructed to possess different properties and release characteristics for the best delivery or encapsulation of the therapeutic agent (Barratt, 2000).

One advantage of nanovectors—nanoparticles is their ability to overcome various biological barriers and to localize into the target tissue. The nanovectors currently used and investigated can be classified into three main groups or "generations" (Sakamoto et al., 2007). The first generation comprises a passive delivery system that localizes into the target site. In case of a tumour, the system reaches the tumour through the fenestrations in the adjacent neovasculature, and is normally decorated by a "stealth" layer in order to avoid their uptake by phagocytic blood cells, thus substantially prolonging their circulation time (Romberg et al., 2008). The unique mechanism of driving systems to the tumour site is the size of particles, not specific recognition of the tumour or neovascular targets. As a case in point, particles based on albumin-paclitaxel have been recently aproved by FDA for their use in metastatic breast cancer (Kratz, 2008). The second generation of nanosystems includes additional functionalities that allow for molecular recognition of the target tissue or for active or triggered release of the payload at the disease site. These include ligands, aptamers and small peptides that bind to specific target-cell surface markers or surface markers expressed in the disease microenvironment (Kang et al., 2008). Responsive systems, such as pH-sensitive polymers, are included in this cathegory. Although the representatives of the second generation have not yet been approved by the FDA, there are numerous ongoing clinical trials involving targeted nanovectors, particularly in cancer applications. Finally, the third generation nanovectors are focused to successfully overcome the natural barriers that the vector needs to bypass to efficiently deliver the drug to the target site. This goal will only be reached by a "multistage" approach, and such a system has been recently reported (Tasciotti et al., 2008).

Polymeric nanoparticles made from natural and synthetic polymers have received the majority of attention due to their stability and ease of surface modification. Polymeric materials used for preparing nanoparticles for drug delivery must be biocompatible at least and biodegradable best. Among natural polymers, proteins or polysaccharides tend to be internalized and degraded rapidly, thus enabling a moderate intracellular release of the drug or gene (Sinha & Trehan, 2003). Polysaccharides have been especially used in the preparation of drug delivery systems.

Polysaccharides are the polymers of monosaccharides. In nature, polysaccharides have various resources from algal origin (e.g.alginate), plant origin (e.g. pectin, guar gum), microbial origin (e.g. dextran, xanthan gum), and animal origin (chitosan, chondroitin) (Sinha & Kumria, 2001). They offer a wide diversity in structure and properties due to their wide range of molecular weight and chemical composition.

Due to the presence of various reactive groups in their structure, polysaccharides can be easily modified chemically and biochemically. Moreover, the presence of hydrophilic groups in their structure, such as hydroxyl, carboxyl and amino groups, enhance bioadhesion with biological tissues, like epithelia and mucous membranes, forming non-covalent bonds, which is an useful strategy to improve bioavailability of drugs included in drug delivery systems (Lee et al., 2000).

One of the main advantages of polysaccharides as natural biomaterials is their availability in natural resources and low cost in their processing, which make them very accessible materials to be used as drug carriers. Furthermore, polysaccharides are highly stable, safe, non-toxic, hydrophilic and biodegradable (Liu et al., 2008).Thus, they have a large variety of composition and properties that cannot be easily mimicked in a chemical laboratory, and the ease of their production makes numerous polysaccharides cheaper than synthetic polymers (Coviello et al., 2007). Therefore, polysaccharides have a promising future as biomaterials.

In recent years, a large number of studies have been conducted on polysaccharides and their derivatives for their potential application as nanoparticle drug delivery systems. The number of polysaccharides that have been investigated for the preparation of nanoparticles suitable as delivery systems is extremely large. As a result, attention has been focused on the latest studies and exploitations related to such systems, including some of the most used polysaccharides, a brief description of their structural features and some of the techniques carried out to prepare polysaccharide-based nanoparticles.

2. Structural features and characteristics of polysaccharides

As the number of polysaccharides used in the preparation of drug delivery systems is very large, some of the most commonly used polymers have been collected, describing their chemical structure, chemical features, and highlighting their applications in different fields, especially, in the preparation of drug delivery systems.

2.1 Alginate

Alginate is a well known polysaccharide obtained from natural sources, such as its extraction from cell walls and intercellular spaces of marine brown algae, and its production by bacteria. It can be characterized as an anionic copolymer whose chemical structure is based on a backbone of (1-4) linked β- D-mannuronic acid (M units) and α-L-guluronic acid (G units) (Fig. 1) of widely varying composition and sequence depending on the source of the alginate, resulting in an irregular blockwise pattern of GG, MG and MM blocks. Alginate has a variable molecular weight, depending on the enzymatic control during its production and the degree of depolymerization caused by its extraction. Typically, commercial alginates have an average molecular weight of approximately 200,000 Da, but alginates with values as high as 400,000-500,000 Da are also available (Rehm, 2009).

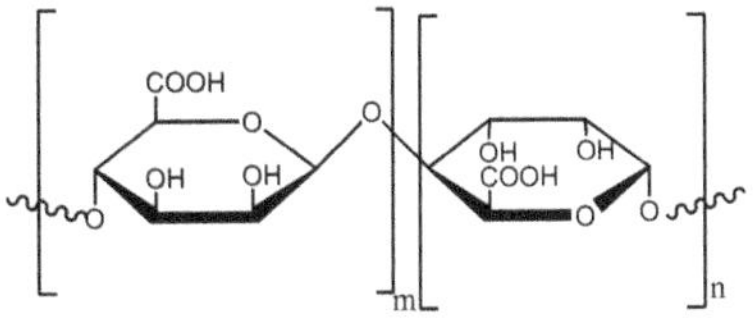

Fig. 1. Chemical structure of alginate

The physico-chemical properties of alginate have been found to be highly affected by the M/G ratio as well as by the structure of the alternating zones, which can be controlled by enzymatic pathways (Coviello et al., 2007). The alginate composition influence on the flexibility of the polysaccharide chain was first reported by Smidsrod (1973), who described that the extension of the alginate chain was dependent on its composition, with the intrinsic flexibility of the blocks decreasing in the order MG>MM>GG. M block segments show linear and flexible conformation because of the β (1→4) linkages. Besides, the guluronic acid gives rise to α (1→4) linkages, which serves to introduce a steric hindrance around the carboxylic groups, and provide folded and rigid structural conformations that are responsible for a pronounced stiffness of the molecular chains (Yang et al., 2011).

Alginate is a biopolymer and a polyelectrolyte considered to be biocompatible, non-inmunogenic, non-toxic and biodegradable, and the composition of the polymer has reported to affect to its applications. Alginate with high content of guluronic acid block can produce, in the form of calcium salts, cross-links stabilizing the structure of the polymer in a rigid gel form. This properly enables alginate solutions to be processed into the form of films, beads and sponges (Sujata, 2002). However, high mannuronic acid alginate capsules are interesting for cell transplantation and for biohybrid organs, because of their less viscosity. In the case of cellular response, some research groups found immunostimulatory activity caused by those alginates with high mannuronic acid content, and immunosuppressive activity caused by alginates with high guluronic acid content. It was concluded that mannuronic acid oligomers would provoke cytokine release by macrophages by a receptor-mediated mechanism, whereas guluronic oligomers should inhibit this reaction (Orive et al., 2002).

Compositional modifications of natural alginates can be obtained by several mannuronan C-5 epimerases produced by alginate-producing bacteria, such as *A. vinelandii*. Recently, the combination of different epimerases has been used as a fundamental tool in order to create specific engineered alginates with any desired block length and composition (Rehm, 2009). Moreover, alginate has a large number of free hydroxyl and carboxyl groups distributed along the backbone, which are highly reactive and turn it into an ideal candidate for being appropriately modified by chemical functionalization. Thus, properties such as solubility, hydrophobicity and physicochemical and biological characteristics may be modified, having proved alginate derivatives to have a lot of potential applications. These chemical modifications of alginate have been achieved using techniques such as oxidation, sulfation, esterification, amidation, or grafting methods (Yang et al., 2011).

Due to its abundance, low price and non-toxicity, alginate has been extensively used in different industries. For instance, it has been used as food additive and thickener in salad dressings and ice-creams in the alimentary industry (Nair & Laurencin, 2007). Moreover, the biocompatibility behavior and the high functionality make alginate a favorable biopolymer material for its use in biomedical applications, such as scaffolds in tissue engineering (Barbosa et al., 2005), immobilization of cells (Lan & Starly, 2011), and controlled drug release devices (Pandey & Ahmad, 2011).

In case of its applications in nanomedicine, alginate has also been extensively investigated as a drug delivery device in which the rate of drug release can be modified by varying the drug polymer interaction, as well as by chemical immobilization of the drug in the polymer backbone using the reactive carboxylate groups (Nair & Laurencin, 2007). Apart from its easy functionalization due to its reactive structure, there are many advantages and favorable properties of alginate for its use in drug delivery. It is a natural polymer compatible with a wide variety of substances, which does not need multiple and complex drug-encapsulation process. Moreover, it is mucoadhesive and biodegradable and, consequently, it can be used in the preparation of controlled drug-delivery systems achieving an enhanced drug bioavailability (Pandey & Ahmad, 2011).

Therefore, the biocompatibility, availability and versatility of this polysaccharide make it an important and hopeful tool in the field of nanomedicine, especially in the preparation of nanoparticulate drug delivery systems.

2.2 Chitosan

Chitosan is a linear polysaccharide composed by units of glucosamine and N -acetyl-glucosamine linked by (1 → 4) β-glycosidic bonds (Fig. 2). It is a hydrophilic biopolymer obtained industrially by hydrolysing the aminoacetyl groups of chitin — which is the main component of the shells of crab, shrimp and krill — by an alkaline deacetylation treatment (Muzzarelli & Muzzarelli, 2005).

The degree of deacetylation (%DD) can be determined by NMR spectroscopy, and generally the %DD in commercial chitosan is in the range 60–100%. On average, the molecular weight of commercially produced chitosan is between 3,800 to 20,000 Da. A commonly used method for the synthesis of chitosan is the deacetylation of chitin, using sodium hydroxide in excess as a reagent and water as a solvent. This reaction pathway, when allowed to go to completion (complete deacetylation), yields up to 98% product (Yuan, 2007). So, once deacetylation happens, chitosan is consisting primarily of repeating units of β-(1,4)-2-amino-deoxy-D-glucose (D-glucosamine).

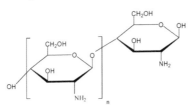

Fig. 2. Chemical structure of chitosan.

This biopolymer is accepted as a biodegradable and non toxic polymer. Despite its biocompatibility, the applications of chitosan are limited due to its insolubility above pH 6. Chitosan is a weak base and it is insoluble in water and organic solvents. However, it is soluble in diluted aqueous acidic solution (pH <6.5), which can convert the glucosamine units into a soluble form with protonated amine groups (Sinha et al., 2004). It is possible to increase the solubility of chitosan in water removing one or two hydrogen atoms from the amino groups of chitosan, and introducing some hydrophilic segments (Srinophakun & Boonmee, 2011).

The non-toxic, biodegradable and biocompatible properties of chitosan provide potential for many applications (Guerrero et al., 2010). Due to its polyelectrolyte nature, chitosan can be used as absorbent for treatment of textile industry effluents as well as for heavy metal ions uptaking from wastewater. It has been also used as template for the preparation of mesoporous metal oxides spheres (Braga et al., 2009). However, it has been more frequently proposed for applications in pharmaceutical and biomedical fields due to its biocompatibility and biodegradability. It has been assayed as biomaterial for wound healing and prosthetic material, since it can be biodegraded by enzyme action (Bernardo et al., 2003). Also it is reported to find applications as an antimicrobial compound, as a drug in the treatment of hyperbilirubinaemia and hypercholesterolaemia and, also, it has been prepared and evaluated for its antitumour activity carrying several antineoplastic agents (Blanco et al., 2000).

In the field of nanomedicine, chitosan has attracted attention as a matrix for controlled release due to its reactive functionalities, polycationic character, easily degradation by enzymes and non-toxic degradation products. Over the years, a variety of natural and synthetic polymers have been explored for the preparation of drug-loaded microparticles and chitosan has been extensively investigated (Davidenko et al., 2009; Muzzarelli & Muzzarelli, 2005). Because of its bioadhesive properties, chitosan has received substantial attention as carrier in novel bioadhesive drug delivery systems which prolong the residence time of the drugs at the site of absorption and increase the drug bioavailability (Varum et al., 2008). Thus, some drugs administered via nasal (Learoyd et al., 2008) or gastrointestinal routes have improved their treatment efficacy when they are included into chitosan-based systems (Guerrero et al., 2010).

Taking all into account, chitosan appears to be a promising matrix for the controlled release of pharmaceutical agents. Experimental *in vitro* and *in vivo* results show chitosan as an ideal carrier for a wide variety of drugs whose efficacy is increased when they are included into these systems.

2.3 Hyaluronic acid

Hyaluronic acid (HA) (also called sodium hyaluronic or hyaluronan) is a polysaccharide with a structure composed of repeating disaccharide units of D-glucuronic acid and N-acetyl D-glucosamine linked by β (1-3) and β (1-4) glycosidic bonds (Fig. 3) (Cafaggi et al., 2011). HA can be modified in many ways to alter the properties of the resulting materials, including modifications leading to hydrophobicity and biological activity. There are three functional groups that can be chemically modified: the glucuronic acid carboxylic acid, the primary and secondary hydroxyl groups, and the N-acetyl group (Burdick & Prestwich, 2011). HA has a molecular weight that can reach as high as 10^7 Da.

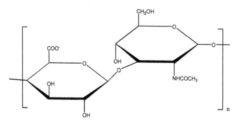

Fig. 3. Chemical structure of disaccharide repeating unit of hyaluronic acid.

It belongs to a group of substances known as glycosaminoglycans. It is the most simple among them, the only one not covalently associated with a core protein, and the only one which is non-sulfated (Kogan et al., 2007). Since hyaluronan is a physiological substance, it is widely distributed in the extracellular matrix of vertebrate tissues. It is mainly synthesized in vertebrate organisms as an essential functional component due to its viscoelastic and rheological properties. It is a major and important component of cartilage, skin and synovial fluid.

HA is usually linked to other biopolymers in the organism, and several separation procedures have to be applied in order to obtain the pure compound, such as protease digestion, HA ion-pair precipitation, membrane ultrafiltration, HA non-solvent precipitation and/or lyophilisation (Mendichi & Soltes, 2002). With these methods HA from several hundred thousand Da up to 2.5 MDa can be obtained. However some microorganisms secreted HA with a molar mass in the range of several MDa, such as attenuated strains of *Streptococcus zooepidemicus* and *S. Equi. Bacillus subtilis* has been recently genetically modified to culture a proprietary formula to yield hyaluronans (Mendichi & Soltes, 2002).

It is a biodegradable, bioactive, non inmunogenic, non cytotoxic and negatively charged polysaccharide (Oh et al., 2010) that has been associated with several cellular processes, including angiogenesis and the regulation of inflammation (Leach & Schmidt, 2005).

Among its applications, it is widely used as a coating for the surface modification of various biomaterials used for prosthetic cartilage, vascular graft, guided nerve regeneration and drug delivery (Li et al., 2006).

Like other glycosaminoglycans, hyaluronan can serve as a targeting vehicle for the delivery of chemotherapeutic agents to cancerous tissues, as many tumours over express the hyaluronan CD44 and RHAMM receptors (Yip et al., 2006). As a drug delivery carrier, HA has several advantages including the negligible non-especific interaction with serum components due to its polyanionic characteristics (Ito et al., 2006) and the highly efficient targeted specific delivery to the liver tissues with HA receptors (Zhou et al., 2003).

More recently, HA has become recognized as an important building block for the creation of new biomaterials with utility in tissue engineering and regenerative medicine (Allison & Grande-Allen, 2006; Prestwich, 2008). Moreover, it has been shown that HA binds to cells and effectively promotes new bone formation. Balazs classified the biomedical applications of the HA and its derivatives in areas as vicosurgery, viscoaugmentation, viscoseparation, viscosupplementation, viscoprotection (Balazs, 2004).

So, in this way, there is a wide number of usages of HA in medicine and cosmetics, such as ophthalmology, orthopaedic surgery and rheumatology, otolaryngology, wound healing, pharmacology and drug delivery (Kogan et al., 2007), which shows HA as a successful biomaterial used in different fields of biomedicine.

2.4 Dextran

Dextran is a polysaccharide made of many glucose molecules composed of chains of varying lengths. It has a substantial number of α (1→6) glucosidic linkages in its main chain (Fig. 4), and a variable amount of α (1→2), α (1→3) and α (1→4) branched linkages (Misaki et al., 1980). The degree and type of branching will be determined by the bacterial strain that synthesizes it. Its average molecular weight is as high as 10^7 - 10^8 Da (Heinze et al., 2006) but can be reduce by acidic hydrolysis obtaining molecular weight fractions that also can interest.

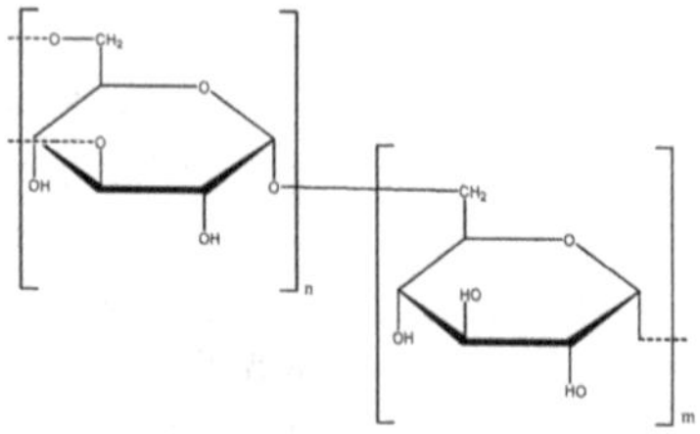

Fig. 4. Chemical structure of dextran with α (1→3) branched linkage.

The natural structure of dextran can be modified by reacting different molecules (such hydrophobic molecules) with its different hydroxyl groups (Lemarchand et al., 2003b). Many amphiphilic dextran derivates have been obtained by varying the nature of the reacting molecules (aromatic rings, aliphatic or cyclic hydrocarbons) and the number of grafted, that is the number of hydrophobic groups per 100 glucopyranose units or the degree of substitution (Rotureau et al., 2004).

Dextran is neutral, water soluble, biocompatible and biodegradable. Its features may vary depending on the molecular mass as well as the distribution, type of branches and the degree of branching, which depend on the bacterial synthesis or post-synthesis reactions to form derivatives.

Dextran is synthetized by a wide variety of bacterial strains. *Leuconostoc mesenteroides* produces dextran from sucrose and *Gluconobacter oxydans* produces dextran from maltodextrin. *Streptococcus mutans* also produces dextran from sucrose (Heinze et al., 2006). It can be also obtained enzimatically using cell-free culture supernatant (Wang et al., 2011). Apart from these methods, dextran can be also produced by chemical synthesis, developing a cationic ring opening polymerisation of levoglucosan (Heinze et al., 2006).

It has wide applications in different areas such as pharmaceutical, chemical, clinical, and food industry. Dextran is used as a drug (as blood plasma volume expander), adjuvant, emulsifier, carrier, stabilizer and thickener of jam and ice cream. Also it is widely used for the separation and purification of proteins (Naessens et al., 2005) based on size exclusion chromatography with a matrix of cross-linked dextran gel layer. Its derivatives also have multiple applications depending on the characteristics that structural modifications give them.

Both dextran and its derivatives have potential application for the preparation of modified drug delivery (Aumelas et al., 2007; Coviello et al., 2007; Chen et al., 2003). Not only has this polysaccharide been used to prepare nanoparticulate systems as a carrier, but also it has been employed to cover these systems (Gavory et al., 2011).

It seems that dextran is a very useful tool in the field of nanomedicine, showing also good availavility, biocompatibility and biodegradability, being selected by a lot of researchers as biomaterial in the preparation of nanosystems.

2.5 Other polysaccharides

2.5.1 Pullulan

Pullulan is a linear bacterial homopolysaccharide produced from starch by the fungus *Aureobasidium pullulans*. The backbone is formed by glycosidic linkages between α-(1→6) D-glucopyranose and α-(1→4) D-glucopyranose units in a 1:2 ratio (Fig. 5). The molecular weight of pullulan range from thousands to 2,000,000 Da depending on the growth conditions (Rekha & Chandra, 2007).

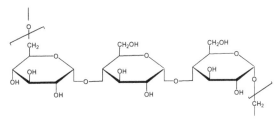

Fig. 5. Chemical structure of pullulan.

The backbone structure of pullulan tends to behave as a random expanded flexible coil in aqueous solution with modelling studies suggesting that this flexibility is imparted by the α-(1→6) linkage. This could be the reason why pullulan is biodegradable and has high adhesion, structural flexibility and solubility (Leathers, 2003). Pullulan can also be easily derivatized in order to impart new physico-chemical properties, e.g. to increase the solubility in organic solvents or to introduce reactive groups.

This polysaccharide has numerous uses: in foods and beverages as a filler; as an edible, mostly tasteless polymer, the chief commercial use of pullulan is in the manufacture of edible films that are used in various breath freshener or oral hygiene products; in pharmaceuticals as a coating agent; in manufacturing and electronics it is used because of its film- and fiber-forming properties. It is worth noting that pullulan films, formed by drying pullulan solutions, are clear and highly oxygen-impermeable and have excellent mechanical properties.

Due to it is hemocompatible, non-immunogenic, non-carcinogenic, FDA approved it for a variety of applications (Coviello et al., 2007). Recently, pullulan has been investigated for being used in various biomedical applications such as drug and gene delivery (Rekha & Chandra, 2007), tissue engineering (Thebaud et al., 2007), and wound healing (Bae et al., 2011).

Numerous papers deal with pullulan hydrogels as drug delivery systems, particularly in the form of micro and nanogels. Despite pullulan is not a natural gelling polysaccharide, an appropriate chemical derivatization of its backbone can actually lead to a polymeric system capable of forming hydrogels. The study of nanogels has been intensified over the last decade due to related potential applications in the development and implementation of new environmentally responsive or smart materials, biomimetics, biosensors, artificial muscles, drug delivery systems and chemical separations (Coviello et al., 2007).

In order to obtain nanostructures that may act as carriers of different drugs, the backbone structure of pullulan is modified with hydrophobic molecules, resulting in a molecule of hydrophobized pullulan that self-assembles in water solutions. Cholesterol, hexadecanol or vitamin H are some molecules that are attached to the structure of pullulan in order to obtain micelles in water solution (Liu et al., 2008).

2.5.2 Guar gum

Guar gum is a water soluble polysaccharide extracted from the seeds of *Cyamopsis tetragonoloba*, which belongs to Leguminosae family. Also called guaran, it is a non-ionic natural polysaccharide derived from the ground endosperm of guar beans. Its backbone consists of linear chains of $(1 \rightarrow 4)$-β-D-mannopyranosyl units with α-D-galactopyranosyl units attached by $(1 \rightarrow 6)$ linkages (Fig. 6), forming short side-branches (Sarmah et al., 2011).

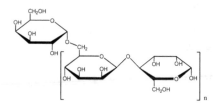

Fig. 6. Chemical structure of guar gum.

Guar gum hydrates in cold water to form a highly viscous solution in which the single polysaccharide chains interact with each other in a complex way (Barbucci et al., 2008). Its nine hydroxyl groups are available for the formation of hydrogen bonds with other molecules, but it remains neutrally charged due to the absence of dissociable functional groups. Extreme pH and high temperature conditions (e.g. pH 3 at 50°C) degrade its structure (Tiraferri et al., 2008). It remains stable in solution over pH range 5-7. Strong acids cause hydrolysis and loss of viscosity, and alkalis in strong concentration also tend to reduce viscosity. It is insoluble in most hydrocarbon solvents.

As the guar gum polymer is a low-cost, easily available and non-toxic polysaccharide, it is widely applied in many industrial fields. Thanks to its property of producing highly viscous aqueous solutions, it is commonly used as a thickening agent in cosmetics and in sauces, salad dressings and ice creams in the food industry (Barbucci et al., 2008). In pharmaceuticals, guar gum is used in solid dosage forms as a binder and disintegrant, and it has also been used as hydrophilic matrix, for designing oral controlled release dosage forms (Sarmah et al., 2011). Guar gum has been extensively used for colon delivery due to its drug release retarding property and susceptibility to microbial degradation in the large intestine (Soumya et al., 2010).

Not only the native guar-gum is used, but also chemically modified products can be used with the objective of changing its intrinsic characteristics of solubility, viscosity and rheological behaviour. For instance, hydrossilalchyl derivatives, which are often used for the formulation of cements and plasters, or carboxymethyl derivatives, which are employed as thickening agents.

In case of biomedical fields or pharmaceutical fields, such as 3D scaffolds for cell culture, fillers for tissue engineering and carriers for drugs, the physically cross-linked product is obtained through a spacer arm between the polymer chains and allows the obtainment of an insoluble compound in a wide range of pH with a good mechanical stability (Barbucci et al., 2008).

Little information is available in the literature for the possibility of using guar gum based nanosized materials as drug carriers due to its solubility in water, what makes difficult to use it as adsorbent in aqueous conditions. Some researchers have incorporated to its structure some compounds like silica, in order to obtain insoluble compounds which could act as adsorbents in aqueous media (Singh et al., 2009). Moreover, guar gum-based nanosystems have been prepared by nanoprecipitation and cross-linking methods (Soumya et al., 2010). A different application of this polysaccharide has been found as stabilizer of nanosuspensions, where the presence of guar gum during the synthesis process allows the achievement of a better stability of the nanoparticles (Tiraferri et al., 2008).

2.5.3 Pectin

Pectin is a structural polysaccharide obtained from the cell wall of all plants, where is implicated in cell adhesion. This natural polymer has a heterogeneous chemical structure based on large amounts of poly (D-galacturonic acid) bonded via α (1 → 4) glycosidic linkage (Fig. 7). Pectin has a few hundred to about one thousand building blocks per molecule, corresponding to an average molecular weight of about 50,000 to about 180,000 Da (Sinha & Kumria, 2001). The carboxyl groups are partially in the methyl ester form with different degree of esterification (DE) and amidation (DA), which determine the content of carboxylic acid in pectin chains.

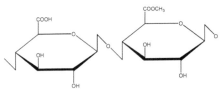

Fig. 7. Chemical structure of pectin

In the beginning, applications of pectin concentrated in food industry, as gelling or thickening agent, but lately it started being also used as an excipient for pharmaceutical purposes (Liu et al., 2003). Nowadays, some of the uses of pectin in biomedical applications include the facilitation of the delivery of specific sequences of amino acids, anti-inflammatory agents, anti-coagulants, and wound healing substances to tissue sites. Also, pectin remains intact in the physiological environment of the stomach and the small intestine, but is degraded by pectinases, which are secreted by the bacteria inhabitants of the human colon. Due to these properties it is highly possible that pectin could function as a delivery vehicle to escort protein and polypeptide drugs from the mouth to the colon (Sinha & Kumria, 2001). To be used as such, pectin based composites can be formed into membranes, microspheres, scaffolds, or injectable gels (Liu et al., 2004).

The most attractive property of pectin for industrial applications is its gelling activity. Parameters such as type and concentration of pectin (DE, DA), modification of hydroxyl groups, pH, temperature and the presence of cations, determine the gel process. For example, a high DE of pectin provides the gel formation, increasing the amounts of hydrophobic areas and reducing the solubility of pectin. In contrast, when the DE is less than 50%, pectin is highly water soluble and gel formation is only at extremely low pH solution or in the presence of divalent cations, which cross-link the galacturonic acids of the main polymer chains (Liu et al., 2003). Also, it is possible to reduce the hydrophilic property with an increasing tendency to form gels by the introduction of amide groups in low DE pectin.

With regard to its use in the preparation of drug delivery systems, pectin is not able to shield its drug load effectively during its passage through the stomach and small intestine due to its high water-solubility (Sinha & Kumria, 2001). Consequently, most of the researching groups focused on looking for water resistant pectin derivatives, which were also enzymatically degradable. For this purpose, calcium salts binding by non-covalent associations with the carbohydrate chains of pectin were investigated, which can reduce the solubility and are stable in low pH solution while resisting extensive hydration *in vivo* in the gastrointestinal tract. Thus, calcium pectinate is a potential candidate as a drug carrier for colon-specific delivery in different formulations such as microspheres, films, gels or droplets (Liu et al., 2003). Another derivative of pectin, amidated pectin cross-linked with calcium, was considered for colonic delivery, with retarding drug release and because of its biodegradability, higher tolerance to pH variations and fluctuations in calcium levels (Sinha & Kumria, 2001).

In addition, combinations of pectin with other polymers, either naturally occurring or synthetic, have been developed in order to obtain useful novel formulations. The combination of pectin and a second polymer into a composite may alter degree of swelling and change mechanical properties (Liu et al., 2003), improving in the most cases the stability of the drug and controlling the drug release. As a case in point, pectin has been combined with 4-aminothiophenol (Perera et al., 2010), chitosan (Fernandez-Hervas & Fell, 1998), hyaluronic acid (Pliszczak et al., 2011) or poly (lactide-co-glycolide) (Liu et al., 2004), showing good results as controlled drug release devices.

3. Preparation methods of polysaccharide-based nanoparticles

As for polysaccharide-based nanoparticles, it can be seen in the literature wide research carried out focusing on the preparation and application of these systems, which enhances their importance and versatility in terms of category and function.

According to the literature and the structural features of the employed polysaccharides, five mechanisms can be mainly applied in order to obtain nanoparticles, namely gelation of emulsion droplets, covalent cross-linking, ionic cross-linking, self-assembling and nanoprecipitation.

3.1 Formation of nanoparticles from an emulsion: nanoparticles obtained by gelation of emulsion droplets

Different methods to prepare emulsified systems have been significantly developed. All of them require two immiscible phases and the presence of a surface active agent, whose nature has been already evolved, replacing the commonly used pluronic or span by new amphiphilic copolymers (Qiu & Bae, 2006). These methods are two-step processes, where the first step consists of the preparation of an emulsified system while nanoparticles are formed during the second step. Generally, the principle of the second step gives its name to the method (Vauthier & Bouchemal, 2009).

Nanoparticles can be obtained from an emulsion method by gelation of the emulsion droplets where the polymer is dissolved, and that have been formed in the first step of the emulsion procedure. Polysaccharides show good gelling properties as well as good solubility in water, which make them ideal candidates to be used for preparing nanoparticles by this method (Vauthier & Couvreur, 2000). Different mechanisms of gelation can be applied depending on the gelling properties of the polymer. Changes in temperature of the emulsion system or gelation induced by covalent or ionic cross-linking are some of these mechanisms which induce the gelation of the pre-formed droplets and allow nanoparticles to be obtained.

Alginate and pectin particles have been obtained by using a modified emulsification/internal gelation method (Opanasopit et al., 2008). The preparation of two different emulsions is required: one containing the gelling polymer in the dispersed phase and the other containing the gelling agent (usually counter-ions) or the pH controlling agent in the dispersed phase. Both emulsions are mixed together under strong stirring conditions in order to achieve collisions between droplets, which are necessary to promote the gelation of the polymer and, consequently, the formation of nanoparticles (Fig. 8). In case of alginate, the size range of particles is greatly dependent on the order of addition of counter-ion to the alginate solution. Some studies show that the addition of a polyelectrolyte complexation step in this procedure shows some benefits in order to obtain a better control of size distribution. Dextran or chitosan can be used as complexing agents in the *in situ* gelation of the droplets obtained in the previous nanoemulsion (Reis et al., 2007). The resulting nanoparticles range in size from 267 nm to 2.76 μm.

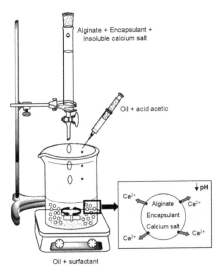

Fig. 8. Schematic representation of the emulsification-internal gelation technique using alginate.

Chitosan particles have been formed from an emulsified system by emulsion cross-linking method or by emulsion-droplet coalescence method. In the first method, the reactive functional amine group of chitosan reacts with aldehyde groups of the cross-linking agent, which usually is glutaraldehyde. A water-in-oil (w/o) emulsion is prepared by emulsifying the chitosan aqueous solution in the oil phase. Aqueous droplets are stabilized using a suitable surfactant. The stable emulsion is cross-linked by glutaraldehyde to harden the droplets. Microspheres are filtered and washed repeatedly with n-hexane, followed by alcohol and then dried (Akbuga & Durmaz, 1994). Particle size can be determined by controlling the size of aqueous droplets, but it is usually ranged in a micrometric scale (Kumbar & Aminabhavi, 2003).

Chitosan nanoparticles with a mean size of 400 nm can be obtained by emulsion-droplet coalescence method. This method, introduced by Tokumitsu and coworkers (Tokumitsu et al., 1999), utilizes the principles of both emulsion cross-linking and precipitation. A stable emulsion containing aqueous solution of chitosan and the encapsulant drug is produced in liquid paraffin. Another emulsion containing NaOH aqueous solution is produced in the same manner and is finally mixed with the other under high speed stirring. Droplets of each emulsion would collide at random and coalesce, precipitating chitosan droplets to give small solid particles.

With a polymer like agarose, gel beads can be formed by cooling down the temperature of the solution which is prepared at high temperature. Thermal gelation results from the formation of helicoidal structures responsible for a three-dimensional network in which large amounts of water can be entrapped. The hydrogel, being hydrophilic, inert, and biocompatible, forms a suitable matrix for macromolecules that can be entrapped in the gel during formation (Vauthier & Couvreur, 2000). Agarose nanoparticles are produced using an emulsion-based technology which requires the preparation of an agarose solution in corn oil emulsion at 408°C. Macromolecules to be encapsulated are initially added to the agarose solution. The small size of the dispersed aqueous nanodroplets is achieved by homogenization. Gelation of agarose is then induced by diluting the emulsion with cold corn oil under agitation at 58°C. The liquid nanodroplets then gel to macromolecule-containing agarose hydrogel nanoparticles (Wang & Wu, 1997). The mean average size of the obtained nanoparticles is 504 nm.

3.2 Polysaccharide-based nanoparticles with covalent cross-links

Among various polysaccharides, chitosan is the early one to be used to prepare nanoparticles based in covalent cross-links. Glutaraldehyde has been usually used as a cross-linker to obtain nanoparticles by emulsion cross-linking method (previously described), but its citotoxicity limits its utility in the field of drug delivery systems. However, some chitosan nanoparticles are still being produced using glutaraldehyde as cross-linker agent (Zhi et al., 2005).

To overcome the problems of toxicity that are presented by glutaraldehyde, some biocompatible cross-linkers, such as natural di- and tricarboxylic acids, including succinic acid, malic acid, tartaric acid and citric acid, are used for intermolecular cross-linking of chitosan nanoparticles (Bodnar et al., 2005). By this method, the pendant amino groups of chitosan react in aqueous media with carboxylic groups of natural acids which were previously activated by a water-soluble carbodiimide, obtaining polycations, polyanions, and polyampholyte nanoparticles with an average size in the range of 270–370 nm depending on the pH.

Hyaluronic acid is another polysaccharide used to prepare nanoparticles by using a carbodiimide method. The preparation of nano-sized particulate systems based on hyaluronic acid takes place by covalently cross-linking via carboxyl groups of the hyaluronic acid chain with a diamine in aqueous media at room temperature. Bodnar and coworkers have obtained spherical nanoparticles whose size varies less than 130 nm (Bodnár et al., 2009).

Recently, nanoparticles based on thiolated alginate and modified albumin have been synthesized and stabilized by the formation of disulphide bonds between both polymers (Martínez et al., 2011). In this case, the covalent interaction is established between the sulfhydryl groups of the albumin, obtained after a reduction process of the protein, and the sulfhydryl groups of the L-cysteine which has been attached to the polysaccharide structure using a carbodiimide reaction. Nanoparticles with a size range of 42-388 nm are obtained by this coacervation method based on a pH change that induces the disulphide bond formation between both structures (Fig. 9).

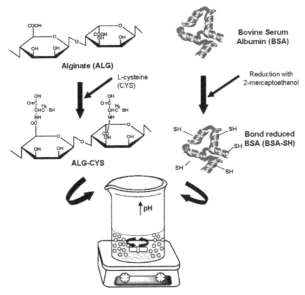

Fig. 9. Schematic representation of preparation method of nanoparticles based on thiolated alginate and modified albumin, stabilized by disulphide bond formation.

3.3 Polysaccharide-based nanoparticles with ionic cross-links

Ionic gelation procedure to obtain nanoparticles is included among the few organic solvent free methods, as nanoparticles are totally synthesized in aqueous media. Compared with covalent cross-linking, this method shows more advantages, such as simple procedures and mild preparation. Nanoparticles can be obtained from aqueous solutions of charged polysaccharides which gel in the presence of small ions of opposite charges. Thus, polyanions and polycations could act as cross-linkers with polycationic and polyanionic polysacchraides, respectively. Very dilute solutions of the polysaccharide are used to perform the gelation process, in which the chains of the polymer reacting with the gelling agent are forming small clusters. These clusters are stabilized by forming complex with opposite charged electrolytes (Vauthier & Bouchemal, 2009).

The cationic nature of chitosan when it is dissolved in an acidic aqueous solution (pH 4-6) can be exploited to form nanoparticles by adding small amounts of tri-polyphosphate (TPP) included in an alkaline phase (pH 7-9), upon mixing of the two phases through inter and intra molecular linkages are created between TPP phosphates and chitosan amino groups (Janes et al., 2001).

Depending on the pH and the ionic strength of the dispersing medium, these nanoparticles are capable of swelling and shrinking, which is used to trigger the release of a drug encapsulated in the nanoparticles upon the action of a pH or an ion concentration variation stimulus. For instance, KCl can be added to the dispersing medium to vary the ionic strength and cause the nanoparticle swelling; or glucosamin groups of chitosan can be deprotonated by raising the pH from acid to basic values causing a shrinking on the gel because the intramolecular electric repulsions inside the particle mesh are reduced. The average size of the obtained chitosan nanoparticles ranges between 20 and 400 nm (Pan et al., 2002).

Some water-soluble chitosan derivatives, like N-(2-hydroxyl) propyl-3-trimethyl ammonium chitosan chloride or N-trimethyl chitosan, have been also ionically cross-linked to prepare nanoparticles. The average size of the obtained systems is between 110 and 350 nm (Amidi et al., 2006).

A slightly modified ionotropic gelation technique was used by de la Fuente and coworkers in order to obtain nanoparticles based on chitosan and hyaluronic acid using TPP as ionic cross-linker. Their results show that hyaluronic acid/chitosan nanoparticles have a small size in the range of 110–230 nm (de la Fuente et al., 2008).

Not only TPP is used as a cross-linker to obtain chitosan naoparticles by ionic gelation method. Some researchers as Kim and co-workers have obtained chitosan-based nanoparticles by this method using the encapsulated drug itself as a cross-linker, establishing electrostatic interactions between amine group of chitosan and hydroxyl group of the drug (Kim et al., 2006).

Among negatively charged polysaccharides, alginate is one of the most used to obtain nanoparticles by ionic gelation. In this case, carboxylic groups on molecular chains of alginate structure can be cross-linked by bivalent calcium ions to form nanoparticles. Then, clusters formed in the pre-gel phase can be stabilized with polications like polylysine and chitosan (De & Robinson, 2003). Polylysine can form polyelectrolyte complexes with alginate without the previous formation of the pre-gel phase with calcium, but more compact nanoparticles are obtained when this previous step is carried out. The size of nanoparticles using polylysine as stabilizer depends, not only on the concentration of alginate, but also on the molecular weight. In fact, an optimal mass balance between sodium alginate: $CaCl_2$: cationic polymer (poly-Llysine or chitosan) has been found to obtain particles with nanometric size (Vauthier & Bouchemal, 2009). Nanoparticles with an average size ranged between 194 nm and 1.6 μm can be obtained by this method (Ahmad et al., 2006; Azizi et al., 2010).

In addition, the interaction of alginate with divalent calcium ions was used to obtain nanoparticles from water-in-oil microemulsions, as it was previously described (Reis et al., 2007).

3.4 Methods based on self assembling macromolecules

3.4.1 Polysaccharide based nanoparticles by polyelectrolyte complexation (PEC)

Polyelectrolyte complexes (PECs) are formed by the interaction between oppositely charged polymers by intramolecular electrostatic interactions. PECs are very interesting materials for different applications because some of their properties, like swelling or permeability, can be easily modified by external stimuli, such as the pH of the medium.

Positively or negatively charged nanoparticles with a core/shell structure can be obtained according to the nature of the polyelectrolyte used in excess. The hydrophobic core is composed by the complexed segments whereas the excess of component not incorporated in the polyelectrolyte complex is segregated in the outer shell ensuring the colloidal stabilization of the nanoparticles against coagulation and conferring the charge of the nanoparticle surface. This charge could affect to the interaction between cells and nanoparticles. Moreover, molecular weight of the two polyelectrolytes influences the size of the nanoparticles (Vauthier & Bouchemal, 2009).

Although any polyelectrolyte could interact with polysaccharides in order to obtain PEC nanoparticles, only water-soluble and biocompatible polymers are used as polyelectrolytes with this goal. Among the existing polyanionic and polycationic polysaccharides to form PEC nanoparticles, chitosan is widely used because it satisfies the needs of safety and solubility. It can be seen in the literature that much research has been carried out on PECs with chitosan as polycation and different negative polymers, such as negative polysaccharides, poly(acrylic acid) (PAA) or nucleic acids.

There is a wide variety of negative polysaccharides that can be attached to chitosan to form PEC nanoparticles. Carboxymethyl cellulose (Cui & Mumper, 2001), dextran-sulfate (Drogoz et al., 2007), alginate (Sarmento et al., 2006) and glucomannan (Alonso-Sande et al., 2006) are just a few examples of PEC combinations that allow the preparation of nanoparticles whose size ranges between 100 and 800 nm.

The formation of the complex between chitosan and poly(acrylic acid) (PAA) has been widely studied. The influence of molecular weight of chitosan and PAA, the ratio of the initial polyelectrolyte concentrations, dropping temperature, pH of the initial solutions and the purification process on the size, stability and morphology of the nanoparticles has been studied by different authors (Chen et al., 2005; Davidenko et al., 2009).

Nucleic acids can also be combined with chitosan to obtain nanospheres. In this case, the drug being the nucleic acid is incorporated in the nanocarrier as part of its structure (de Martimprey et al., 2009). The N/P ratio, which is defined as the ratio number of amine groups of the polycation (N) divided by the number of phosphate groups of the nucleic acid (P), has to be taken into account to obtain the desired size of nanospheres.

Apart from chitosan, polyelectrolyte complexes with nanometric size can be formed using alginate, a negatively charged polysaccharide, combined with polylysine, a positively charged peptide (George & Abraham, 2006). Although both structures could interact without the previous formation of the alginate pre-gel phase with calcium, more compact nanoparticles are obtained when this previous step is carried out. Nanoparticles with a mean size of 250–850 nm are obtained using very well defined concentrations of both electrolytes.

3.4.2 Nanoparticles obtained from self-assembling of hydrophobically modified polysaccharides

Amphiphilic copolymers are synthesized when hydrophobic segments are added to chains of hydrophilic polymers. In aqueous solutions, amphiphilic molecules orientate themselves in order to achieve a state of minimum free energy and the hydrophobic blocks are removed from the aqueous environment. Consequently, polymeric micelles with core/shell structure are formed. Thanks to their hydrophobic domain, surrounded by a hydrophilic outer shell, they can serve as reservoir for various hydrophobic drugs (Letchford & Burt, 2007).

The synthesis and application of polysaccharide-based-self-aggregate nanoparticles as drug delivery systems have been recently investigated. There are various hydrophobic molecules that can be attached to polysaccharides in order to obtain these kind of systems, such as poly(ethylene glycol) derivatives, long chain fatty acids, poly(ε-caprolactone), pluronic copolymers, cholesterol and poly(isobutilcyanoacrylate) (PIBCA).

Poly(ethylene glycol) (PEG) has been often used in pharmaceutical and biomedical fields as soluble polymeric modifier in organic synthesis, and as a pharmacological polymer with high hydrophilicity, biocompatibility and biodegradability. PEG and its derivatives can be attached to the polysaccharide structure to form micelles directly in an aqueous medium by adjusting the hydrophobicity/hydrophilicity of the polysaccharide chain. Chitosan has been grafted with different molecules of poly(ethylene glycol), obtaining nanoparticles with an average size ranged between 80-260 nm (Yang et al., 2008; Yoksan et al., 2004).

Some long-chain fatty acids like hexanoic acid, linoleic acid, linolenic acid, palmitic acid or stearic acid have been used for modifying polysaccharides and obtaining polymeric micelles. Nanoparticles based on linoleic acid-chitosan have been obtained through a carbodiimide-mediated reaction, and their size ranged between 200-600 nm. (Chen et al., 2003). Hu and coworkers employed a similar methodology in order to obtain stearic acid-chitosan nanoparticles. To increase the stability of the micelle *in vivo* and controlled drug release, the shells of micelles were cross-linked by glutaraldehyde (Hu et al., 2006). Dextran has been also employed to obtain nanoparticles by coupling lipoic acid to the structure of dextran and forming nanoparticles in water, whose size varied from 145 to 221 nm (Li et al., 2009).

Poly(ε-caprolactone) (PCL) is a well known biodegradable polyester, frequently used as implantable carrier for drug delivery systems, and a promising molecule which allows the formation of nanometric micelles when it is attached with a polysaccharide structure. The combination of the hydroxyl groups of dextran with the carboxylic function present on preformed PCL monocarboxylic acid results in the formation of nanoparticles of less than 200 nm (Lemarchand et al., 2003a).

Hyaluronic acid and pluronics form the polymeric shell of nanoparticles obtained by Han and coworkers (Han et al., 2005). Pluronics self-assemble to form a spherical micellar structure above the lower critical solution temperature by hydrophobic interaction of the poly(propylene oxide) middle block in the structure. Depending on the composition and molecular weight, they show at high concentration a sol-gel transition behaviour when raising the temperature above the lower critical solution temperature, which means a swelling or de-swelling behaviour temperature-dependent (Liu et al., 2003).

Among different cyclic hydrophobic molecules, cholesterol is one of the most used to give hydrophobic character to polysaccharide structures and form self-assembly nanoparticles in aqueous solution. Cholesterol attached to pullulans of different molecular weights, or to chitosan, allow to obtain nanoparticles of very small sizes: 20-30 nm in case of cholesterol-pullulans based nanoparticles (Akiyoshi et al., 1998) and 417 nm in case of cholesterol-chitosan systems (Wang et al., 2007). Apart from cholesterol, 5β-Cholanic acid is a non-toxic bile acid present in humans that can be chosen as the hydrophobic moiety to be attached to hyaluronic acid in order to form amphiphilic conjugates. The mean diameters of the obtained nanoparticles are in the range of 237–424 nm (Choi et al., 2010).

Poly(isobutilcyanoacrylate) (PIBCA), which belongs to polyacrylate family, can form amphiphilic copolymers in combination with polysaccharide structures due to the hydrophobic nature of this molecule that shows carboxylic esters on its structure. Among the different existing combinations of PIBCA with polysaccharides, chitosan and dextran are good examples of hydrophilic molecules that have been used to obtain nanometric micelles though an emulsion polymerization of IBCA in the presence of polysaccharides. In case of chitosan, the obtained micelles show very small sizes, lower than 35 nm (Bertholon et al., 2006).

3.5 Nanoprecipitation

Nanoprecipitation is one of the most used methods to produce nanoparticles owing to the reproducibility, the simplicity and the economy of the technique. Systems of three components are needed to perform nanoprecipitation process: the polymer, the polymer solvent and the non-solvent of the polymer. To produce nanoparticles, the polymer solution is mixed with the non-solvent. Nanoparticles are instantaneously formed during the fast diffusion of the polymer solution in the non-solvent. This technique has been up to date mainly applied for poly(lactic acid), poly(lactic-co-glycolic acid) (Lassalle & Ferreira, 2007) as well as for polysaccharide derivatives (Hornig & Heinze, 2008). Chitosan and amphiphilic cycodextrins are some examples of polysaccharide derivatives employed to obtain nanoparticles by this method.

In case of chitosan, sodium sulfate is commonly used as a precipitating agent to form chitosan particles. When sodium sulfate is slowly added into a solution of chitosan and polysorbate 80 under both stirring and ultrasonication, desolvated chitosan in a particulate form is obtained. The precipitated particles are at micro/nano interface (900 ± 200 nm) (Berthold et al., 1996).

Skiba and coworkers synthesised nanoparticles based on amphiphilic cyclodextrines (Skiba et al., 1996). Nanospheres are prepared by progressive dispersion of an organic solution of modified β-cyclodextrin in an aqueous phase with or without surfactant. Various physicochemical parameters were studied such as the effect of the chain length of acyl groups and type of surfactant on the size and physicochemical properties and stability of the nanospheres. Systems with a mean diameter varying between 90 and 150 nm were obtained by this method.

4. Application as drug delivery systems

In the design of drug carriers, issues of safety, toxicity and availability have to be taken into account, and the application of polysaccharides simplifies some of these issues. Thus, polysaccharides are emerging as new promising materials to be applied in the design of drug delivery systems. Some examples of drug delivery systems that have been prepared during the last years using polysaccharides in their composition have been collected below (Table 1).

4.1 Peptides and proteins included in polysaccharide-based delivery systems

Some of the most significant advances in biotechnology in recent years are related with the discovery of some therapeutic and antigenic peptides and proteins (Vila et al., 2002). Despite these great advances, some problems like low stability, short biological half-life and the need to cross biological barriers limit the use and the *in vivo* application of the most of biologically derived drugs. The inclusion of the biologically derived drug into drug delivery systems based on polysaccharides have shown good results and some of the problems of systemic administration can be overcome.

Insulin is one of the most widely used therapeutic peptides. It is a 5.8 kDa protein used exogenously to treat insulin-dependent diabetes mellitus when normal pancreatic production is insufficient. Orally administration of insulin normally shows low bioavailability due to acidic gastric pH, the enzymatic barrier of the intestinal tract and the physical barrier made up of the intestinal epithelium. The inclusion of insulin into nano- and microparticulate systems potentially provides gastric protection, controlled release and enhanced absorption by mucosal adhesion and nanoparticle direct uptake (Tiyaboonchai et al., 2003). There are various studies in which insulin has been successfully included into polysaccharide-based nanoparticles prepared by different methods, that have shown good results in loading efficacy of the drug, good release control and, in some of them, *in vivo* efficacy of the systems.

Nanoemulsion/in situ triggered gelation is one of the methods employed to prepare insulin-loaded nanoparticles. Pinto Reis and co-workers have designed nanoparticles based on alginate and alginate coated with chitosan that have been loaded with insulin obtaining an encapsulation efficiency between 76-93% due to the interaction of amino groups of insulin with carboxylic groups of alginate. They observed that insulin was strongly retained into both systems at low pH, but when the release medium was changed by one at near neutral pH, up to 40-70% of the insulin was released almost immediately. Alginate-chitosan nanospheres demonstrated the best controlled insulin profile release in simulated intestinal conditions (Reis et al., 2008). The same method has been used to prepare insulin-loaded nanoparticles based on alginate-dextran, achieving an encapsulation efficency of 82.5%. At low gastric pH, insulin was fully retained likely due to alginate polymer forming a compact acid-gel structure reducing permeability and potentially stabilizing insulin from acid attack. Up to 89% of the insulin was released almost immediately after changing the medium to near neutral pH, and full release was observed after 1h. Nanoencapsulated insulin was bioactive, which was demonstrated through both *in vivo* and *in vitro* bioassays (Reis et al., 2007).

Insulin-loaded nanoparticles have been prepared by ionic cross-link methods using chitosan as the selected polysaccharide and TPP anions as cross-linker agent. Pan and co-workers used this method in order to prepare chitosan nanoparticles which enhanced the intestinal absorption of insulin *in vivo* (Pan et al., 2002). The association between the drug and the system was up to 80%, and the drug release was very dependent on the pH of the release medium. *In vivo* studies revealed that insulin was released in its active form, and the dosages of insulin loaded into chitosan nanoparticles were found orally effective, prolonging the hypoglycemia over 15 h.

Group of drugs	Entrapped Drug	Method of synthesis	Polysaccharide	NP size (nm)	References
Peptides and proteins	Insulin	Emulsification /ionic gelation	Alginate	564	Reis et al., 2008;
			Alginate-chitosan	1280	
			Alginate-dextran	267-2760	Reis et al., 2007
		Ionic cross-link	Chitosan	300-400	Pan et al., 2002
				250-400	
		Polyelectrolyte complexation (PEC)	Chitosan-dextran	423-850	Sarmento et al., 2006
			Chitosan-alginate		
			Chitosan-glucomannan	200-700	Alonso-Sande et al., 2006
		Self-assembling	Pullulan-cholesterol	20-30	Akiyoshi et al., 1998
	Growth factors	Ionic cross-link	Chitosan	301-424	Cetin et al., 2007
	Antisenses	Ionic cross-link	Alginate-chitosan	196-430	Azizi et al., 2010
Anticancer drugs	Tamoxifen	Covalent cross-link	Alginate-albumin	42-388	Martínez et al., 2011
		Emulsification	Guar gum	200-300	Sarmah et al., 2009
	Mitoxantrone	Ionic cross-link	Chitosan	75	Lu et al., 2006
	Methotrexate	Self-assembling	PEG-chitosan	262	Yang et al., 2008
	Doxorubicin	Emulsification	Oleoyl-chitosan	255	Zhang et al., 2007
	Paclitaxel	Self-assembling	Stearic acid-chitosan	28-175	Hu et al, 2006
	Epirubicin	Self-assembling	Chitosan-cholesterol	417	Wang et al., 2007
Nucleic acids and genetic material	Nucleic acids	Ionic cross-link	Chitosan	200-700	Katas & Alpar, 2006
		Polyelectrolyte complexation (PEC)	Chitosan-alginate	323-1600	Douglas & Tabrizian, 2005
			Chitosan-CMC	180-200	Cui & Mumper, 2001

Group of drugs	Entrapped Drug	Method of synthesis	Polysaccharide	NP size (nm)	References
			Chitosan-nucleic acids	129-450	de Martimprey et al., 2009
	Antisenses	Polyelectrolyte complexation (PEC)	Alginate-poly-L-lysine	28.000-10.000	González Ferreiro et al., 2002
	Genes	Emulsification /ionic gelation	Alginate	80	You & Pen, 2005
Other drugs	Furosemide	Emulsification /ionic gelation	Chitosan	30-150	Zhi et al., 2005
	Cyclosporine A	Ionic cross-link	Chitosan	250-400	De la Fuente et al., 2010
	Rertinol	Nanoprecipitation	Chitosan	50-200	Kim et al., 2006
	Antituber-cular drugs	Ionic gelation	Alginate	236	Ahmad et al., 2006
	Anfortericin B	Polyelectrolyte complexation (PEC)	Chitosan-dextran sulfate	600-800	Tiyaboonchai & Limpeanchob, 2007

Table 1. Some groups of drugs included in polysaccharide-based nanoparticles.

Ionotropic complexation (PEC) between chitosan and different polyanions (alginate, dextran sulfate and glucomannan) has been described in different works as another method that allowed the preparation of insulin-loaded nanoparticles. Insulin association efficiency from 63 to 94% and loading capacity from 5 to 13% were obtained with chitosan/alginate and chitosan/dextran nanoparticles, providing dextran sulfate combination the highest insulin association efficiency. It might be seen that insulin release occurred very rapidly from both systems, but a significant increment of insulin retention when using dextran sulfate in the formulation compared with alginate was observed in simulated gastric conditions (Sarmento et al., 2006). Alonso-Sande and co-workers prepared insulin-loaded chitosan/glucomannan nanoparticles reaching association efficiency values of 89%. They analyzed the influence of the pH release medium and the TPP presence on the release rate, obtaining more restricted diffusion of the peptide through a highly cross-linked system with TPP and at high pH of the release medium (Alonso-Sande et al., 2006).

Finally, nanoparticles based on self-assembling methods were also obtained and loaded with insulin. In this case, insulin was easily complexed with the hydrogel nanoparticle of hydrophobized cholesterol-bearing pullulan in water. The presence of different concentrations of bovine serum albumin (BSA) in the release medium was analyzed, and approximately 15% of insulin still remained in the complex, even at concentrations of BSA comparable to its physiological value in the blood. The physiological activity of complexed insulin was preserved in vivo after i.v. injection, showing an excellent behaviour of the complex as a possible protein drug carrier (Akiyoshi et al., 1998).

Basic fibroblast growth factor (bFGF) is a protein with molecular mass of 18 kDa that is a potent mitogen which regulates angiogenesis during growth and development. This growth factor stimulates the proliferation of a wide variety of cells, including mesenchymal, neuroectodermal, and endothelial cells, and it is effective on protecting neurons from oxidative stress processes (Li et al., 2006). However, bFGF has minimal pharmacological effects in the central nervous system (CNS) because of the presence of bloodbrain barrier that reduces its transport to CNS. In order to enhance the amount of bFGF that could reach the CNS, Cetin and co-workers have developed bFGF-loaded chitosan nanoparticles according to a ionotropic gelation process, using TPP as cross-linker agent. Nanoparticles displayed a low bFGF-association efficiency (27.4%) leading to final bFGF-loading values as low as 0.021%, but since bFGF exerts its biological activity in the concentration range of 0.1 to 10.0 ng/mL, they concluded that the bFGF loading was acceptable. *In vitro* release studies showed that around 30% of the loaded protein was immediately released into PBS and that the highest extent of release (68%) was observed at 24h. Moreover, their results showed that integrity of the encapsulated bFGF was not affected by the entrapment procedure and release conditions (Cetin et al., 2007).

Epidermal growth factor receptor (EGFR) antisense (AS) is used to reduce the expression of EGFR, a receptor tyrosine kinase proto-oncogene that plays a central role in the initiation and development of several human malignancies, notably of the breast, brain, and lung. The problem is that the macromolecular drug can degrade during preparation of the nanoparticles, during storage, and also during *in vitro* or *in vivo* release. For this reason, the ability of nanoparticles to protect the antisense molecules from degradation in aqueous medium is necessary to be investigated. Azizi and co-workers prepared EGFR-AS-loaded chitosan/alginate nanoparticles with different composition ratios, obtaining that nanoparticles can release antisense over 45-50 hours and 67-96% loading efficacy. Thus, they concluded that the nanoparticles could retain and stabilize the content of antisense in their hydrogel matrix (Azizi et al., 2010).

4.2 Anticancer drugs included in polysaccharide-based delivery systems

One of the major problems facing cancer chemotherapy is the achievement of the required concentration of the drug at the tumour site for a desired period of time, since tumours usually present resistances to treatment, and high dosages are frequently toxic. Thus, one of the main goals of nanomedicine is to develop safe and effective drug carriers that are systemically applied but will selectively deliver cytotoxic drugs to tumour cells without harming normal cells (Gullotti & Yeo, 2009). Among the available potential drug carrier systems in this size range, polysaccharide-based nanoparticles play an important role and their use with some anticancer drugs show promising results.

Tamoxifen has been successfully used since 1970s in treatment of hormone dependent breast cancer (Ameller et al., 2004). However, tamoxifen shows low water solubility, which limits the administration of this drug only to the oral route. Furthermore, following a long-term therapy, tamoxifen has some side effects, such as endometrial cancer and development of drug resistance. To overcome the undesirable side effects of tamoxifen, and to increase the concentration at the tumour site, tamoxifen could be entrapped into polymeric nanoparticles, which may provide better means of delivery in terms of enhanced uptake by the tumour and increased local concentration of the drug at the receptor site. Tamoxifen-loaded nanoparticles were prepared by Sarmah and co-workers based on guar gum, which is commonly used for colon specific drug delivery in the pharmaceutical industry. Nanoparticles were obtained by o/w emulsification and in situ polymer cross-linking, using dichloromethane as the best solvent of the drug and glutaraldehyde as cross-linker agent during the process. An efficiency loading of 15% was obtained when dichloromethane was used as selected solvent (Sarmah et al., 2009). Recently, nanoparticles based on thiolated alginate (ALG-CYS) and disulfide bond reduced albumin (BSA-SH) have been synthesized by coacervation method and stabilized by disulfide bond formation between both polymers. *In vitro* studies revealed that total release of the drug was not achieved in any case; only the 23–61% of the drug was released. Maximum release took place between 7 and 75 h. According to the results, it was concluded that the presence of alginate in the nanoparticle composition allowed the modulation of the amount of released TMX (Martínez et al., 2011).

Mitoxantrone is often used to treat breast cancer clinically, but the prolonged treatment with this drug results in some side-effects, such as heart toxicity and myelosupression, which are often a problem. Mitoxantrone is positively charged and it can be absorbed by negatively charged polysaccharides, such as chitosan. Nanospheres can be obtained by ion gelation method using sodium TPP as gelation agent, and obtaining an encapsulation efficacy of 98%. Tests for *in vitro* release in physiological saline or physiological saline containing 0.5% (w/v) ascorbic acid by a dialysis bag showed sustained release and little burst effect (Lu et al., 2006).

Methotrexate (MTX) is a folate antimetabolite and has been used in the treatment of various malignancies, including childhood acute lymphocytic leukemia, osteosarcoma, non-Hodgkin's lymphoma, Hodgkin's disease, head and neck cancer, lung cancer, and breast cancer. However, it may cause some adverse effects such as bone marrow suppression, acute and chronic hepatotoxicity, interstitial pneumonitis and chronic interstitial obstructive pulmonary disease. Therefore, in order to reduce its toxic and side effects and to improve specificity and selectivity, it is a good candidate to be included in drug delivery systems. Yang and co-workers have designed methoxy poly(ethylene glycol)-grafted-chitosan (mPEG-g-CS) self-aggregated nanoparticles being used as a carrier of MTX . Depending on the formulation, loading efficiency varied from 21 to 95%. *In vitro* release studies showed that the MTX appeared to be released in a biphasic way, which characterized by an initial release or rapid release period followed by a step of slower release. A fast release was observed in 4 h, in which 40% of the drug was released from nanoparticles. After this initial effect, MTX was released in a continuous way for up to 48 h, reaching percentage of cumulative release close to 60%. Therefore, the self-aggregated nanoparticles delayed the drug release in the release process when these results were compared with the release of the free drug (Yang et al., 2008).

Doxorubicin (DOX) is a member of the anthracycline ring antibiotics, with a broad spectrum of antitumor activity, including a variety of human and animal solid tumors. Despite advances of this antitumor agent, it is hydrophobic and possesses inevitable, serious side effects such as nonspecific toxicity that limit the dose and use of the drug. Therefore, a lot of studies have been carried out in order to entrap this drug into different drug carriers. DOX has been successfully entrapped into oleoyl-chitosan nanoparticles prepared by o/w emulsification method. Nanoparticles had a high encapsulation efficiency of 53% in loading doxorubicin, being drug incorporation into nanoparticles generally limited by the large surface area of the latter, as well as by the solubility of the drug in water. The drug was completely released from nanoparticles in the buffer medium of Na_2HPO_4– citric acid (pH 3.8), whereas in PBS (pH 7.4) 65% of doxorubicin was released after 6 hours, followed by a sustained release until 72 hours. Approximately 72% of DOX was released for 3 days, showing the potential of the nanoparticles as a sustained drug delivery system, suggesting that the nanoparticles might act as a barrier against the release of entrapped DOX. This result indicated that the nanoparticles contributed to an extended circulation of DOX and thus an improvement in therapeutic efficacy (Zhang et al., 2007).

Paclitaxel is an anticancer drug which is used in different types of malignancies, such as ovarian and breast cancer. Despite its multiple applications, it has important problems of solubility in water and, consequently, its administration is carried out including an oily component (Cremophor) and dehydrated alcohol within its formulation. Thus, relevant side effects such as hypersensibility, hypotension or breast pain are frequently observed after its intravenous administration. In the last years, a lot of studies have included this drug as good candidate to be included in drug delivery systems to overcome its administration problems. Hu and coworkers have obtained paclitaxel-loaded nanoparticles by self-cross-linking stearic-grafted chitosan and using glutaraldehyde as cross-linker in order to stabilize the systems. These formulations showed high encapsulation efficiency which ranged from 95% to 99%. When the surfaces of micelles were cross-linked by glutaraldehyde, the burst release of the micelles at earlier stage was highly improved and the drug release time was prolonged. By controlling the amino substitution of stearic-grafted chitosan and the cross-link degree, the prolonged and controlled release could be achieved (Hu et al., 2006).

As an anthracycline anticancer agent, **epirubicin** has a wide range of antitumor activity and is used to treat various carcinomas. However, this therapy may cause some serious side effects such as allergic reactions, cardiotoxicity and blood problems. Therefore, nanoparticles being used as a carrier of epirubicin are hoped to sustain its release, prolong its circulation time, enhance its therapeutic index and decrease its toxic effects. Cholesterol-modified nanoparticles have been prepared by self-assembling method by Wang and co-workers (Wang et al., 2007). They synthesized cholesterol-modified chitosan conjugate with succinyl linkages, investigated its self-aggregation behaviour, and prepared self-aggregated nanoparticles by sonication in aqueous media. Nanoparticles loading efficiency varied from 25 to 71% depending on the weight ratio of epirubicin and the nanoparticles and exhibited the release profile relating to the pH of the release media. When the pH of the release media increased, an evident decrease of drug release rate was observed owing to the solubility of epirubicin is greatly influenced by the pH of the aqueous solution, and self-aggregated nanoparticles were also pH sensitive because of the presence of many amino groups in their molecules. Thus, the drug release from self-aggregated nanoparticles was very slow in PBS (pH 7.4) and the total release amount was about 25% in 48 h, which suggested self-aggregated nanoparticles had a potential as a sustained-release carrier of epirubicin.

4.3 Nucleic acids and genetic material included in polysaccharide-based delivery systems

Small interfering RNAs (siRNAs) have proven to be versatile agents for controlling gene expression in mammalian cells. They have been employed as a novel tool since they can block the expression of genes, such as those expressed in infectious diseases and cancers.

However, siRNA suffers particular problems including poor cellular uptake, rapid degradation as well as limited blood stability. Therefore, effective systems which can protect and transport siRNA to the cytoplasm of the targeted cells are needed to exploit the promising potential applications offered by successful delivery of siRNA. Chitosan has been used to prepare nanoparticles based on modified ionic gelation with TPP as cross-linker agent, and they have been applied as possible vector that should be able to be taken up by the cells and escape the endosomal vesicle to avoid lysosomal degradation. *In vitro* study revealed the transfection efficiency of siRNA depends on the method of siRNA association to the chitosan, and entrapping siRNA using ionic gelation has shown to yield a better biological effect than simple complexation or siRNA adsorption onto the chitosan nanoparticles (Katas & Alpar, 2006).

Ionotropic complexation (PEC) between chitosan and different polyanions has been used as an useful method of preparation of nanoparticles in order to include nucleic acids, since chitosan-**DNA** nanoparticles demonstrated low transfection efficiencies and the incorporation of secondary polymers improved the characteristics of these systems (Kaul & Amiji, 2002). Among the chitosan–polyanion complexes investigated, the combination of chitosan and alginate is considered to be among the most interesting for delivery systems. The method used to prepare the nanoparticles is a two-step method where the first step is the formation of a calcium–alginate pre-gel. Various concentrations of chitosan solutions were then added with continuous stirring. High loading efficacy was obtained (26-60%), while maximum mass loading was 60 μg DNA/mg nanoparticles and 6% and 3.5% of the adsorbed DNA was released. Therefore, the high encapsulation of DNA from alginate-chitosan nanoparticles is encouraging for application in the field of gene therapy (Douglas & Tabrizian, 2005). The complexation between chitosan and carboxymethylcellulose (CMC) is another PEC combination employed in the inclusion of nucleic acid into delivery systems. Cui and Mumper coated the surface of this pre-formed nanoparticles with plasmid-DNA, obtaining a final plasmid DNA concentration of up to 400 mg/ml. Chitosan/CMC based nanoparticles containing plasmid DNA were applied topically to the skin of shaved mice and resulted in detectable and quantifiable levels of luciferase expression in skin after 24 h (Cui & Mumper, 2001). Nucleic acids can be also combined with chitosan in order to obtain nanospheres, being part of the structure of the system at the same time they are the encapsulant of the nanospheres. Chitosan showed a very high protection efficiency for siRNA. Polyplexes prepared with this polysaccharide are able to protect a siRNA for 7 h and an efficient transfection of the cells may be obtained after the intranasal administration of siRNA/chitosan polyplexes (de Martimprey et al., 2009).

Antisense oligonucleotides are therapeutic agents known to selectively modulate gene expression. The development of non-parenteral dosage forms for these compounds is desirable. However, the high molecular weight, the hydrophilicity and multiple negative charges result in a poor absorption of antisense oligonucleotides. Moreover, the oral administration faces additional problems such as degradation in the acidic gastric environment, enzymatic metabolism in the lumen and at the gastrointestinal epithelium and first-pass hepatic clearance. To achieve a successful non-parenteral delivery of antisense therapeutics, it is necessary to solve the specific problems of the oral administration route, together with general concerns of correct time-space targeting, improved cellular uptake and nuclear localization to exert gene transfection (Akhtar et al., 2000). Using a ionotropic gelation method, microparticles made of alginate cross-linked with calcium ions and poly-L-lysine have been reported to effectively act as transfecting agents. Encapsulation efficacies of 32-95% depending on the composition of the formulation have been obtained and the *in vitro* release behaviour depended mostly on the medium composition. The importance of competitive anions and ionic strength on the mechanism of dissociation of the oligonucleotide from the polymeric matrix was observed. Thus, the presence of phosphate anions preferably displaced alginate from the structure, resulting in the release of complexed oligonucleotide. Rat *in vivo* studies showed promising oligonucleotide bioavailability for microparticles after intrajejunal administration in the presence of a mixture of permeation enhancers to achieve a successful intestinal application (Gonzalez Ferreiro et al., 2002).

Natural biopolymers are also widely used in the field of **gene delivery**. In fact, alginate nanoparticles have been prepared using w/o microemulsion as a template followed by calcium cross-linking of guluronic acid units of alginate polymer. Ca-alginate nanoparticles were loaded with GFP-encoding plasmids in order to study their potency as carriers for gene delivery. The degree of endocytosis by NIH 3T3 cells and ensuing transfection rate were investigated. Results showed that Ca-alginate nanoparticles were very efficient gene carriers (You & Peng, 2004).

4.4 Other drugs included in polysaccharide-based delivery systems

Furosemide is a loop diuretic used in the treatment of congestive heart failure and edema, and it has been recently incorporated into chitosan nanoparticles (Zhi et al., 2005). Since the chitosan molecule has strong interaction with the organic compounds, it can be applied to adsorb diuretics from the water samples. Zhi and coworkers prepared chitosan nanoparticles by a nanoemulsion system, in which the chitosan nanoparticles were prepared by adding NaOH solution or glutaraldehyde as the solidification solution. Size was smaller in case of using NaOH as cross-linker agent (30 - 150 nm). The adsorptive efficiency of furosemide on the nanoparticle was 51.9%, and the presence of less cross-linking agent could be in favour of the adsorption ability. The furosemide adsorption capacity on the chitosan nanoparticles was affected not only by the –NH2 content, but also by the molecular weight of chitosan.

Ophthalmic drug delivery, probably more than any other route of administration, may benefit from the characteristics of nanotechnology-based drug delivery systems, mainly because of their capacity to protect the encapsulated molecule while facilitating its transport to the different compartments of the eye (Raju & Goldberg, 2008). For this purpose, the common use of chitosan is justified by its mucoadhesive and penetration enhancing properties, as well as by its good biocompatibility with the ocular structures. **Cyclosporine A (CyA)** is an immunosuppressant drug widely used in post-allogeneic organ transplant to reduce the activity of the immune system, and therefore the risk of organ rejection. The entrapment of the hydrophobic polypeptide CyA was achieved by ionic gelation technique using TPP and a previous dissolution of the peptide in an acetonitrile: water mixture, and a further nanoprecipitation into the nanoparticles in the form of small nanocrystals (de la Fuente et al., 2010). Entrapment efficiencies were reported to be as high as 73.4%. The *in vivo* evaluation of this new prototype, in rabbits, evidenced the capacity of these nanoparticles providing a selective and prolonged delivery of CyA to the cornea and conjunctiva. More importantly, it was observed that CyA-loaded chitosan nanoparticles provided therapeutic levels of CyA in the conjunctiva and the cornea for up to 24 and 48 h postadministration, respectively, while reducing the access of CyA to the blood circulation. This positive behavior of chitosan nanoparticles was attributed to their improved interaction with the corneal and conjunctival and the prolonged delivery of the CyA molecules associated to them (de la Fuente et al., 2010).

Retinol and its derivatives are extensively used in the pharmaceutical and cosmetic area. Especially, retinoids are recognized as being important for modern therapy of dermatological treatment of wrinkled skin. The stability and solubility problems of these compouns make them ideal candidates to be included into carriers. Nanoparticles are reported to be useful formulation to solve the poor aqueous solubility of retinoids and are able to use it by intravenous injection. Chitosan has been used by Kim and coworkers in order to obtain retinol-encapsulated chitosan nanoparticles for application of cosmetic and pharmaceutical applications (Kim et al., 2006). Retinol-encapsulated nanoparticles were completely reconstituted into aqueous solution as same as original aqueous solution.

The requirement of multidrug administration daily or several times a week for at least 6 months is the main cause of the failure of antitubercular chemotherapy. The dose-dependent side effects with high incidence and the need of daily-dosing results in discontinuation of medication, relapse of symptoms and an alarming increase in the prevalence of multidrug-resistant strains (Pandey & Khuller, 2005). Consequently, reductions in dose and dosing frequency are the major goals of tuberculosis research and the application of drug delivery systems constitutes an important therapeutic strategy. **Isoniazid, pyrazinamide and rifampicin**, which are important antitubercular drugs, were encapsulated into alginate nanoparticles prepared by cation-induced gelification, reaching drug encapsulation efficiencies of 70–90% for isoniazid and pyrazinamide and 80–90% for rifampicin (Ahmad et al., 2006). When alginate nanoparticles were administered through nebulisation, the drugs were detected in plasma from 3 h onwards. Encapsulated drugs were observed up to 14, 10 and 14 days, respectively, in contrast with free drugs that were cleared from the circulation within 12–24 h. The levels of drugs in various organs remained above the minimum inhibitory concentration at both doses for equal periods, demonstrating their equiefficiency. Alginate nanoparticles hold great potential in reducing dosing frequency of antitubercular drugs.

Amphotericin B (AmB) is a polyene macrolide antifungal agent and the drug of choice for systemic fungal infection. Unfortunately, it is poorly absorbed from the gastrointestinal tract due to its low aqueous solubility and it must be given parenterally to treat systemic fungal infections. Currently, two types of drug formulations for AmB are available. The first one is a micellar solution of AmB and the second one are lipid-based nanoparticulate formulations. Despite these formulations reduce nephrotoxicity of the treatment, they are still quite expensive. Thus, much effort has been spent to develop cheaper delivery systems with reduced amphotericin B toxicity. Polysaccharide based nanoparticle delivery systems are one approach that has been investigated by Tiyaboonchai and coworkers (Tiyaboonchai & Limpeanchob, 2007). They prepared AmB-chitosan-dextrane sulfate nanoparticles by polyelectrolyte complexation at room temperature, showing an an association efficiency of 50–65%. They observed a fast release characteristic of AmB independent of the processing conditions, with most of AmB released from particles within 5 min. This suggested that

AmB exhibited only moderate interaction with the weakly cross-linked polymers of the nanoparticles. A reduction of nephrotoxicity was observed in an *in vivo* renal toxicity study.

5. Conclusion

As pointed out throughout this review, polysaccharides show variability and versatility, due to their complex structure, which is difficult to be reproduced with synthetic polymers. Thus, native polysaccharides and their derivatives are emerging in the last years as one of the most used biomaterials in the field of nanomedicine, especially being chosen by a lot of researchers as carriers to be used in the preparation of nanoparticulate drug delivery systems.

A wide variety of preparation methods of nanoparticles has been developed, and three aspects have marked the evolution of these methods: need for lees toxic agents, simplification of the procedures and optimization to improve yield and entrapment efficiency. Now it is possible to choose the best method of preparation and the best suitable polymer to achieve an efficient encapsulation of the drug, taking into account the drug features in this selection.

As reviewed above, so many nanoparticle drug delivery systems have been prepared using various polysaccharides as carriers combined with different groups of drugs, and they have been investigated in terms of physicochemical features, drug -loading efficiency, *in vitro* toxicity and comparative *in vivo* test. Deeper studies, such as evaluation of interaction between cells, tissues and organs, as well as how the administration of these systems can affect to the metabolism, need to be carried out. In fact, more and more nanoparticle systems are emerging nowadays, and these necessary studies will be focused on in the near future, completing the evaluation of these hopeful polysaccharide-based systems.

6. Acknowledgements

The financial support of the Ministerio de Ciencia e Innovación of Spain FIS(PS09/01513) and MAT2010-21509-C03-03, and the FPI grant from UCM to A. Martínez are gratefully acknowledged.

7. References

Ahmad, Z.; Pandey, R.; Sharma, S. & Khuller, G. K. (2006). Pharmacokinetic and pharmacodynamic behaviour of antitubercular drugs encapsulated in alginate nanoparticles at two doses. *Int J Antimicrob Agents*, 27, 5, pp. (409-416), 0924-8579 (Print) 0924-8579 (Linking).

Akbuga, J. & Durmaz, G. (1994). Preparation and evaluation of crosslinked chitosan microspheres containing furosemide. *International Journal of Pharmaceutics*, 11, pp. (217-222).

Akhtar, S.; Hughes, M. D.; Khan, A.; Bibby, M.; Hussain, M.; Nawaz, Q.; Double, J. & Sayyed, P. (2000). The delivery of antisense therapeutics. *Adv Drug Deliv Rev*, 44, 1, pp. (3-21), 0169-409X (Print) 0169-409X (Linking).

Akiyoshi, K.; Kobayashi, S.; Shichibe, S.; Mix, D.; Baudys, M.; Kim, S. W. & Sunamoto, J. (1998). Self-assembled hydrogel nanoparticle of cholesterol-bearing pullulan as a carrier of protein drugs: complexation and stabilization of insulin. *J Control Release*, 54, 3, pp. (313-320), 0168-3659 (Print) 0168-3659 (Linking).

Alonso-Sande, M.; Cuna, M. & Remunan-Lopez, C. (2006). Formation of new glucomannan–chitosan nanoparticles and study of their ability to associate and deliver proteins. *Macromolecules*, 39, pp. (4152-4158).

Allison, D. D. & Grande-Allen, K. J. (2006). Review. Hyaluronan: a powerful tissue engineering tool. *Tissue Eng*, 12, 8, pp. (2131-2140), 1076-3279 (Print) 1076-3279 (Linking).

Ameller, T.; Legrand, P.; Marsaud, V. & Renoir, J. M. (2004). Drug delivery systems for oestrogenic hormones and antagonists: the need for selective targeting in estradiol-dependent cancers. *J Steroid Biochem Mol Biol*, 92, 1-2, pp. (1-18), 0960-0760 (Print) 0960-0760 (Linking).

Amidi, M.; Romeijn, S. G.; Borchard, G.; Junginger, H. E.; Hennink, W. E. & Jiskoot, W. (2006). Preparation and characterization of protein-loaded N-trimethyl chitosan nanoparticles as nasal delivery system. *J Control Release*, 111, 1-2, pp. (107-116), 0168-3659 (Print) 0168-3659 (Linking).

Aumelas, A.; Serrero, A.; Durand, A.; Dellacherie, E. & Leonard, M. (2007). Nanoparticles of hydrophobically modified dextrans as potential drug carrier systems. *Colloids Surf B Biointerfaces*, 59, 1, pp. (74-80), 0927-7765 (Print) 0927-7765 (Linking).

Azizi, E.; Namazi, A.; Haririan, I.; Fouladdel, S.; Khoshayand, M. R.; Shotorbani, P. Y.; Nomani, A. & Gazori, T. (2010). Release profile and stability evaluation of optimized chitosan/alginate nanoparticles as EGFR antisense vector. *Int J Nanomedicine*, 5, pp. (455-461), 1178-2013 (Electronic) 1176-9114 (Linking).

Bae, H.; Ahari, A. F.; Shin, H.; Nichol, J. W.; Hutson, C. B.; Masaeli, M.; Kim, S. H.; Aubin, H.; Yamanlar, S. & Khademhosseini, A. (2011). Cell-laden microengineered pullulan methacrylate hydrogels promote cell proliferation and 3D cluster formation. *Soft Matter*, 7, 5, pp. (1903-1911), 1744-6848 (Electronic) 1744-683X (Linking).

Balazs, E. A. (2004). *Viscoelastic properties of hyaluronan and its therapeutics use*, Elsevier, Amsterdam

Barbosa, M.; Granja, P.; Barrias, C. & Amaral, I. (2005). Polysaccharides as scaffolds for bone regeneration. *ITBM-RBM*, 26, pp. (212-217).

Barbucci, R.; Pasqui, D.; Favaloro, R. & Panariello, G. (2008). A thixotropic hydrogel from chemically cross-linked guar gum: synthesis, characterization and rheological behaviour. *Carbohydr Res*, 343, 18, pp. (3058-3065), 1873-426X (Electronic) 0008-6215 (Linking).

Barratt, G. M. (2000). Therapeutic applications of colloidal drug carriers. *Pharm Sci Technolo Today*, 3, 5, pp. (163-171), 1461-5347 (Electronic) 1461-5347 (Linking).

Bernardo, M. V.; Blanco, M. D.; Sastre, R. L.; Teijon, C. & Teijon, J. M. (2003). Sustained release of bupivacaine from devices based on chitosan. *Farmaco*, 58, 11, pp. (1187-1191), 0014-827X (Print) 0014-827X (Linking).

Berthold, A.; Cremer, K. & Kreuter, J. (1996). Preparation and characterization of chitosan microspheres as drug carrier for prednisolone sodium phosphate as model for anti-inflammatory drugs. *journal of Controlled Release*, 39, pp. (17-25).

Bertholon, I.; Lesieur, S.; Labarre, D.; Besnard, M. & C., V. (2006). Characterization of Dextran-Poly(isobutylcyanoacrylate) Copolymers Obtained by Redox Radical and Anionic Emulsion Polymerization. *Macromolecules*, 39, pp. (3559-3567).

Blanco, M. D.; Gomez, C.; Olmo, R.; Muniz, E. & Teijon, J. M. (2000). Chitosan microspheres in PLG films as devices for cytarabine release. *Int J Pharm*, 202, 1-2, pp. (29-39), 0378-5173 (Print) 0378-5173 (Linking).

Bodnár, M.; Daróczi, L.; Batta, G.; Bakó, J.; Hartmann, J. F. & Borbély, J. (2009). Preparation and characterization of cross-linked hyaluronan nanoparticles. *Colloid & Polymer Science*, 287, pp. (991-1000).

Bodnar, M.; Hartmann, J. F. & Borbely, J. (2005). Preparation and characterization of chitosan-based nanoparticles. *Biomacromolecules*, 6, 5, pp. (2521-2527), 1525-7797 (Print) 1525-7797 (Linking).

Braga, T. P.; Chagas, E. C.; Freitas de Sousa, A.; Villarreal, N. L.; Longhinotti, N. & Valentini, A. (2009). Synthesis of hybrid mesoporous spheres using the chitosan as template. *Journal of Non-Crystalline Solids*, 355, pp. (860-866).

Burdick, J. A. & Prestwich, G. D. (2011). Hyaluronic acid hydrogels for biomedical applications. *Adv Mater*, 23, 12, pp. (H41-56), 1521-4095 (Electronic) 0935-9648 (Linking).

Cafaggi, S.; Russo, E.; Stefani, R.; Parodi, B.; Caviglioli, G.; Sillo, G.; Bisio, A.; Aiello, C. & Viale, M. (2011). Preparation, characterisation and preliminary antitumour activity evaluation of a novel nanoparticulate system based on a cisplatin-hyaluronate complex and N-trimethyl chitosan. *Invest New Drugs*, 29, 3, pp. (443-455), 1573-0646 (Electronic) 0167-6997 (Linking).

Cetin, M.; Aktas, Y.; Vural, I.; Capan, Y.; Dogan, L. A.; Duman, M. & Dalkara, T. (2007). Preparation and in vitro evaluation of bFGF-loaded chitosan nanoparticles. *Drug Deliv*, 14, 8, pp. (525-529), 1071-7544 (Print) 1071-7544 (Linking).

Coviello, T.; Matricardi, P.; Marianecci, C. & Alhaique, F. (2007). Polysaccharide hydrogels for modified release formulations. *J Control Release*, 119, 1, pp. (5-24), 1873-4995 (Electronic) 0168-3659 (Linking).

Cui, Z. & Mumper, R. J. (2001). Chitosan-based nanoparticles for topical genetic immunization. *J Control Release*, 75, 3, pp. (409-419), 0168-3659 (Print) 0168-3659 (Linking).

Chen, Q.; Hu, Y.; Chen, Y.; Jiang, X. & Yang, Y. (2005). Microstructure formation and property of chitosan-poly(acrylic acid) nanoparticles prepared by macromolecular complex. *Macromol Biosci*, 5, 10, pp. (993-1000), 1616-5187 (Print) 1616-5187 (Linking).

Chen, X. G.; Lee, C. M. & Park, H. J. (2003). O/W emulsification for the self-aggregation and nanoparticle formation of linoleic acid-modified chitosan in the aqueous system. *J Agric Food Chem*, 51, 10, pp. (3135-3139), 0021-8561 (Print) 0021-8561 (Linking).

Choi, K. Y.; Chung, H.; Min, K. H.; Yoon, H. Y.; Kim, K.; Park, J. H.; Kwon, I. C. & Jeong, S. Y. (2010). Self-assembled hyaluronic acid nanoparticles for active tumor targeting. *Biomaterials*, 31, 1, pp. (106-114), 1878-5905 (Electronic) 0142-9612 (Linking).

Davidenko, N.; Blanco, M. D.; Peniche, C.; Becherán, L.; Guerrero, S. & Teijón, J. M. (2009). Effects of different parameters on characteristics of chitosan-poly(acrilic acid) nanoparticles obtained by the method of coacervation. *Journal of Applied Polymer Science*, 111, pp. (2362-2371).

de la Fuente, M.; Ravina, M.; Paolicelli, P.; Sanchez, A.; Seijo, B. & Alonso, M. J. (2010). Chitosan-based nanostructures: a delivery platform for ocular therapeutics. *Adv Drug Deliv Rev*, 62, 1, pp. (100-117), 1872-8294 (Electronic) 0169-409X (Linking).

de la Fuente, M.; Seijo, B. & Alonso, M. J. (2008). Design of novel polysaccharidic nanostructures for gene delivery. *Nanotechnology*, 19, 7, pp. (075105), 0957-4484 (Print) 0957-4484 (Linking).

de Martimprey, H.; Vauthier, C.; Malvy, C. & Couvreur, P. (2009). Polymer nanocarriers for the delivery of small fragments of nucleic acids: oligonucleotides and siRNA. *Eur J Pharm Biopharm*, 71, 3, pp. (490-504), 1873-3441 (Electronic) 0939-6411 (Linking).

De, S. & Robinson, D. (2003). Polymer relationships during preparation of chitosan-alginate and poly-l-lysine-alginate nanospheres. *J Control Release*, 89, 1, pp. (101-112), 0168-3659 (Print) 0168-3659 (Linking).

Douglas, K. L. & Tabrizian, M. (2005). Effect of experimental parameters on the formation of alginate-chitosan nanoparticles and evaluation of their potential application as DNA carrier. *J Biomater Sci Polym Ed*, 16, 1, pp. (43-56), 0920-5063 (Print) 0920-5063 (Linking).

Drogoz, A.; David, L.; Rochas, C.; Domard, A. & Delair, T. (2007). Polyelectrolyte complexes from polysaccharides: formation and stoichiometry monitoring. *Langmuir*, 23, 22, pp. (10950-10958), 0743-7463 (Print) 0743-7463 (Linking).

Fernandez-Hervas, M. & Fell, J. (1998). Pectin/chitosan mixtures as coatings for colon-specific drug delivery: an in vitro evaluation. *International Journal of Pharmaceutics*, 169, pp. (115-119).

Gavory, C.; Durand, A.; Six, J. L.; Nouvel, C.; Marie, E. & Leonard, M. (2011). Polysaccharide-covered nanoparticles prepared by nanoprecipitation. *Carbohydrate Polymers*, 84, pp. (133-140).

George, M. & Abraham, T. E. (2006). Polyionic hydrocolloids for the intestinal delivery of protein drugs: alginate and chitosan--a review. *J Control Release*, 114, 1, pp. (1-14), 0168-3659 (Print) 0168-3659 (Linking).

Gonzalez Ferreiro, M.; Tillman, L.; Hardee, G. & Bodmeier, R. (2002). Characterization of alginate/poly-L-lysine particles as antisense oligonucleotide carriers. *Int J Pharm*, 239, 1-2, pp. (47-59), 0378-5173 (Print) 0378-5173 (Linking).

Guerrero, S.; Teijón, C.; Muñiz, E.; Teijón, J. M. & Blanco, M. D. (2010). Characterization and in vivo evaluation of ketotifen-loaded chitosan microspheres. *Carbohydrate Polymers*, 79, pp. (1006-1013).

Gullotti, E. & Yeo, Y. (2009). Extracellularly activated nanocarriers: a new paradigm of tumor targeted drug delivery. *Mol Pharm*, 6, 4, pp. (1041-1051), 1543-8384 (Print) 1543-8384 (Linking).

Gurwitz, D. & Livshits, G. (2006). Personalized medicine Europe: health, genes and society: Tel-Aviv University, Tel-Aviv, Israel, June 19-21, 2005. *Eur J Hum Genet*, 14, 3, pp. (376-380), 1018-4813 (Print) 1018-4813 (Linking).

Han, S. K.; Lee, J. H.; Kim, D.; Cho, S. H. & Yuk, S. H. (2005). Hydrophilized poly(lactide-coglycolide) nanoparticles with core/shell structure for protein delivery. *Science and Technology of Advanced Materials*, 6, pp. (468-474).

Heinze, T.; Liebert, T.; Heublein, B. & Hornig, S. (2006). Functional Polymers Based on Dextran. *Advances in Polymer Science*, 205/2006, pp. (199-291).

Hornig, S. & Heinze, T. (2008). Efficient approach to design stable water-dispersible nanoparticles of hydrophobic cellulose esters. *Biomacromolecules*, 9, 5, pp. (1487-1492), 1526-4602 (Electronic) 1525-7797 (Linking).

Hu, F. Q.; Ren, G. F.; Yuan, H.; Du, Y. Z. & Zeng, S. (2006). Shell cross-linked stearic acid grafted chitosan oligosaccharide self-aggregated micelles for controlled release of paclitaxel. *Colloids Surf B Biointerfaces*, 50, 2, pp. (97-103), 0927-7765 (Print) 0927-7765 (Linking).

Ito, T.; Iida-Tanaka, N.; Niidome, T.; Kawano, T.; Kubo, K.; Yoshikawa, K.; Sato, T.; Yang, Z. & Koyama, Y. (2006). Hyaluronic acid and its derivative as a multi-functional gene expression enhancer: protection from non-specific interactions, adhesion to targeted cells, and transcriptional activation. *J Control Release*, 112, 3, pp. (382-388), 0168-3659 (Print) 0168-3659 (Linking).

Janes, K. A.; Calvo, P. & Alonso, M. J. (2001). Polysaccharide colloidal particles as delivery systems for macromolecules. *Adv Drug Deliv Rev*, 47, 1, pp. (83-97), 0169-409X (Print) 0169-409X (Linking).

Kang, J.; Lee, M. S.; Copland, J. A., 3rd; Luxon, B. A. & Gorenstein, D. G. (2008). Combinatorial selection of a single stranded DNA thioaptamer targeting TGF-beta1 protein. *Bioorg Med Chem Lett*, 18, 6, pp. (1835-1839), 1464-3405 (Electronic) 0960-894X (Linking).

Katas, H. & Alpar, H. O. (2006). Development and characterisation of chitosan nanoparticles for siRNA delivery. *J Control Release*, 115, 2, pp. (216-225), 0168-3659 (Print) 0168-3659 (Linking).

Kaul, G. & Amiji, M. (2002). Long-circulating poly(ethylene glycol)-modified gelatin nanoparticles for intracellular delivery. *Pharm Res*, 19, 7, pp. (1061-1067), 0724-8741 (Print) 0724-8741 (Linking).

Kim, D. G.; Jeong, Y. I.; Choi, C.; Roh, S. H.; Kang, S. K.; Jang, M. K. & Nah, J. W. (2006). Retinol-encapsulated low molecular water-soluble chitosan nanoparticles. *Int J Pharm*, 319, 1-2, pp. (130-138), 0378-5173 (Print) 0378-5173 (Linking).

Kogan, G.; Soltes, L.; Stern, R. & Gemeiner, P. (2007). Hyaluronic acid: a natural biopolymer with a broad range of biomedical and industrial applications. *Biotechnol Lett*, 29, 1, pp. (17-25), 1573-6776 (Electronic) 0141-5492 (Linking).

Kratz, F. (2008). Albumin as a drug carrier: design of prodrugs, drug conjugates and nanoparticles. *J Control Release*, 132, 3, pp. (171-183), 1873-4995 (Electronic) 0168-3659 (Linking).

Kumbar, S. G. & Aminabhavi, T. M. (2003). Synthesis and characterization of modified chitosan microspheres: effect of the grafting ratio on the controlled release of nifedipine through microspheres. *Journal of Applied Polymer Science*, 89, pp. (2940-2949).

Lan, S. F. & Starly, B. (2011). Alginate based 3D hydrogels as an in vitro co-culture model platform for the toxicity screening of new chemical entities. *Toxicol Appl Pharmacol*, pp.), 1096-0333 (Electronic) 0041-008X (Linking).

Lassalle, V. & Ferreira, M. L. (2007). PLA nano- and microparticles for drug delivery: an overview of the methods of preparation. *Macromol Biosci*, 7, 6, pp. (767-783), 1616-5187 (Print) 1616-5187 (Linking).

Leach, J. B. & Schmidt, C. E. (2005). Characterization of protein release from photocrosslinkable hyaluronic acid-polyethylene glycol hydrogel tissue engineering scaffolds. *Biomaterials*, 26, 2, pp. (125-135), 0142-9612 (Print) 0142-9612 (Linking).

Learoyd, T. P.; Burrows, J. L.; French, E. & Seville, P. C. (2008). Chitosan-based spray-dried respirable powders for sustained delivery of terbutaline sulfate. *Eur J Pharm Biopharm*, 68, 2, pp. (224-234), 0939-6411 (Print) 0939-6411 (Linking).

Leathers, T. D. (2003). Biotechnological production and applications of pullulan. *Appl Microbiol Biotechnol*, 62, 5-6, pp. (468-473), 0175-7598 (Print) 0175-7598 (Linking).

Lee, J. W.; Park, J. H. & Robinson, J. R. (2000). Bioadhesive-based dosage forms: the next generation. *J Pharm Sci*, 89, 7, pp. (850-866), 0022-3549 (Print) 0022-3549 (Linking).

Lemarchand, C.; Couvreur, P.; Besnard, M.; Costantini, D. & Gref, R. (2003a). Novel polyester-polysaccharide nanoparticles. *Pharm Res*, 20, 8, pp. (1284-1292), 0724-8741 (Print) 0724-8741 (Linking).

Lemarchand, C.; Couvreur, P.; Vauthier, C.; Costantini, D. & Gref, R. (2003b). Study of emulsion stabilization by graft copolymers using the optical analyzer Turbiscan. *Int J Pharm*, 254, 1, pp. (77-82), 0378-5173 (Print) 0378-5173 (Linking).

Letchford, K. & Burt, H. (2007). A review of the formation and classification of amphiphilic block copolymer nanoparticulate structures: micelles, nanospheres, nanocapsules and polymersomes. *Eur J Pharm Biopharm*, 65, 3, pp. (259-269), 0939-6411 (Print) 0939-6411 (Linking).

Li, Y.; Nagira, T. & Tsuchiya, T. (2006). The effect of hyaluronic acid on insulin secretion in HIT-T15 cells through the enhancement of gap-junctional intercellular communications. *Biomaterials*, 27, 8, pp. (1437-1443), 0142-9612 (Print) 0142-9612 (Linking).

Li, Y. L.; Zhu, L.; Liu, Z.; Cheng, R.; Meng, F.; Cui, J. H.; Ji, S. J. & Zhong, Z. (2009). Reversibly stabilized multifunctional dextran nanoparticles efficiently deliver doxorubicin into the nuclei of cancer cells. *Angew Chem Int Ed Engl*, 48, 52, pp. (9914-9918), 1521-3773 (Electronic) 1433-7851 (Linking).

Liu, L.; Fishman, M. L.; Kost, J. & Hicks, K. B. (2003). Pectin-based systems for colon-specific drug delivery via oral route. *Biomaterials*, 24, 19, pp. (3333-3343), 0142-9612 (Print) 0142-9612 (Linking).

Liu, L.; Won, Y. J.; Cooke, P. H.; Coffin, D. R.; Fishman, M. L.; Hicks, K. B. & Ma, P. X. (2004). Pectin/poly(lactide-co-glycolide) composite matrices for biomedical applications. *Biomaterials*, 25, 16, pp. (3201-3210), 0142-9612 (Print) 0142-9612 (Linking).

Liu, Z.; Jiao, Y.; Wang, Y.; Zhou, C. & Zhang, Z. (2008). Polysaccharides-based nanoparticles as drug delivery systems. *Adv Drug Deliv Rev*, 60, 15, pp. (1650-1662), 1872-8294 (Electronic) 0169-409X (Linking).

Lu, B.; Xiong, S. B.; Yang, H.; Yin, X. D. & Zhao, R. B. (2006). Mitoxantrone-loaded BSA nanospheres and chitosan nanospheres for local injection against breast cancer and its lymph node metastases. I: Formulation and in vitro characterization. *Int J Pharm*, 307, 2, pp. (168-174), 0378-5173 (Print) 0378-5173 (Linking).

Martínez, A.; Iglesias, I.; Lozano, R.; Teijón, J. M. & Blanco, M. D. (2011). Synthesis and characterization of thiolated alginate-albumin nanoparticles stabilized by disulfide bonds. Evaluation as drug delivery systems. *Carbohydrate Polymers*, 83, 3, pp. (1311-1321).

Mendichi, R. & Soltes, L. (2002). Hyaluronan molecular weight and polydispersity in some commercial intra-articular injectable preparations and in synovial fluid. *Inflamm Res*, 51, 3, pp. (115-116), 1023-3830 (Print) 1023-3830 (Linking).

Misaki, A.; Torii, M.; Sawai, T. & Goldstein, I. J. (1980). Structure of the dextran of Leuconostoc mesenteroides B-1355. *Carbohydrate Research*, 84, pp. (273-285).

Moghimi, S. M.; Hunter, A. C. & Murray, J. C. (2005). Nanomedicine: current status and future prospects. *FASEB J*, 19, 3, pp. (311-330), 1530-6860 (Electronic) 0892-6638 (Linking).

Muzzarelli, R. A. A. & Muzzarelli, C. (2005). Chitosan chemistry: Relevance to the biomedical sciences Polysaccharides 1: Structure, characterization and use. *Advances in Polymer Science*, 186, pp. (151-209).

Naessens, M.; Cerdobbel, A.; Soetaert, W. & Vandamme, E. J. (2005). Leuconostoc dextransucrase and dextran: production, properties and applications. *Journal of Chemical Technology & Biotechnology*, 80, pp. (845-860).

Nair, L. S. & Laurencin, C. T. (2007). Biodegradable polymers as biomaterial. *Progress in Polymer Science*, 6, pp. (762-798).

Oh, E. J.; Park, K.; Kim, K. S.; Kim, J.; Yang, J. A.; Kong, J. H.; Lee, M. Y.; Hoffman, A. S. & Hahn, S. K. (2010). Target specific and long-acting delivery of protein, peptide, and nucleotide therapeutics using hyaluronic acid derivatives. *J Control Release*, 141, 1, pp. (2-12), 1873-4995 (Electronic) 0168-3659 (Linking).

Opanasopit, P.; Apirakaramwong, A.; Ngawhirunpat, T.; Rojanarata, T. & Ruktanonchai, U. (2008). Development and characterization of pectinate micro/nanoparticles for gene delivery. *AAPS PharmSciTech*, 9, 1, pp. (67-74), 1530-9932 (Electronic) 1530-9932 (Linking).

Orive, G.; Ponce, S.; Hernandez, R. M.; Gascon, A. R.; Igartua, M. & Pedraz, J. L. (2002). Biocompatibility of microcapsules for cell immobilization elaborated with different type of alginates. *Biomaterials*, 23, 18, pp. (3825-3831), 0142-9612 (Print) 0142-9612 (Linking).

Pan, Y.; Li, Y. J.; Zhao, H. Y.; Zheng, J. M.; Xu, H.; Wei, G.; Hao, J. S. & Cui, F. D. (2002). Bioadhesive polysaccharide in protein delivery system: chitosan nanoparticles improve the intestinal absorption of insulin in vivo. *Int J Pharm*, 249, 1-2, pp. (139-147), 0378-5173 (Print) 0378-5173 (Linking).

Pandey, R. & Ahmad, Z. (2011). Nanomedicine and experimental tuberculosis: facts, flaws, and future. *Nanomedicine*, 7, 3, pp. (259-272), 1549-9642 (Electronic) 1549-9634 (Linking).

Pandey, R. & Khuller, G. K. (2005). Antitubercular inhaled therapy: opportunities, progress and challenges. *J Antimicrob Chemother*, 55, 4, pp. (430-435), 0305-7453 (Print) 0305-7453 (Linking).

Perera, G.; Barthelmes, J. & Bernkop-Schnurch, A. (2010). Novel pectin-4-aminothiophenole conjugate microparticles for colon-specific drug delivery. *J Control Release*, 145, 3, pp. (240-246), 1873-4995 (Electronic) 0168-3659 (Linking).

Pliszczak, D.; Bourgeois, S.; Bordes, C.; Valour, J. P.; Mazoyer, M. A.; Orecchioni, A. M.; Nakache, E. & Lanteri, P. (2011). Improvement of an encapsulation process for the preparation of pro- and prebiotics-loaded bioadhesive microparticles by using experimental design. *Eur J Pharm Sci*, 44, 1-2, pp. (83-92), 1879-0720 (Electronic) 0928-0987 (Linking).

Prestwich, G. D. (2008). Engineering a clinically-useful matrix for cell therapy. *Organogenesis*, 4, 1, pp. (42-47), 1547-6278 (Print) 1547-6278 (Linking).

Qiu, L. Y. & Bae, Y. H. (2006). Polymer architecture and drug delivery. *Pharm Res*, 23, 1, pp. (1-30), 0724-8741 (Print) 0724-8741 (Linking).

Raju, H. B. & Goldberg, J. L. (2008). Nanotechnology for ocular therapeutics and tissue repair. *Expert Review of Ophthalmology*, 3, pp. (431-436).

Rehm, B. H. A. (Ed). (2009). *Alginates: Biology and Applications*, Springer, 978-3-540-92678.

Reis, C. P.; Ribeiro, A. J.; Houng, S.; Veiga, F. & Neufeld, R. J. (2007). Nanoparticulate delivery system for insulin: design, characterization and in vitro/in vivo bioactivity. *Eur J Pharm Sci*, 30, 5, pp. (392-397), 0928-0987 (Print) 0928-0987 (Linking).

Reis, C. P.; Ribeiro, A. J.; Veiga, F.; Neufeld, R. J. & Damge, C. (2008). Polyelectrolyte biomaterial interactions provide nanoparticulate carrier for oral insulin delivery. *Drug Deliv*, 15, 2, pp. (127-139), 1071-7544 (Print) 1071-7544 (Linking).

Rekha, M. R. & Chandra, P. S. (2007). Pullulan as a promising biomaterial for biomedical applications: A perspective. *Trends in Biomaterials & Artificial Organs*, 20, pp. (116-121).

Romberg, B.; Hennink, W. E. & Storm, G. (2008). Sheddable coatings for long-circulating nanoparticles. *Pharm Res*, 25, 1, pp. (55-71), 0724-8741 (Print) 0724-8741 (Linking).

Rotureau, E.; Leonard, M.; Dellacherie, E. & Durand, A. (2004). Amphiphilic derivatives of dextran: adsorption at air/water and oil/water interfaces. *J Colloid Interface Sci*, 279, 1, pp. (68-77), 0021-9797 (Print) 0021-9797 (Linking).

Sakamoto, J.; Annapragada, A.; Decuzzi, P. & Ferrari, M. (2007). Antibiological barrier nanovector technology for cancer applications. *Expert Opin Drug Deliv*, 4, 4, pp. (359-369), 1742-5247 (Print) 1742-5247 (Linking).

Sarmah, J. K.; Bhattacharjee, S. K.; Mahanta, R. & Mahanta, R. (2009). Preparation of cross-linked guar gum nanospheres containing tamoxifen citrate by single step emulsion in situ polymer cross-linking method. *Journal of Inclusion Phenomena and Macrocyclic Chemistry*, 65, pp. (329-334).

Sarmah, J. K.; Mahanta, R.; Bhattacharjee, S. K. & Biswas, A. (2011). Controlled release of tamoxifen citrate encapsulated in cross-linked guar gum nanoparticles. *Int J Biol Macromol*, 49, 3, pp. (390-396), 1879-0003 (Electronic) 0141-8130 (Linking).

Sarmento, B.; Martins, S.; Ribeiro, A.; Veiga, F.; Neufeld, R. & Ferreira, D. (2006). Development and comparison of different nanoparticulate polyelectrolyte complexes as insulin carriers. *International Journal of Peptide Research and Therapeutics*, 12, pp. (131-138).

Singh, V.; , P., S.; Singh, S. K. & Sanghi, R. (2009). Removal of cadmium from aqueous solutions by adsorption using poly(acrylamide) modified guar gum–silica nanocomposites. *Separation and Purification Technology*, 67, pp. (251-261).

Sinha, V. R. & Kumria, R. (2001). Polysaccharides in colon-specific drug delivery. *Int J Pharm*, 224, 1-2, pp. (19-38), 0378-5173 (Print) 0378-5173 (Linking).

Sinha, V. R.; Singla, A. K.; Wadhawan, S.; Kaushik, R.; Kumria, R.; Bansal, K. & Dhawan, S. (2004). Chitosan microspheres as a potential carrier for drugs. *Int J Pharm*, 274, 1-2, pp. (1-33), 0378-5173 (Print) 0378-5173 (Linking).

Sinha, V. R. & Trehan, A. (2003). Biodegradable microspheres for protein delivery. *J Control Release*, 90, 3, pp. (261-280), 0168-3659 (Print) 0168-3659 (Linking).

Skiba, M.; Wouessidjewe, D.; Puisieux, F.; Duchène, D. & Gulik, A. (1996). Characterization of amphiphilic fl-cyclodextrin nanospheres. *International Journal of Pharmaceutics*, 142, pp. (121-124).

Soumya, R. S.; Ghosh, S. & Abraham, E. T. (2010). Preparation and characterization of guar gum nanoparticles. *Int J Biol Macromol*, 46, 2, pp. (267-269), 1879-0003 (Electronic) 0141-8130 (Linking).

Srinophakun, T. & Boonmee, J. (2011). Preliminary Study of Conformation and Drug Release Mechanism of Doxorubicin-Conjugated Glycol Chitosan, via cis-Aconityl Linkage, by Molecular Modeling. *Int J Mol Sci*, 12, 3, pp. (1672-1683), 1422-0067 (Electronic) 1422-0067 (Linking).

Sujata, V. B. (2002). *Biomaterials*, Springer, 0-7923-7058-9, India.

Tasciotti, E.; Liu, X.; Bhavane, R.; Plant, K.; Leonard, A. D.; Price, B. K.; Cheng, M. M.; Decuzzi, P.; Tour, J. M.; Robertson, F. & Ferrari, M. (2008). Mesoporous silicon particles as a multistage delivery system for imaging and therapeutic applications. *Nat Nanotechnol*, 3, 3, pp. (151-157), 1748-3395 (Electronic) 1748-3387 (Linking).

Thebaud, N. B.; Pierron, D.; Bareille, R.; Le Visage, C.; Letourneur, D. & Bordenave, L. (2007). Human endothelial progenitor cell attachment to polysaccharide-based hydrogels: a pre-requisite for vascular tissue engineering. *J Mater Sci Mater Med*, 18, 2, pp. (339-345), 0957-4530 (Print) 0957-4530 (Linking).

Tibbals, H. F. (2010). *Medical Nanotechnology and Nanomedicine*, CRC Press, 978-1-4398-0876-4, USA.

Tiraferri, A.; Chen, K. L.; Sethi, R. & Elimelech, M. (2008). Reduced aggregation and sedimentation of zero-valent iron nanoparticles in the presence of guar gum. *J Colloid Interface Sci*, 324, 1-2, pp. (71-79), 1095-7103 (Electronic) 0021-9797 (Linking).

Tiyaboonchai, W. & Limpeanchob, N. (2007). Formulation and characterization of amphotericin B-chitosan-dextran sulfate nanoparticles. *Int J Pharm*, 329, 1-2, pp. (142-149), 0378-5173 (Print) 0378-5173 (Linking).

Tiyaboonchai, W.; Woiszwillo, J.; Sims, R. C. & Middaugh, C. R. (2003). Insulin containing polyethylenimine-dextran sulfate nanoparticles. *Int J Pharm*, 255, 1-2, pp. (139-151), 0378-5173 (Print) 0378-5173 (Linking).

Tokumitsu, H.; Ichikawa, H. & Fukumori, Y. (1999). Chitosan-gadopentetic acid complex nanoparticles for gadolinium neutron-capture therapy of cancer: preparation by novel emulsion-droplet coalescence technique and characterization. *Pharm Res*, 16, 12, pp. (1830-1835), 0724-8741 (Print) 0724-8741 (Linking).

Varum, F. J.; McConnell, E. L.; Sousa, J. J.; Veiga, F. & Basit, A. W. (2008). Mucoadhesion and the gastrointestinal tract. *Crit Rev Ther Drug Carrier Syst*, 25, 3, pp. (207-258), 0743-4863 (Print) 0743-4863 (Linking).

Vauthier, C. & Bouchemal, K. (2009). Methods for the preparation and manufacture of polymeric nanoparticles. *Pharm Res*, 26, 5, pp. (1025-1058), 1573-904X (Electronic) 0724-8741 (Linking).

Vauthier, C. & Couvreur, P. (2000). Development of nanoparticles made of polysaccharides as novel drug carrier systems, In: *Handbook of Pharmaceutical Controlled Release Technology*, D. L. Wise, (Ed), pp. (413-429), Marcel Dekker, New York.

Vila, A.; Sanchez, A.; Tobio, M.; Calvo, P. & Alonso, M. J. (2002). Design of biodegradable particles for protein delivery. *J Control Release*, 78, 1-3, pp. (15-24), 0168-3659 (Print) 0168-3659 (Linking).

Wang, N. & Wu, X. S. (1997). Preparation and characterization of agarose hydrogel nanoparticles for protein and peptide drug delivery. *Pharm Dev Technol*, 2, 2, pp. (135-142), 1083-7450 (Print) 1083-7450 (Linking).

Wang, S.; Mao, X.; Wang, H.; Lin, J.; Li, F. & Wei, D. (2011). Characterization of a novel dextran produced by Gluconobacter oxydans DSM 2003. *Appl Microbiol Biotechnol*, 91, 2, pp. (287-294), 1432-0614 (Electronic) 0175-7598 (Linking).

Wang, Y. S.; Liu, L. R.; Jiang, Q. & Zhang, Q. Q. (2007). Self-aggregated nanoparticles of cholesterol-modified chitosan conjugate as a novel carrier of epirubicin. *European Polymer Journal*, 43, pp. (43-51).

Yang, J. S.; Xie, Y. J. & He, W. (2011). Research progress on chemical modification of alginate: A review. *Carbohydrate Polymers*, 84, pp. (33–39).

Yang, X.; Zhang, Q.; Wang, Y.; Chen, H.; Zhang, H.; Gao, F. & Liu, L. (2008). Self-aggregated nanoparticles from methoxy poly(ethylene glycol)-modified chitosan: synthesis; characterization; aggregation and methotrexate release in vitro. *Colloids Surf B Biointerfaces*, 61, 2, pp. (125-131), 0927-7765 (Print) 0927-7765 (Linking).

Yip, G. W.; Smollich, M. & Gotte, M. (2006). Therapeutic value of glycosaminoglycans in cancer. *Mol Cancer Ther*, 5, 9, pp. (2139-2148), 1535-7163 (Print) 1535-7163 (Linking).

Yoksan, R.; Matsusaki, M.; Akashi, M. & Chirachanchai, S. (2004). Controlled hydrophobic/hydrophilic chitosan: colloidal phenomena and nanosphere formation. *Colloid & Polymer Science*, 282, pp. (337-342).

You, J. O. & Peng, C. A. (2004). Calcium-alginate nanoparticles formed by reverse microemulsion as gene carriers. *Macromolecular Symposia*, 219, pp. (147-153).

Yuan, Z. (2007). Study on the synthesis and catalyst oxidation properties of chitosan bound nickel(II) complexes. *Journal of Agricultural and Food Chemistry*, 21, 5, pp. (22-24).

Zhang, J.; Chen, X. G.; Li, Y. Y. & Liu, C. S. (2007). Self-assembled nanoparticles based on hydrophobically modified chitosan as carriers for doxorubicin. *Nanomedicine*, 3, 4, pp. (258-265), 1549-9642 (Electronic) 1549-9634 (Linking).

Zhi, J.; Wang, Y. J. & Luo, G. S. (2005). Adsorption of diuretic furosemide onto chitosan nanoparticles prepared with a water-in-oil nanoemulsion system. *Reactive and Functional Polymers*, 249-257, pp.),

Zhou, B.; McGary, C. T.; Weigel, J. A.; Saxena, A. & Weigel, P. H. (2003). Purification and molecular identification of the human hyaluronan receptor for endocytosis. *Glycobiology*, 13, 5, pp. (339-349), 0959-6658 (Print) 0959-6658 (Linking).

9

Aptamer-Nanoparticle Bioconjugates for Drug Delivery

Veli C. Özalp and Thomas Schäfer
University of the Basque Country
Spain

1. Introduction

Drug delivery systems traditionally relied on passive diffusion mechanisms for targeting and releasing of therapeutically active molecules. The major problems associated with traditional delivery are poor specificity and dose-limited toxicity. Nanoparticles have found applicability in the development of novel drug delivery systems by easily overcoming toxicity problem. However, specificity of delivery has remained as a challenge. Developments in the methods of reaching to targeted tissue have lead to new and improved drug delivery platforms. Recently, active targeting has been incorporated by cell specific ligands such as antibodies, lectins, growth factor receptors. More recently, aptamers gained popularity in construction of novel actively targeted drug delivery systems (Ozalp *et al.*, 2011). Considerable proportions of aptamer-based delivery systems have been incorporated to a variety of nanomaterials in order to improve their specific targeting properties (Chen *et al.*, 2011; Zhou *et al.*, 2011).

A successful therapy starts with diagnosis and then continues with application of right dose of therapeutic molecules to the site of diseased tissues or cells. Specific detection of cancer cells is essential for early diagnosis. Cancer cell types are diverse even within the same cancer category. Fast and accurate diagnosis requires multiple specific probes which can profile more than one type of cancer cell at a given time on the same sample. In that respect, cell-based aptamer selection can supply multiple biorecognition probe molecules for any desired number of target cell type. Early diagnosis of cancer relies largely on the sensitivity of detection methods. Nanoparticle-bioconjugates has been designed to develop high sensitivity sensors. In this chapter, we will summarize the recent developments in selecting cell specific aptamer selection and development of aptamer-targeted nanoparticles for medical applications.

2. Selection of cell specific aptamers

Aptamers are increasingly recognized as the future affinity molecules that can readily incorporated in nanodevices for medical applications. The recent progress in cell-specific selection methodology reconfirms this premise by supplying highly specific and high affinity aptamers for any type of diseased cells. Aptamers have generally been shown to have affinities and specificity that are comparable to antibodies. They have the advantage of high stability at a variety of extreme physical conditions such as high temperature, high salt and ionic conditions with reversible denaturation properties (Deng et al., 2001; Liss et al., 2002). Cell-specific apramers have been obtained by following two types of Selex procedure; by using the traditional approach i) against purified protein targets or by new approaches ii) against whole cells. Using purified proteins as targets has the advantage of technical control over assessments of enrichment during Selex procedure. However, the second approach has a clear advantage when the marker target is unknown.

2.1 Aptamers selected for cell-specific proteins

Aptamers are selected through a combinatorial procedure to identify short nucleic acid sequences that recognize specific targets. The procedure is based on a repeated cycle of affinity and separation. First target molecules are brought together with a library of nucleic acids. The usual number of members in such libraries is 10^{14}-10^{18} random sequences. The library members will bind to target molecules if there is any recognition. The next step is the separation of bound and unbound library members through various chromatography techniques. It is commonly required to a cycle of 8 to 20 to enrich enough aptamer sequences which can be identified through sequencing since unspecific or common binder sequences will be favoured under the binding conditions. Detailed procedures on SELEX methods can be found elsewhere (Mayer, 2009). The simplest and direct way to obtain cell-specific aptamers is to isolate a marker protein that exists on the surface of target cell type and to use that protein as a target molecule in SELEX procedure. Many cell-specific aptamers have been selected using this strategy (Janas, 2011). The critical parameters for a successful selection are to identify cell-specific markers and to obtain the marker as pure starting material. Furthermore, the target that will be used in selection should be processed in such way that cell surface exposed regions of the molecule can be used during selection procedure. Several aptamers against cancer-related cells have been selected by using prufied proteins, including platelete-drived growth factor (PDGF) (Green *et al.*, 1996), vascular endothelial growth factor (VEGF) (Ruckman *et al.*, 1998), epidermal growth factor receptor 3 (HER3) (Chen *et al.*, 2003), transcription factor NFkB (Cassiday *et al.*, 2001), tenascin-C (Hicke *et al.*, 2001), prostate-specific membrane antigen (PSMA) (Lupold *et al.*, 2002), and ErbB in breast cancer (Kim *et al.*, 2011).

The aptamer against ErbB is a typical selex procedure for selecting cell specific aptamers by using purified surface proteins. Epidermal growth factor receptor family (ErbB family) of receptor tyrosine kinases plays major roles in formation and progression of human cancers. Under normal physiological, ErbB receptors play important roles in some basic cellular processes such as proliferation, differentiation, motility and apoptosis. Upon ligand binding, ErbB receptors form homo- and heterodimers leading to the activation of their tyrosine kinase domain and subsequent autocatalytic phosphorylation of specific tyrosine residues in the cytoplasmic region. Amplification and overexpression of ErbB tyrosine kinases is associated with many types of solid tumors including breast, ovarian, head and neck, gastric, bladder, colorectal, salivary, renal, prostate and lung cancers (Salomon et al., 1995). ErbB receptors consist of an extracellular region, a single transmembrane region and a cytoplasmic tyrosine kinase domain (Cho et al., 2002). Thus, ErbB has been a target in cancer treatment. 2'-fluorine-modified RNA aptamers for the extracellular region of the receptor has been recently selected for their recognition of breast cancer cells (Kim and Jeong, 2011).

2.2 Aptamers selected for whole cells (cell-Selex)

Cell specific aptamers can also be selected by using whole cells as target. Instead of selecting for a purified and defined target, aptamers can be selected by subtracting the background binding sequences after a selection cycle against the target. Whole-cell approach has been increasingly adapted more commonly for selecting new aptamers. Selection using live cell target means that the cell surface proteins will be in their native conformations by the right exposure as it would be in a real application. A panel of aptamers were selected for leukemia, myeloid leukemia, liver cancer, and lung cancer by using cell-Selex procedures. The aptamers selected by cell-selex resulted in high affinity specific aptamers, but more selection rounds compared to traditional selex were required (Fang et al., 2010). Figure 1 is a schematic representation of the cell-Selex approach. Complex targets such as tumor cells have been demonstrated to be compatible with the cell-Selex approach. The cell-selex procedure is easy to implement and faster than traditional Selex based methodologies. Most cell targets like cancer cells usually lack a clear marker protein which is highly specific for the type of application. Another important advantage of cell-selex is that the target

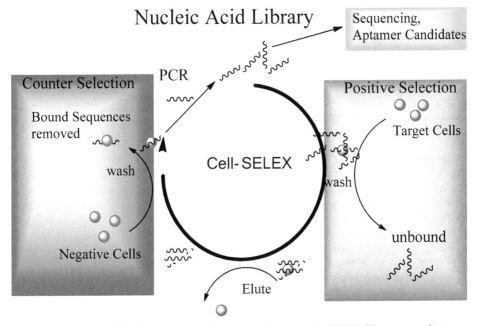

Fig. 1. Steps in a typical cell-Selex procedure. A single-stanraded DNA library was first incubated with target cells. Nano-binding sequences were washed away and the sequences bound on target cells were eluted and isolated. The recovered sequences were incubated with control cells (negative cells) to eliminate the sequences which bind to common epitopes present on both target and control cell surfaces. This step ensures the enrichment of target cell specific aptamers. The sequences specifically binding to target cells were then amplified to obtain the pool for the next round of selex. The enrichment of target cell binding sequences were evaluated using a binding assay and the sequences from last round were cloned and sequenced.

molecules are at their optimum exposed form in vivo. A traditional selex procedure using purified marker protein can select aptamers recognizing any subregion of the protein, which can be cytoplasmic site, transmembrane regions or the regions of target proteins can be hidden when they are sitting in membrane. Therefore, cell-Selex allows a comparative strategy to identify differences at molecular levels. In this section, cell-Selex will be explained by using specific examples. However, recent reviews on cell-selex can be referred for a more detailed information (Guo et al., 2008; Fang and Tan, 2010).

2.2.1 Aptamers targeting tumor cells

Cancer cell types have been commonly a target for cell-Selex procedures in multiple example. An aptamer against RET receptors was selected by using cell line engineered for expressing RET tyrosine kinase receptors and background subtraction against non-expressing cell line (Cerchia et al., 2005). Another advantage is that cell-selex can be carried out without knowing identity of surface proteins in a target cell. Moreover, it is possible to select a panel of aptamers recognizing surface proteins which does not exist in counter selection target. Cultured T cell acute lymphoblastic leucemia (CCRF-CEM) cells was used as target and human Burkitt's lymphoma B-cells for counter-selex (Shangguan et al., 2006). Flow cytometry has been used to monitor the enrichment of aptamers after each cycle and several high affinity aptamers (Kd in nanomolar ranges) have been obtained with specific recognition properties for CCRF-CEM cells in a mixture of cells. The same procedure was also adapted to adherent liver tumor cells by first detaching the cells with a non-enzymatic solution (Shangguan et al., 2008). Six aptamers for Toledo and CCRF-CEM cell lines were tested for their ability to differentiate between real samples from leucemia patients (Shangguan et al., 2007). The results showed the ability of these aptamers to specifically detect molecular differences between patients.

The identity of target on the surface of cell to which an aptamer are selected using cell-selex procedures can be determined by using post-Selex procedures. TD05 is an aptamer selected against Ramos cells by using cell-Selex methodology. The aptamer was determined to bind to Ramos cell through immunoglobulin heavy mu chain which is the heavy portion of the IgM protein. B-cell receptor complex are expressed in Ramos cells and IgM is the major component of this complex. The identification of identity of aptamer target was achieved by chemically modifying the aptamers with a photoactive uracil derivative for covalent binding of aptamer with their targets. Once aptamers are bound to targets covalently, they are isolated through magnetic extraction and the target protein was determined through mass spectroscopy techniques (Mallikaratchy et al., 2007).

2.2.2 Aptamers targeting stem cells

Aptamers can be used for recognition of any given cell population. The aptamer-based capture of circulating cells has been proposed for applications in regenerative medicine and tissue engineering. Mesechymal stem cells from whole bone marrow were used to generate aptamers for isolation purposes by local injection methods (Guo et al., 2006; Schaefer et al., 2007). The aptamers recognizing osteoblasts were used to enrich them quickly and efficiently on titanium surfaces (Guo et al., 2007). Stents coated with capturing agents for endothelial proginator cells (EPC) from circulating blood stream have been proposed to achieve fast enthelialization of the implanted stents and hence acceleration of the healing process of vascular tissue (Aoki *et al.*, 2005; Szmitko *et al.*, 2006). The capturing techniques was demonstrated with antibodies, peptides and aptamers (Wendel *et al.*, 2010). In one example, aptamers for high affinity for circulating porcine EPC was selected for EPC fishing (Hoffmann *et al.*, 2008). The application of EPC capture stents is a promising therapeutic approach to rapidly create an in vivo endothelialization of artificial surfaces. However, antibody based application has the problems due to the heterogeneity of EPC populations and the poor definition of EPC surface molecules. Therefore, the success of this approach depends on a better characterization and understanding of EPC biology and aptamers can offer the needed biorecognition molecules for the desired cell population for using in fishing approach as a successful therapy. Aptamer-based biological coatings which mimic the self healing potential of the patients' stem cell pool for vascular wall regeneration offer exciting new therapy strategies (Wendel, Avci-Adali et al., 2010). Aptamers recognizing EPC cells have also been demonstrated as one of the cell sources useful in cardiac stem cell therapy because they can contribute to neoangiogenesis and regenerate infarcted myocardium. This new therapeutic method based on aptamers was demonstrated that the transplantation of aptamer-isolated EPCs after myocardial infarction improves angiogenesis (Sobolewska *et al.*, 2010).

2.2.3 Aptamers targeting microbial cells

Another group of cell specific aptamers has been selected for microbial cells. Aptamer-based detection of pathogens is an effective diagnostic tool in food industry and medicine. Aptamers targeting viruses, protozoa, or bacteria have been selected for biosensor development (Karkkainen *et al.*, 2011). A subtype of human influenza virus has been detected by aptamers demonstrating the feasibility of using aptamers in detecting pathogenic viruses in contaminated environmental and food matrixes (Gopinath *et al.*, 2006). African trypanosomes are a specific class of protozoan organisms responsible for the parasitic disease sleeping sickness. Aptamers specific for the African trypanosomes were reported for a surface protein located within the flagellar pocket of the parasite (Hoffmann, Paul et al., 2008) and for a surface glycoprotein of the Trypnosoma brucei subsp. Brucei (Lorger *et al.*, 2003). Aptamers can target against some of the surface proteins of micro-organism, and the growth of the bacteria can be inhibited. The only example of therapeutic aptamer with antimicrobial properties has been demonstrated with virulent *Mycobacterium tuberculosis* (Chen *et al.*, 2007).

3. Drug delivering nanoparticles

Nanoparticles have many potential applications in medicine ranging from bioimaging to drug delivery. They have been synthesized from a variety of materials that could be degradable or stable, depending on the type of application. Although developing new types of materials is an exciting research area for future applications, pharmaceutical applications mostly rely on materials with history in human use to reduce the risks involved (Grama et al., 2011). Exposing malignant tumour cells that migrate to adjacent tissues or circulate in bloodstream is critical for early detection and effective therapy. Passive delivery refers to nanoparticles loaded with drugs are administered into body without any specific targeting. The improvement over conventional methods has been obtained by adjusting the release properties of the nanoparticles by selecting special materials during the synthesis of particles. On the other hand, increasing surface area to volume ratio for a delivery vehicle has huge benefits in improving solubility of biologically available drugs. Simple entrapment of bioactives inside nanoparticles has been shown to result in significantly higher bioavailability compared to their conventional administration methods (Mittal et al., 2007). Designed formulations for adjusting release rate are pursued with success in multiple examples (Cozar-Bernal et al., 2011; Tsai et al., 2011). For example, PLGA nanoparticles are biocompatible, synthesized at different grades, and safe (used for long time with humans). The release, degradation and elimination rates of PLGA nanopartiles can be tailored to desired parameters. On the other hand, nanoparticles can be functionalized for targetd delivery. Active delivery refers to incorporating a specific targeting mechanism in drug-loaded nanoparticle delivery vehicle.

Aptamers have been considered as potentially interesting targeting molecules for nanoparticles. The main reason is that conjugation of aptamers to other molecules or nanoparticles is a straight-forward procedure and the attachment does not change the affinity or specificity properties of aptamers. Early diagnosis can be achieved by systems of aptamer-nanoparticle conjugates because high affinity aptamers can be obtained for tumour cells. There have been many approaches in for detection of tumor cells that is based on mechanical forces, immunohistochemistry, magnetic cell sorting, and flow cytometry. The detection based on affinity interactions is expected to yield higher efficiency and sensitivity. Antibody based affinity sensors are often found to be high levels of cross-reactivity. There are also technical challenges in crosslinking antibodies on the surface of nanodevices (Wan et al., 2010). Thus, there is an increasing recognition for aptamers as promising utility in cancer diagnosis.

Three critical steps should be considered in designing an efficient aptamer-targeted nanoparticle drug therapy; targeting, internalization and degradation of the targeted cells by release of loaded drugs (Fig. 2).

4. Aptamer-based targeting of nanoparticles

Aptamers are biorecognition molecules with multiple properties that make them attractive targeting elements. Aptamers can bind to their targets with high affinity and specificity, comparable to antibodies. The *in vitro* selection methodology is the major advantage ensuring availability for any desired application. Moreover, nucleic acids have unique properties that are frequently exploited in construction of nanomachines . A combination of all of the above-mentioned properties makes aptamers a desirable targeting agent for novel nanoparticle-based drug delivery systems. Recent developments in cell based Selex technology have raised expectations for medical applications required targeting of whole living cells. Two major medical applications of aptamer-based targeting of nanoparticles are bioimaging and drug delivery.

4.1 Bioimaging applications

In vivo imaging is a valuable tool in clinical diagnostics and critical to develop ultrasensitive methods (Soontornworajit et al., 2011). Many techniques of imaging from whole organism (e.g. NMR) to specific molecular imaging (e.g. Fluorescence) are available and they are envisioned to be on the focus for developing cancer diagnosis, drug delivery, guided stem

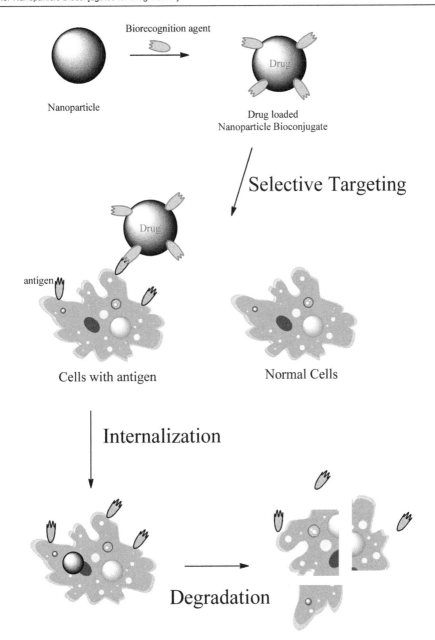

Fig. 2. A scheme of targeted Nanoparticle drug delivery. Drugs are loaded inside the nanoparticles that are conjugated to antigen-binding recognition element which bind specifically to antigen expressing cells. The nanoparticles are rapidly internalized once bound on the surface of diseased cells. The drugs encapsulated in the nanoparticles are released inside the cells and destroys the integrity of cells.

cell therapies, imaging of gene expression to monitor disease development, gene therapy, image guided surgery (Hahn et al., 2011). Nanoparticles for bioimaging should fullfill several requirements; being dispersable and stable in vivo, low non-specific binding, and high selectivity for their targets. There are four types of nanoparticles that are commonly used in conjugation with aptamers for biological imaging applications; 1) gold nanoparticles, 2) quantum dots, 3) silica nanoparticles and 4) magnetic nanoparticles.

A nonoparticle imaging and drug delivery system has been developed by conjugating A10 aptamers to quantum dots (Bagalkot et al., 2007). Fig. 3 shows a schematic representation of aptamer-quantum dot bioconjugate for drug delivery and imaging of delivered drugs as reported. The chemotherapeutic drug doxorubicin is an nucleic acid intercalating molecules. A bi-FRET system was prepared by attaching PSMA aptamers on the surface of quantum dot nanoparticles and doxorubicin was immobilized on the aptamer molecules by intercalation. Doxorubicin is a fluorescent molecule, but quenched by gold and nucleic acids. Therefore, the system targets the doxorubicin in non-fluorescent form to the surface of prostate cell. When the drug carrier system is internalized by the cells, doxorubicin is free from aptamer and quantum dots and become fluorescent, signalling the delivery of drug into cancer cells. Another example of quantum dot-aptamer conjugates in bioimaging was to identify and isolate subpopulations of tumor cells with high specificity. In another example, quantum dots were conjugated to liver hepatoma cell line aptamers through a sterptavidin-biotin binding. The quantum-dor-aptamer complexes were shown to specifically recognize MEAR cells specifically comapered to HeLa cells and BNL cells (Zhang et al., 2010).

Aptamer conjugated magnetic nanoparticles are proposed for in vivo isolation of specific cells and also for magnetic resonance imaging (MRI) techniques. Aptamer-magnetic nanoparticles were used to isolate mesenchymal stem cells and endothelian progenitor cells for isolating "ready to transplant" cells as promising methods (Schaefer, Wiskirchen et al., 2007; Sobolewska, Avci-Adali et al., 2010). Prostate cancer specific A10 aptamers conjugated to magnetic nanoparticles were used as magnetic resonance contrast agents for lower toxicity and stronger binding properties (Wang et al., 2008).

Silica nanoparticles can be readily loaded with fluorescent molecules and highly fluorescent, stable nanoparticles can be obtained. They are stable, hydrophilic, biocompatible, chemically inert. Another advantage is that organic dyes can be encapsulated in the silica with advantages of less photobleaching. Furthermore, aptamers can be easily attached to the surface of silica. Up to 100 fold increase in sensitivity has been obtained in flow cytometry applications with dye-doped silica nanopartiles (Estevez et al., 2009). In an effort to develop sensitive monitoring of multiple cancer cell lines, three different aptamers were conjugated to fluorescent doped silica nanoparticles (Chen et al., 2009). Human acute lymphoblastic leukemia cells specific aptamer, Ramos cells specific aptamer, and Toledo cells specific aptamer were used in developing multiple assays. Nanomaterials are often used for multivalent ligand binding purposes to increase the signal and sometimes enhance the binding affinity. For example, gold-silver nanorods were used as a platform for multiple aptamer immobilization. Up to 80 aptamers could be attached to rods for obtaining 26-fold higher affinity and 300-fold higher fluorescence (Huang et al., 2008).

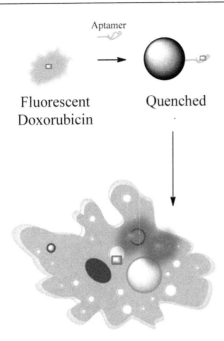

Fig. 3. Represantation of nanoparticle-aptamer based bi-FRET system for drug delivery and bioimaging. Doxorubicin is a fluophore and intercalates into nucleic acids. The fluorescence of doxorubicin is quenched by both gold and the aptamers. The quantum dot-aptamer-doxorubicin complex is targeted to tumor cells and internalized, and doxorubicin is released inside the cell, removing the quenching. The system targets the drug into diseased cells and the release of the drug can be monitored once it is delivered into desired site.

4.2 Aptamer-targeted drug delivery in nanoparticles

Aptamer-targeted delivery of drugs by using nanoparticles has similar advantages to bioimaging applications. The nanoparticles are specifically targeted to cell-types that are required for therapeutic molecules to act on. The drug delivery through nanoparticle encapsulation provides a vehicle for controlling the precise level and location of drugs in the body, reducing unwanted toxicity to healthy cells, lowering the doses needed. For example, drug encapsulated nanoparticles with a controlled release system can be developed and targeting to diseased cells can maximize the therapeutic efficacy and reduce undesired side effects (Langer, 2001).

Proof-of-concept nanoparticle-based vehicles have been developed for targeted drug delivery to prostate cancer cell surfaces with high specificity and efficiency. Once bound to tumor antigens, nanoparticles will be internalized through endocytosis. The toxic cargo of the nanoparticles will be released inside the tumour cells. Targeted delivery achieves a safe vehicle to destroy selectively the cancer cells. During the research of similar proof-of-concept research, teh nanoparticles commonly loaded with fleorescent molecules to study the pathways, the success of targeting and release rate of cargo.

4.2.1 Prostate cancer as model system

Aptamers potentially can direct nanoparticles to cancer cell antigens present on the surface of tumour cells. Therapeutic nanoparticles are specifically designed to encapsulate drugs and release the drugs in preregulated and predetermined way. Prostate cancer has been the first model disease for aptamer mediated nanoparticle targeting. Prostate specific membrane antigen protein (PMSA) is a well-known marker antigen which is over expressed on tumorigenic prostatic epithelial cells. Two RNA aptamer binding to extracellular portions of PSMA has been selected and shown to adhere selectively to PMSA positive cells (Lupold, Hicke et al., 2002). Later, an in vitro analysis device was developed in order to study the specific targeting properties of PMSA aptamer-nanoparticle conjugates (Farokhzad et al., 2005). The device was a model of prostate cancer targeting using a microfluidic cell under physiological fluid conditions similar to microvasculature. In this experiments, the microchannels were seeded with protate cancer epithelial cells expressing PMSA or non-expressing cells lines. Pegylated poly(lactic acid) (PLA) particles conjugated to A10 PMSA aptamer were tested for various experimental parameters mimicking the intraveneous administration of nanoparticles. The results confirmed that the particles were rapidly internalized into PMSA positive cells. The optimized aptamer-nanoparticle vehicle loaded with the chemotherapeutic drug docetaxel was later shown to reduce tumors upon injection into mice (Farokhzad et al., 2006; Cheng et al., 2007). This example was the first targeted delivery using nanoparticle-aptamer bioconjugates. The drug delivery was achieved to be highly specific for PMSA positive cells under *in vivo* contions. The same system of nanoparticle-aptamer bioconjugate was shown to deliver cisplatin drug to prostate cancer cells by improving simultaneously both tolerance and efficacy in mice (Kolishetti et al., 2010).

Nanomaterials for drug delivery are required to be stable and biocompatible. Current technology can supply materials that are capable of partially fulfilling these properties. Thus, combinations of surface coatings with complementary properties are proposed for developing better functioning delivery systems. In one example, gold nanorods were encapsulated within thin and uniform layer of silica shell for enhanced stability and and a layer of PEG was used for biocompatibility. The composite nanoparticles were functionalized with PSMA aptamers and tested for their ability to target the nanopartciles to prostate cancer cells (Hu et al., 2011).

4.2.2 Other examples

Vasculature targeted delivery is an attractive strategy due to the phenotypic changes on the endothelial cell surface associated with pathological conditions such as inflammation and angiogenesis. E-selectin is expressed in inflammed vaculator in advanced tumors and it can be used as marker for pathological vasculator. A high affinity thio-backbone modified aptamers against E-selectin were recently selected by using recombinant-expressed protein (Mann et al., 2010). E-selection aptamers were incorporated in liposomes, the targeting efficiency and pharmokinetic properties of delivery were investigated (Mann et al., 2011). CCRF-CEM recognizing aptamers (sgc8c) were conjugated to doxorubucin-doped silica nanoparticles and delivered leukemia cells with high specificity and efficiency (He et al., 2011).

4.2.3 Aptamer-based gating systems

The efficacy of drug therapy is determined by two parameters; efficient targeting of drug to the diseased site and the controlled release of the drug. The potential of aptamer-based targeting have been demonstrated in multiple examples as presented in above sections in order to improve specific delivery properties of drug carrying nanoparticles. The bioactive molecules in the nanoparticles were released with traditional mechanism such as internalization, endocytosis and degradation of nanomaterial once the nanoparticles attach to the surface of diseased cell surface. However, aptamers can be also exploited in controlling the release of drugs from nanoparticles. Two recent reports indicate to future potential of aptamers as targeting and gating agents in nanoparticle-based drug delivery. A snap-top type molecular gate using aptamer-gold nanoparticle complex to control the release of cargo of silica nanoparticles was shown to respond to aptamer target for stimuli-responsive release (Zhu *et al.*, 2011). In another report, switchable control of the gating in nanocarriers through aptamers was shown to control the release rate of active molecules in response to taptamer target stimuli (Ozalp *et al.*, 2011).

5. Conclusion

Nanoparticle-aptamer bioconjugates provide exciting prospects in medical nanotechnology for future disease treatments. The advancements in nanomaterial field together with cell-Selex procedures offer the controlled release polymer systems conjugated to aptamers tweaked to the any target diseased cell. Therefore, it is possible to produce a diverse range of specific and selective nanoparticle-aptamer bioconjugates. Drug delivery vehicles can improve the therapy of a myriad of important human diseases by using aptamer-nanoparticle bioconjugates. Aptamer-based targeting has already achieved progress in demonstrating the usability nanoparticle-aptamer bioconjugate systems to improve nanomaterial drug delivery vehicles in model animals. Next critical step for realization of potentials of bioconjugated nanoparticles will be clinical realization studies in near future. It is worth to indicate that with aptamers being in principle available for any kind of target (e.g., small metabolites, proteins, lipids, ions), it can be anticipated a straightforward transfer of proof-of-concept results to other applications when developing efficient drug-delivery systems. Finally, aptamer-based molecular gating of nanoparticle cargo can bring new approaches in drug delivery field by combining targeting and controlled release in the same mechanism.

6. Acknowledgment

A starting-grant from the European Research Council (No. 209842; "MATRIX") is gratefully acknowledged.

7. References

Aoki, J., P. W. Serruys, H. van Beusekom, A. T. Ong, E. P. McFadden, G. Sianos, W. J. van der Giessen, E. Regar, P. J. de Feyter, H. R. Davis, S. Rowland and M. J. Kutryk (2005). "Endothelial progenitor cell capture by stents coated with antibody against

CD34: the HEALING-FIM (Healthy Endothelial Accelerated Lining Inhibits Neointimal Growth-First In Man) Registry. J Am Coll Cardiol 45(10): 1574-9.

Bagalkot, V., L. Zhang, E. Levy-Nissenbaum, S. Jon, P. W. Kantoff, R. Langer and O. C. Farokhzad (2007). "Quantum dot - Aptamer conjugates for synchronous cancer imaging, therapy, and sensing of drug delivery based on Bi-fluorescence resonance energy transfer. Nano Letters 7(10): 3065-3070.

Cassiday, L. A. and L. J. Maher, 3rd (2001). "In vivo recognition of an RNA aptamer by its transcription factor target. Biochemistry 40(8): 2433-8.

Cerchia, L., F. Duconge, C. Pestourie, J. Boulay, Y. Aissouni, K. Gombert, B. Tavitian, V. de Franciscis and D. Libri (2005). "Neutralizing aptamers from whole-cell SELEX inhibit the RET receptor tyrosine kinase. Plos Biology 3(4): 697-704.

Cozar-Bernal, M. J., M. A. Holgado, J. L. Arias, I. Munoz-Rubio, L. Martin-Banderas, J. Alvarez-Fuentes and M. Fernandez-Arevalo (2011). "Insulin-loaded PLGA microparticles: flow focusing versus double emulsion/solvent evaporation. Journal of Microencapsulation 28(5): 430-441.

Chen, C. H. B., G. A. Chernis, V. Q. Hoang and R. Landgraf (2003). "Inhibition of heregulin signaling by an aptamer that preferentially binds to the oligomeric form of human epidermal growth factor receptor-3. Proceedings of the National Academy of Sciences of the United States of America 100(16): 9226-9231.

Chen, F., J. Zhou, F. Luo, A. B. Mohammed and X. L. Zhang (2007). "Aptamer from whole-bacterium SELEX as new therapeutic reagent against virulent Mycobacterium tuberculosis. Biochem Biophys Res Commun 357(3): 743-8.

Chen, T., M. I. Shukoor, Y. Chen, Q. A. Yuan, Z. Zhu, Z. L. Zhao, B. Gulbakan and W. H. Tan (2011). "Aptamer-conjugated nanomaterials for bioanalysis and biotechnology applications. Nanoscale 3(2): 546-556.

Chen, X., M. C. Estevez, Z. Zhu, Y.-F. Huang, Y. Chen, L. Wang and W. Tan (2009). "Using Aptamer-Conjugated Fluorescence Resonance Energy Transfer Nanoparticles for Multiplexed Cancer Cell Monitoring. Analytical Chemistry 81(16): 7009-7014.

Cheng, J., B. A. Teply, I. Sherifi, J. Sung, G. Luther, F. X. Gu, E. Levy-Nissenbaum, A. F. Radovic-Moreno, R. Langer and O. C. Farokhzad (2007). "Formulation of functionalized PLGA-PEG nanoparticles for in vivo targeted drug delivery. Biomaterials 28(5): 869-876.

Cho, H. S. and D. J. Leahy (2002). "Structure of the extracellular region of HER3 reveals an interdomain tether. Science 297(5585): 1330-1333.

Deng, Q., I. German, D. Buchanan and R. T. Kennedy (2001). "Retention and separation of adenosine and analogues by affinity chromatography with an aptamer stationary phase. Analytical Chemistry 73(22): 5415-5421.

Estevez, M. C., M. B. O'Donoghue, X. Chen and W. Tan (2009). "Highly Fluorescent Dye-Doped Silica Nanoparticles Increase Flow Cytometry Sensitivity for Cancer Cell Monitoring. Nano Research 2(6): 448-461.

Fang, X. H. and W. H. Tan (2010). "Aptamers Generated from Cell-SELEX for Molecular Medicine: A Chemical Biology Approach. Accounts of Chemical Research 43(1): 48-57.

Farokhzad, O. C., J. J. Cheng, B. A. Teply, I. Sherifi, S. Jon, P. W. Kantoff, J. P. Richie and R. Langer (2006). "Targeted nanoparticle-aptamer bioconjugates for cancer chemotherapy in vivo. Proceedings of the National Academy of Sciences of the United States of America 103(16): 6315-6320.

Farokhzad, O. C., A. Khademhosseini, S. Y. Yon, A. Hermann, J. J. Cheng, C. Chin, A. Kiselyuk, B. Teply, G. Eng and R. Langer (2005). "Microfluidic system for studying nanoparticles and microparticles the interaction of with cells. Analytical Chemistry 77(17): 5453-5459.

Gopinath, S. C., T. S. Misono, K. Kawasaki, T. Mizuno, M. Imai, T. Odagiri and P. K. Kumar (2006). "An RNA aptamer that distinguishes between closely related human influenza viruses and inhibits haemagglutinin-mediated membrane fusion. J Gen Virol 87(Pt 3): 479-87.

Grama, C. N., D. D. Ankola and M. N. V. R. Kumar (2011). "Poly(lactide-co-glycolide) nanoparticles for peroral delivery of bioactives. Current Opinion in Colloid & Interface Science 16(3): 238-245.

Green, L. S., D. Jellinek, R. Jenison, A. Ostman, C. H. Heldin and N. Janjic (1996). "Inhibitory DNA ligands to platelet-derived growth factor B-chain. Biochemistry 35(45): 14413-24.

Guo, K.-T., G. Ziemer, A. Paul and H. Wendel (2008). "CELL-SELEX: Novel Perspectives of Aptamer-Based Therapeutics. International Journal of Molecular Sciences 9(4): 668-678.

Guo, K. T., R. SchAfer, A. Paul, A. Gerber, G. Ziemer and H. P. Wendel (2006). "A new technique for the isolation and surface immobilization of mesenchymal stem cells from whole bone marrow using high-specific DNA aptamers. Stem Cells 24(10): 2220-31.

Guo, K. T., D. Scharnweber, B. Schwenzer, G. Ziemer and H. P. Wendel (2007). "The effect of electrochemical functionalization of Ti-alloy surfaces by aptamer-based capture molecules on cell adhesion. Biomaterials 28(3): 468-74.

Hahn, M. A., A. K. Singh, P. Sharma, S. C. Brown and B. M. Moudgil (2011). "Nanoparticles as contrast agents for in-vivo bioimaging: current status and future perspectives. Analytical and Bioanalytical Chemistry 399(1): 3-27.

He, X. X., L. Hai, J. Su, K. M. Wang and X. Wu (2011). "One-pot synthesis of sustained-released doxorubicin silica nanoparticles for aptamer targeted delivery to tumor cells. Nanoscale 3(7): 2936-2942.

Hicke, B. J., C. Marion, Y. F. Chang, T. Gould, C. K. Lynott, D. Parma, P. G. Schmidt and S. Warren (2001). "Tenascin-C aptamers are generated using tumor cells and purified protein. J Biol Chem 276(52): 48644-54.

Hoffmann, J., A. Paul, M. Harwardt, J. Groll, T. Reeswinkel, D. Klee, M. Moeller, H. Fischer, T. Walker, T. Greiner, G. Ziemer and H. P. Wendel (2008). "Immobilized DNA aptamers used as potent attractors for porcine endothelial precursor cells. J Biomed Mater Res A 84(3): 614-21.

Hu, X. and X. Gao (2011). "Multilayer coating of gold nanorods for combined stability and biocompatibility. Physical Chemistry Chemical Physics 13(21): 10028-10035.

Huang, Y.-F., H.-T. Chang and W. Tan (2008). "Cancer cell targeting using multiple aptamers conjugated on nanorods. Analytical Chemistry 80(3): 567-572.

Janas, T. (2011). "The Selection of Aptamers Specific for Membrane Molecular Targets. Cellular & Molecular Biology Letters 16(1): 25-39.

Karkkainen, R. M., M. R. Drasbek, I. McDowall, C. J. Smith, N. W. G. Young and G. A. Bonwick (2011). "Aptamers for safety and quality assurance in the food industry: detection of pathogens. International Journal of Food Science and Technology 46(3): 445-454.

Kim, M. Y. and S. Jeong (2011). "In Vitro Selection of RNA Aptamer and Specific Targeting of ErbB2 in Breast Cancer Cells. Oligonucleotides 21(3): 173-178.

Kolishetti, N., S. Dhar, P. M. Valencia, L. Q. Lin, R. Karnik, S. J. Lippard, R. Langer and O. C. Farokhzad (2010). "Engineering of self-assembled nanoparticle platform for precisely controlled combination drug therapy. Proceedings of the National Academy of Sciences of the United States of America 107(42): 17939-17944.

Langer, R. (2001). "Perspectives: Drug delivery - Drugs on target. Science 293(5527): 58-59.

Liss, M., B. Petersen, H. Wolf and E. Prohaska (2002). "An aptamer-based quartz crystal protein biosensor. Analytical Chemistry 74(17): 4488-4495.

Lorger, M., M. Engstler, M. Homann and H. U. Goringer (2003). "Targeting the variable surface of African trypanosomes with variant surface glycoprotein-specific, serum-stable RNA aptamers. Eukaryot Cell 2(1): 84-94.

Lupold, S. E., B. J. Hicke, Y. Lin and D. S. Coffey (2002). "Identification and characterization of nuclease-stabilized RNA molecules that bind human prostate cancer cells via the prostate-specific membrane antigen. Cancer Research 62(14): 4029-4033.

Mallikaratchy, P., Z. Tang, S. Kwame, L. Meng, D. Shangguan and W. Tan (2007). "Aptamer directly evolved from live cells recognizes membrane bound immunoglobin heavy mu chain in Burkitt's lymphoma cells. Molecular & Cellular Proteomics 6(12): 2230-2238.

Mann, A. P., R. C. Bhavane, A. Somasunderam, B. L. Montalvo-Ortiz, K. B. Ghaghada, D. Volk, R. Nieves-Alicea, K. S. Suh, M. Ferrari, A. Annapragada, D. G. Gorenstein and T. Tanaka (2011). Thioaptamer Conjugated Liposomes for Tumor Vasculature Targeting.

Mann, A. P., A. Somasunderam, R. Nieves-Alicea, X. Li, A. Hu, A. K. Sood, M. Ferrari, D. G. Gorenstein and T. Tanaka (2010). "Identification of Thioaptamer Ligand against E-Selectin: Potential Application for Inflamed Vasculature Targeting. Plos One 5(9).

Mayer, G. (2009). Nucleic Acid and Peptide Aptamers, Humana Press.

Mittal, G., D. K. Sahana, V. Bhardwaj and M. N. V. R. Kumar (2007). "Estradiol loaded PLGA nanoparticles for oral administration: Effect of polymer molecular weight and copolymer composition on release behavior in vitro and in vivo. Journal of Controlled Release 119(1): 77-85.

Ozalp, V. C., F. Eyidogan and H. A. Oktem (2011). "Aptamer-Gated Nanoparticles for Smart Drug Delivery. Pharmaceuticals 4(8): 1137-1157.

Ozalp, V. C. and T. Schafer (2011). "Aptamer-Based Switchable Nanovalves for Stimuli-Responsive Drug Delivery. Chemistry - A European Journal: DOI: 10.1002/chem.201101403.

Ruckman, J., L. S. Green, J. Beeson, S. Waugh, W. L. Gillette, D. D. Henninger, L. Claesson-Welsh and N. Janjic (1998). "2'-fluoropyrimidine RNA-based aptamers to the 165-amino acid form of vascular endothelial growth factor (VEGF(165)) - Inhibition of receptor binding and VEGF-induced vascular permeability through interactions requiring the exon 7-encoded domain. Journal of Biological Chemistry 273(32): 20556-20567.

Salomon, D. S., R. Brandt, F. Ciardiello and N. Normanno (1995). "Epidermal Growth Factor-Related Peptides and Their Receptors in Human Malignancies. Critical Reviews in Oncology/Hematology 19(3): 183-232.

Schaefer, R., J. Wiskirchen, K. Guo, B. Neumann, R. Kehlbach, J. Pintaske, V. Voth, T. Walker, A. M. Scheule, T. O. Greiner, U. Hermanutz-Klein, C. D. Claussen, H. Northoff, G. Ziemer and H. P. Wendel (2007). "Aptamer-based isolation and subsequent imaging of mesenchymal stem cells in ischemic myocard by magnetic resonance imaging. Rofo-Fortschritte Auf Dem Gebiet Der Rontgenstrahlen Und Der Bildgebenden Verfahren 179(10): 1009-1015.

Shangguan, D., Z. C. Cao, Y. Li and W. Tan (2007). "Aptamers evolved from cultured cancer cells reveal molecular differences of cancer cells in patient samples. Clinical Chemistry 53(6): 1153-1155.

Shangguan, D., Y. Li, Z. Tang, Z. C. Cao, H. W. Chen, P. Mallikaratchy, K. Sefah, C. J. Yang and W. Tan (2006). "Aptamers evolved from live cells as effective molecular probes for cancer study. Proceedings of the National Academy of Sciences of the United States of America 103(32): 11838-11843.

Shangguan, D., L. Meng, Z. C. Cao, Z. Xiao, X. Fang, Y. Li, D. Cardona, R. P. Witek, C. Liu and W. Tan (2008). "Identification of liver cancer-specific aptamers using whole live cells. Analytical Chemistry 80(3): 721-728.

Sobolewska, B. W., M. Avci-Adali, B. Neumann, T. O. Greiner, A. Stolz, D. Bail, T. Walker, A. Scheule, G. Ziemer and H. P. Wendel (2010). "A novel method for isolation of endothelial progenitor cells for cardiac stem cell therapy. Kardiochirurgia I Torakochirurgia Polska 7(1): 61-65.

Soontornworajit, B. and Y. Wang (2011). "Nucleic acid aptamers for clinical diagnosis: cell detection and molecular imaging. Analytical and Bioanalytical Chemistry 399(4): 1591-1599.

Szmitko, P. E., M. J. Kutryk, D. J. Stewart, M. H. Strauss and S. Verma (2006). "Endothelial progenitor cell-coated stents under scrutiny. Can J Cardiol 22(13): 1117-9.

Tsai, Y.-M., W.-C. Jan, C.-F. Chien, W.-C. Lee, L.-C. Lin and T.-H. Tsai (2011). "Optimised nano-formulation on the bioavailability of hydrophobic polyphenol, curcumin, in freely-moving rats. Food Chemistry 127(3): 918-925.

Wan, Y. A., Y. T. Kim, N. Li, S. K. Cho, R. Bachoo, A. D. Ellington and S. M. Iqbal (2010). "Surface-Immobilized Aptamers for Cancer Cell Isolation and Microscopic Cytology. Cancer Research 70(22): 9371-9380.

Wang, A. Z., V. Bagalkot, C. C. Vasilliou, F. Gu, F. Alexis, L. Zhang, M. Shaikh, K. Yuet, M. J. Cima, R. Langer, P. W. Kantoff, N. H. Bander, S. Jon and O. C. Farokhzad (2008). "Superparamagnetic iron oxide nanoparticle-aptamer bioconjugates for combined prostate cancer imaging and therapy. Chemmedchem 3(9): 1311-1315.

Wendel, H. P., M. Avci-Adali and G. Ziemer (2010). "Endothelial progenitor cell capture stents - hype or hope? International Journal of Cardiology 145(1): 115-117.

Zhang, J., X. Jia, X.-J. Lv, Y.-L. Deng and H.-Y. Xie (2010). "Fluorescent quantum dot-labeled aptamer bioprobes specifically targeting mouse liver cancer cells. Talanta 81(1-2): 505-509.

Zhou, J. H. and J. J. Rossi (2011). "Cell-Specific Aptamer-Mediated Targeted Drug Delivery. Oligonucleotides 21(1): 1-10.

Zhu, C. L., C. H. Lu, X. Y. Song, H. H. Yang and X. R. Wang (2011). "Bioresponsive Controlled Release Using Mesoporous Silica Nanoparticles Capped with Aptamer-Based Molecular Gate. Journal of the American Chemical Society 133(5): 1278-1281.

10

Microbubble Therapies

Ajay Sud[1,2,3] and Shiva Dindyal[1,2]
[1]Whipps Cross University Hospital, London,
[2]The Royal London Hospital, London,
[3]The Royal Liverpool University Hospital, Liverpool,
United Kingdom

1. Introduction

Microbubble technologies have recently gained acceptance as radiological molecular imaging or perfusion agents. These intravenous ultrasound contrast agents consist of gas-filled microspheres stabilized by a 'shell' layer which interfaces with a surrounding solvent pool. Development in the form of encapsulation with complex biocompatible polymer shell consisting of proteins, lipids and surfactants has improved their stability and half-life significantly. These shell modifications have optimised the mechanical index of microbubbles, making them more resistant to compression rather than expansion.

The relatively inert nature of microbubbles contributes to a favourable safety profile and 'stealth' delivery profile. These indolent properties of microspheres are fundamental to their ability to function as targeted delivery vehicles for potential therapeutic modalities. Their small size (1-8 micrometre) enables unobstructed diffusion between the vascular and capillary compartments. Complete isolation of the active therapeutic components from normal native tissues and liberal exposure to the target disease tissue is the panacea of drug delivery.

Incorporation of lipopolysaccharide ligands adhering to selective cell surface epitopes, to the microbubble surface interface, is the principal method by which these vehicles can be precisely targeted. Selective bioactive material delivery into the target cell significantly improves therapeutic efficacy and can completely negate the adverse side-effects.

Externally applied focused ultrasound directed at the ligand bound microbubbles and target cells complex is the crucial final step ensuring optimal delivery. Application of sonication results in the oscillation followed by destruction of the bubbles and liberation of the bioactive compound; the high intensity focused acoustic energy disrupts the microbubbles. Passage of the active product into the intracellular space is facilitated through cellular sonoporation, at least in part as a result of acoustic cavitation of the target cell's membrane.

It has been demonstrated that microbubble technology can accommodation active constituents, hence be 'drug loaded'. Microbubble delivery technology has been utilised as a feasible drug delivery modality for gene therapies, cytotoxics, drugs and dyes. Co-administration of the microbubbles with the active product, without drug loading, has also been utilised to improve targeted drug delivery.

The premise of the clinically applicable microbubble delivered therapeutic agent delivery is an attractive proposition, but the clinical experience is still limited. Establishing the clinical benefits of microbubble technologies in the field of diagnosis has opened the door to a precise and accurate drug delivery vector. The proliferation in interest surrounding microbubbles has yet to meet its climax; the scientific, translational and clinical challenges of this exciting technology are still to be fully elucidated.

In this chapter we will consider the current understanding of the elucidated theory underlying this technology, and the potential clinical applications with drug and gene delivery. We also endeavour to summaries uncertainties and controversies.

2. The evolution of microbubble

High-intensity focused ultrasound is used routinely for the lithotripsy of the renal calculi. The ablation of the solid tumours and regions of tissue with high-intensity focused ultrasound is a less proven application. Since the inception of the microbubbles their application has grown exponentially with incremental advances in their formulations (Tinkov et al., 2009).

The dawn of microbubble technologies began with an innocuous *in vitro* observation that simple agitated saline enhanced ultrasound signal. This observation was demonstrated in cardiological and aortic ultrasound echocardiogram signal within the aortic root and chambers of the heart (Gramiak & Shah, 1968). The first generation microbubble contrast ultrasound agents were largely only applicable to diagnostic situations where additional opacification of the myocardium and coronary vessels is was necessary. The limited spectrum of clinical application was the result of the inherent characteristics of these first generation microbubbles agents (Kaul. 1997).

The fragile *in vitro* composition of these microbubbles is the result of unstable gas fluid interface and large bubble size. First generation microbubble agents consist of simple air bubbles dispersed in aqueous solution with an absence of a stabilising shell. The large bubble diameter ensured these agents do not to pass through the small capillaries, including the intrapulmonary circulation. This with combined with the exceptionally short life-span beds means that these first generation agents will never become detectable in within the left ventricle after intravenous injection (Kaul. 1997).

Later generations of microbubble agents have increasingly sophisticated colloidal gaseous/aqueous shells, which results in enhanced stability characteristics. The improved modalities of the later generations of microbubbles contrast dyes are now are recognised part imaging adjunct. The smaller size distribution of second generation enabled transpulmonary passage. Following intravenous administration of this generation of microbubble agents with stabilised solvent-gaseous interface they can become detectable within the left heart chambers. This generation is limited by unstable gaseous cores. Despite the composite shells the air cores have a tendency to dissolve in the blood within minutes. Subsequent generations replaced these highly soluble gaseous air nucleuses to overcome this inherent characteristic which limited their half-life (Voci et al, 1994).

Third generation microbubble compounds contained physiologically inert gas cores propagated their stability almost tripling half-life to at least 15 minutes. This generation has extended the application of microbubble agents to any ultrasound amenable organs. Further manipulation of the agents will ensure further broadening of their scope of application (Tinkov et al., 2009).

The trends in the developments in later generations of microbubbles agents have moved away from the diagnostic efficacy and towards therapeutic modalities with drug carrying capacities. Precise formulation and engineering is necessary to ensure the efficacious delivery of specific active products (Pitt et al., 2004).

3. Drug targeting

The essential facet of effective targeted drug delivery is the ability to improve the therapeutic efficacy and minimise the quantity of circulating application of unbind active drug. The implications of this include reduction of the adverse effects and a significant minimisation of the effective required dosage necessary. The second property is paradoxically essential for the premise of the drug-loaded microbubbles to be efficacious because of the shell drug-loading capacity is limited. Because of this, only extremely potent agents with a low toxicity threshold would be amenable to this form of delivery. Poly/oligonucleotide gene therapy, cytotoxic, steroid, antimicrobials and active protein compounds may benefit from microbubble compounds.

It is necessary to expose the systemic circulation to the active compound bound to microbubble to ensure close enough proximity to the target organ. It is undeterminable and unpredictable to accurately the proportion of the active product liberated by focused ultrasound destruction at the desired target organ site. Of the remaining non-utilised microbubble complexes some will be eliminated or be dispersed remotely from the target organ. Well perfused organs, such as the renal (Koike et al 2005) and myocardial (Vannan et al 2002; Shohet et al 2000) systems will witness higher microbubble agent concentrations, making an ultrasound liberation of the active products more effective. Organs with inferior blood perfusion levels may not witness the necessary concentrations to make the targeted ultrasound microbubble destruction efficacious. Hence, the extent of blood supply alone may significantly ameliorate any possible projections of the efficacy of targeted therapeutic microbubbles (Tinkov et al., 2009).

The bodies' innate mechanisms can be harnessed to preferentially channel microbubbles to the desire target tissues. This approach of passive targeting relies on the innate non-specific macro- or microcellular characteristic. Reliance on these non-specific in vitro mechanisms means that these agents will be amenable to the influences of pharmacodynamics forces. Two relevant mechanisms include the particle clearance, aspects of the innate and even adaptive immune systems. An example of such a modality includes the effects of immuno-receptive microbubbles typically include those with phospholipid microbubbles containing phosphatidylserine and those which possess polymer-shells. Phagocytic immune white blood cells consume these antigenic microbubbles and convey these microbubbles to sites of active inflammation. Spontaneous non-macrophage mediate passive accumulation of ligand-containing microbubble agents via pathological or physiological enhancement in endothelial permeability and then retention has been found not to be possible (Bloch et al., 2004).

Active molecular targeting is essentially the incorporation of intentionally designed microbubble manipulations to ensure preferential affinity to the desire target tissues. Cell targeting can be achieved with a series of complementary immunoliposome (Siwak et al., 2002) and nanoparticle ligands (Shi et al., 2007) to specific cell surface receptors attached to the microbubble shells. Isolated or preferential expression of particular diseased cell-specific receptors on their surface provides a means by which targeting discrimination can be achieved. Complex formation between the complementary cell surface and attached microbubble shell protein ensure both selectivity and persistence of the vector to the target.

A major area of research has been into the active molecular targeting of microbubble agents are that they wholly reside within the intravascular compartment and will only come into contact with potential target markers situated on the endothelial wall. The shear force of vessel Blood-flow does significantly weaken microbubble ligand complexes-receptor binding complexes. Increasing the ligand density present on the microbubble surface has been found to be a reliable method by which to target binding can be enhanced. No further improvement in receptor binding is exhibited above a certain ligand concentration (Tinkov et al 2009). Supple ligand-microbubble spacer arms are a promising method by which endothelial wall receptor binding can be stabilised against shear blood-vessel forces. It has been theorised that the formation of the ligand endothelial wall receptor complexes induces microbubble core pressurization. The resultant deformation in the shape of the microbubble shell produces limited gaseous leakage and wrinkling microbubbles with outward protrusions, which enhance adhesion characteristics (Kilbanov et al 1999).

The laminar flow patterns present in larger vessels may ameliorate the possibility for significant target binding. The application of ultrasound beams can deflect circulating microbubbles on to the endothelial vessel walls enabling improved contact targeting. This property can help overcome the problem of sclerotic plaque targeting in larger vessels and the microbubble targeting inflammatory deep venous areas (Dayton et al 1999).

4. Therapeutic microbubble vectors

Specificity of drug delivery and targeting is the panacea pharmaceutical technologies. The biological barrier forms a worthy barrier to entry of the therapeutic compounds. These biological barriers include the cell membrane, capillary endothelium, and blood - brain barrier and vessel walls. These barriers can selectively transport molecules of varying molecular sizes. Ultrasound targeted microbubble destruction will enable the transport of 2-3 MDa, 6-8 μm, ~ 9 nm,~100 nm through the cell membrane, capillary endothelium (vessel diameter <7 μm), vessel wall diameter ~55 μm) and blood brain barrier respectively (Schlicher et al., 2006; Skyba et al., 1998).

A dichotomy exists between the desired characteristics necessary for the delivery of compounds to intracellular target sites. A simple innately lipophilic natured compound will preferentially access the intracellular compartment but, lacks the ability to freely transit in the aqueous haematological system to the cellular compartment. Uncomplicated hydrophilic agents can easily around the circulation, but without the benefit of selective specialised transport proteins are unable to enter intracellular compartments. Interiorisation of active compounds may be augmented through the application of ultrasound to cell-membrane barrier. Small molecular compounds, including polynucleotides and proteins can pass into the intracellular compartment through the process of cavitation when facilitated with focused ultrasound. High energy ultrasound cellular damage is an undesirable effect of focused high-energy ultrasound. The acoustic cellular damage induced at these energies required to induce significant gases cavitation in vitro make this prospect clinically unreasonable (O'Brien 2007).

The application of ultrasound resonance field to microbubbles induces their frequency dependent oscillation. Microbubbles destabilisation and destruction can be induced with targeted high-energy ultrasound. Local delivery can be achieved through the liberation of the active product at the target site with the microbubbles acting cavitation nuclei. Significantly lower levels of the ultrasound energy are necessary when used in combination with microbubble agents.

If insonated microbubbles pass in the near vicinity of cellular membranes, the characteristics of the cell membranes may be altered and demonstrate the presence of sonopores. Their appearance may be the result of the proposed phenomena specific to microbubbles, these include microstream swirling, micro-jetting, the impact of enhanced ion-channel conductance, the formation of hydrodynamic shock waves or controversially free radical formation (12-24)(Miller & Quddus 2000; Miller & Pislaru 2002; Marmottant & Hilgenfeldt 2003; Barnett 1998; Guzman et al., 2003; Wang et al., 1999; Wei et al., 2004; Juffermans et al., 2006; Miller & Thomas 1993). There remain dissenters as to the whole validity of the cavitation theory, whilst others propose alternative ionic channel conductance mechanisms (Lawrie et al., 2003; Bouakaz A et al., 2006).

Bioactive substances may be able to infiltrate into the intracellular compartment through these sonate induced pores. These induced cellular membrane transit pores are transient (seconds to minutes) in their nature because of an endogenous vesicle-based healing responses resulting in these channels resealing. These healing processes are dependent calcium and ATP based processes. Their estimated sizes range from 30-100nm in size up to a maximum of a couple of millimetres (Pan et al., 2005).

It is questionable whether the formation membrane pores results in proportional changes in the nuclear membrane. Some studies have found that the viscous cytoplastic nature meant that the formation of these pores do not affect the nucleus. In contradiction to this one in vivo study has demonstrated that targeted ultrasound microbubble destruction resulted in nuclear uptake of rhodamine-labelled (Duvshani-Eshet et al., 2006).

It has been proposed that the ultrasound focused destruction of microbubbles increases the capillary permeability at the cellular microvascular level. The implication is that permeability can be enhanced active compounds can be enhanced. The reciprocal undesirable effect is that the application of focused-ultrasound microbubble destruction can result in local haemorrhage (Tinkov et al., 2009)

5. Drug loading

Manipulation to the microbubble formulation alters their physiochemical characteristics significantly, particularly the shell volumes and widths. There are four classes of microbubble structures with the capacity as gene and drug carriers. Potentially higher drug loads can be loaded into acoustically-active lipospheres (shell thickness [triacetin layer 500-

1000 nm], [soya-bean oil layer 300-700nm]), microcapsules(usually emulsification method, shell thickness 50-200 nm) and protein-shelled microbubbles (HSA-shelled microbubbles, probe-type sonication method, thickness 200-300 nm) possess larger shell volumes to which active products can be embedded. Whereas, Phospholipid-microbubbles (shell thickness 2-3 nm), because they have smaller shell volumes have better acoustic properties (Tinkov et al., 2009).

The most common approach for drug-loading in phospholipid and protein microbubbles is for shell surface loading. The loading process of the active agents within the entire volume of the shell is appropriate for microcapsules and on occasion protein composite microbubbles.

Microbubble drug coupling can be achieved through a number of the modalities. In essence microbubbles are gas-filled colloidal particle, which comprises a surfactant wrapping a flexible protein and polymer shell.

The outer shell surface can be used as a foundation by which active products can be attached. Drug compound attachment of the microbubble composite polymer shell can be utilise van-der-waal, hydrophobic or electrostatic forces. The association of active agents can be achieved through the association of smaller secondary shell surface anchor/carrier particles proteins or complexation to adhesive human serum albumin molecules. The nucleic acids, DNA and RNA can be attached to the outer microbubble shells with charge coupling.

Physical encapsulation in biodegradable polymeric shells offers the advantage of facilitation of the appropriate elimination from the systemic circulation. These biocompatible shells can consist of compounds such as gelatin, and may reduce immunogenicity and unfortunately ligand affinity.

Hydrophobic drugs molecules incorporated within can be submerged within the oily layer of lipospheres or intercalated between the monolayer of phospholipids.

A 'phase-shift' colloid emulsion vehicle is an elaborate system consisting of a perfluoropentane microemulsion system stabilised by biodegradable surfactants. This colloid emulsion is then loaded with the active compound possessing the appropriate physiochemical properties. The droplet-colloid composition is phase-shifted to bubble form through the application of heated sonication (Rapport et al., 2007).

6. Co-administration drug targeting

The drug loading of microbubbles is not essential to glean the benefits of their potential drug targeting properties. Administration of microbubble agents can be accompanied by the active product. Ultrasonic beam application at the target sites may facilitate the entrance of otherwise circulating active product into the interstitial compartment and into the intracellular compartment. Further study is necessary to study the therapeutic modality of this approach (Tinkov et al., 2009).

7. Targeted microbubble production

The process production and formulation begins with the identification of an amenable disease and correlated target disease specific receptor. A therapeutic mechanism and active compound are prerequisites to ensure the efficacy of any potential enhancement with microbubble targeting. Comprehensive understanding of pharmacological, pharmacodynamics, physiochemical and dosages are essential to ensure realistic and cost-effective production of clinically applicable microbubble compounds. The active compounds must be amenable to the microbubble formulation process, which in many instances may denature and degrade biological compounds (e.g peptide, nucleic acid and monoclonal antibodies) and many synthetic compounds. Highly thermodynamic and immunogenical components be they active compounds or ligands are unsuitable for incorporation into microbubble formulations. For several molecular therapeutic compounds the impact of the formulation and manufacture process makes targeted microbubble modalities non-feasible.

The production process must incorporate uncompromising quality control based on sound reproducible manufacturing practices and in-depth understanding of both components and the required end-product.

Complexation of the active products to the shell utilises the relative physiochemical characteristics of the shell and the active product. Non-covalent bonding to the shell ensures appropriate strength to their loading and for ultrasonic liberation. Electrostatic charged albumin and phospholipids can be coupled to charged active products.

7.1 Phospholipids

Semi-synthetic flexible phospholipid thin monolayer possesses suitable acoustic properties. When present in solution the microbubble the hydrophobic portion orientated internally towards the gaseous centre and the electrostatic hydrophilic elements coming in continuity with the aqueous solvent. The condensation of the saturated fatty acid chains portions of the phospholipids structure imparts integral stability to the gel micelle shells.

There is a wide spectrum of the phospholipid compositions, including unsaturated fatty acid chains. Amalgamation of certain phospholipids and excipients can drastically alter the microbubble stability. An example, includes the integration of unsaturated chained phospholipids into the shell significantly reduces the stability of the microbubble shells. Alternatively, the steric stabilisation can be achieved with the inclusion of PEGylated phospholipid.

Other phospholipid compounds or phospholipid adjuvants impart an extended range of characteristics to these microbubble vectors. The inclusion of PEGylated or non-bilayer phospholipids can prolong circulation and gene payload transfection, respectively. Alteration of the condensed and disordered phase domains within the phospholipid layers changes the miscibility of components within the shell, with a result in its characteristics.

The heterogeneity of constituent components changes the degree of phase separation, which itself alters the capacity for drug/ligand loading, stability, and acoustic characteristics. Heterogeneity of phase separation can confer desirable characteristics to the microbubble formulations. One such advantage imparted by the incorporation of brush moieties is through the steric protection they can provide. The major disadvantage is the threat to formulation quality, which may be the result of phase separation in the lateral orientation. It has been suggested that the degree lateral phase separation may be the limited through adjustment of the fabrication process and through manipulation of the shells composition.

The condensation state is a crucial in determining the *in vivo* stability and is reliant upon on the temperature. To ensure that an agent has sufficient in vivo stability the physiological body temperature must be below the threshold phase-transition temperature (Borden et al., 2005).

When considering phospholipid constructs the stability is very dependent upon the hydrophobic carbon chain lengths. Longer carbon chained microbubble phospholipid shells possess less surface tension and an increased tendency to resist gaseous permeation. Increased viscosity with enhanced durability, but reduced acoustic characteristics with diminished echogenicity (Duncan P.B. & Needham D. 2004).

The ideal compromise to ensure optimal phospholipid microbubble stability and the existence of the minimal phase-shift characteristics, which are achieved by ensuring the equivalence of the various constituents phase-transition temperatures and phospholipid fatty-acid residue lengths. Avoidance of unsaturated phospholipid components possessing a large degree of conformational liberty would also improve stability. These 'loose' phospholipids destabilise the order of the otherwise densely packed monolayer. Microbubble monolayer destabilising and disorganising abundant surface electrostatic charges should also be avoided because they induce electrostatic repulsions within a layer orientation (Borden et al., 2006).

Non-conventional production of the phospholipid can be undertaken using a flow-focused method, which encompasses the passage of the inert gaseous core and shell material through a fine nozzled device and into a water reservoir.

7.2 Polymeric

In terms of polymeric microcapsule formulations characteristics are dependent upon material biocompatibility polymeric component weight, which possesses a direct relation to the shell thickness. Variability to the polymer constituents influences the stability, echogenic, time to destruction and acoustic characteristics (Forsberg et al., 2004).

Inappropriate selection of the incorporated of Proteinaceous excipients into the microbubble shell can also embarrass the functional stability of therapeutic microbubble formulations. Exposural factors, which may disrupt the self-association forces between shell components, which must be mitigated against, include heat and free-radicals. Adhesive shell building factors include thiol-rich proteins including HSA. These forces are imparted through the sonication of the shell constituents' results in intensive cavitation and free-radical production, which result in the production of HSA linking thiol bridges. These bridges can also be formed with the addition of glutaraldehyde and formaldehyde during the shell formation process (Tinkov et al., 2009).

7.3 Mechanical agitation

The most frequently utilised method for the production of acoustically active lipsomes and phospholipid-shelled microbubbles. The initial phase of the process involves the formation of a thin hydrated phospholipid sheet, which is then either infiltrated with an alcoholic component or phase inverted by an alternate approach. A gaseous component is then pressed in the space above the liposomal dispersion and then agitated at least several thousand oscillations every second. Loading of the active agents can take place at initial microbubble production or after this second gaseous core forming stage. To ensure a predictable desire loading pattern it essential to take steps to ensure loading of the active molecular compounds are bound to the liposomal shell instead of encapsulated within the water soluble liposomal centre (Tinkov et al., 2009).

The key to the production of the acoustically active liposomes revolves around the production of the appropriate micro-emulsion. This can be undertaken through the combination of the perpetual aqueous phase and injectable oil phase. Suitable constituents of the injectable oil phase include castor or triacetin oil. Successful emulsion formation requires the incorporation of co-solubilizers, such as synthetic block copolymers, because of the defoaming characteristics of the oil phase. The adequately lipophilic agent can then be dissolved and remain persistently encapsulated within the oily phase and acoustically active liposomes (Fang et al 2007)

This versatile modus operandi for the production of therapeutic microbubbles is advantageous when integrating delicate targeting ligands and drugs molecules. Successful utilisation of this drug-loading approach requires meticulous control of important variable production parameters. Components concentrations, viscosity, Agitation time and temperature must be all thoroughly controlled to ensure predictable characteristics.

7.4 Emulsification

Some phospholipid-shelled microbubble and many polymer shelled microcapsule formulations are produced with this oil-in-water freeze-drying emulsification production method. Some of the commonly used later generations of contrast enhanced ultrasound agents have been manufactured in this manner. The inner emulsion phase consists of a volatile solid compound (such as camphor), an organic solvent phase, which has the characteristic of being lyophilizable water-immiscibility and polymeric shell material. The water-water-insoluble benzene compound based solvents, such as toluene or p-xylene. The polymeric shell material consists of compounds similar to the highly biodegradable and biocompatibile co-polymer poly-lactide-co-glycolide (PLGA). The emulsion matrix is the residual after the removal of the volatile aqueous and organic phases remaining after the freeze-drying process. The gaseous cores are formed after reconstitution of the gaseous vial and emulsion matrix with the injection solvent medium. Coverage of these microbubbles with gelatin or albumin compounds included in the aqueous phase of this emulsification production method may significantly improve their biocompatibility profile (Schneider et al., 1991; El-Sherif et al, 2003; Short 2005).

Lipophilic active products can be easily accommodated within organic emulsion phase of polymeric shelled microbubbles compounds. Water-soluble drugs and biological agents can be accommodated in these formulations through encapsulation. Incorporation of hydrophilic compounds within the mixed-phase emulsion is undertaken through their dispersion within the shell containing the organic phase. The next step of the process involves the production of a double emulsion, which consists of water in organic in water emulsion. The lyophilization of the double emulsion is the final synthetic mechanism resulting in hollow shell drug-loaded microbubbles. The important factors that need to be controlled include the emulsion droplet sizes; quantity and molecular weight of the shell material monomers (Tinkov et al, 2009).

7.5 Probe-type sonication

Low frequency and high intensity ultrasound is the commonest technique for the production of protein-shelled microbubbles and less commonly phospholipid microbubbles. Ultrasound induces exuberant convection streams to disperse the inert gaseous phase into the aqueous phase and produces the microbubble gaseous core. The process generates free radicals and temperature rises in excess of seventy degrees centigrade, which itself produces insoluble-shell microbubble shells. The proteins can be denatured with unveiling of their thiol-groups with the subsequent formation of covalent and non-covalent bridging bonds. Phospholipid microbubbles do not benefit from higher temperatures as their exceeding their phase-transition temperature can be deleterious to the formulation.

The most commonly utilised shell building component was HSA because of its exceptional affinity to a wide spectrum of active agents, including nucleic acid products. Drug loading to only the outer shell can be undertaken with the cell surface loading approach. This involves the incubation of the active consignment compounds to composed microbubbles. The disadvantage of this approach is a relatively low concentration of active product can be achieved through this process of surface absorption. This limitation can be overcome by the inclusion of the compound throughout the depth of the shell. This can be achieved through the application of sonication to the shell active product mixture prior to the microbubble formulation. Fragile active compounds may not be amenable to the entire shell loading approach as they will be damaged by the harsh cavitation and chemical and thermal stresses.

7.6 Spray-drying method

This versatile approach is essentially the formation of hollow particles through a spray-drying process. Encapsulation with the spraying technique is commonly utilised for the production conventional microbubbles. It provides an opportunity for a reliable method by which drug-loaded polymeric- and protein-microbubbles can be manufactured. This spray-drying method provides drug volume loading, a stable dried product and a final composition produced in favourable conditions.

Phospholipid-, protein- and polymer- microbubble formation utilises volatile ammonium derivatives and organic liquids (e.g. halogenated hydrocarbons) to induce pores and cavities in spray-dried particles. The commonly utilised HSA components of microbubble shells can be thermally stabilised and undergo chemical, but not free-radical cross- processes. The premise of this approach is the application of an evaporation process to a solvent mixture containing the dissolved shell material. The resultant saturation of the air-water interface with solvent-mixture forms an elastic film layer. On the droplets surface the shell components within the film solidifies, whilst the associated solvent completely evaporates. The particle formation is determined by the pressure induced fluid conversion into solid. The optimal acoustic characteristics are exhibited by microbubbles with the largest possible core cavity volumes. This spray-drying technique does not result in a uniform population of microbubble particle characteristics. The core cavity sizes can vary dramatically from exceptionally large cavities with fragile shells to multiple small voids (Porter 1997; Lentacker et al., 2007).

8. Stability

The shell is fundamental for microbubble stability is maintains the integrity of the gaseous core by preventing the leaching of the gaseous core and minimising interface tensions. There is a complex gaseous balance reached between the solvent saturated gases and the gaseous core content. The passage of gas occurs from the core into the plasma solvent and simultaneously from the solvent into the gaseous core. This continues until the gaseous partial pressure equilibrium has been achieved. When the net gaseous movement is static the equilibrium osmotic gradient is zero. The aqueous insoluble Perfluorocarbon microbubbles vivo have negative osmotic gradient, so initially tend to increase in size because the net passage of aqueous dissolved gases into the gaseous core. Formulation techniques to improve stability include the incorporation of mixed gaseous cores, such core have been termed osmotically stabilized. Specifically engineered shells may reduce the degree of gaseous passage, but will be unlikely completely ameliorate it. Alteration of the gaseous core constituents will prove a more robust modification (Tinkov et al., 2009).

Air-filled microbubbles consisting of proteineous shells possess inferior echogenicity characteristics than fluorinated gas core microbubbles. The water insoluble perfluorobutane and sulphur hexafluoride consisting gaseous core microbubbles confer some superior properties to more advanced microbubble generations. Perfluorocarbon containing third generation microbubbles are disadvantaged by their tendency to absorb blood containing gases resulting in an inconsistent size profile and potentially reduced stability due to swelling. Perfluorocarbon and nitrogen mixtures may fare better and possess inherently superior pharmacodynamics properties (Tinkov et al 2009).

9. Quality control

Quality assurance requires a detailed understanding of microbubble properties. The formulation characteristics, which complicate analysis, revolve around the dynamic nature of microbubbles and their delicate quiescent phases. Their complicated characteristics include their unpredictable buoyancy and excessive sensitivity to shear-stress, pressure, temperature. These drug-loaded microbubbles are relatively complex composite mixture of molecular interacting structures, which can behave in a broad manner at any one time. Despite many of these unique properties many conventional assays can be utilised in an efficacious manner. Specific analytic tests are necessary to characterise therapeutic microbubble destructibility and echogenicity (Tinkov et al 2009).

9.1 Size

The terminological dogma underlying therapeutic microbubbles dictates size limitations of between 500 nm and 1µm in size. This size ranges enables optimal compromise between drug-loading in the gaseous core and echo signal. Nanobubble preparations have a greater predilection to take advantage of the permeability and retention effect. Nanoparticles more easily cross leaky capillary walls. Nanobubble units can have a tendency to coalescence into a microbubble conformation, thus ameliorating this advantageous property of the smaller bubbles (Tinkov et al 2009).

Acoustic activity, pharmacodynamics and pharmacokinetics of microbubble preparations are highly dependent upon their size distrubtion. A Broad, mult- or bimodal distribution rather than a more conventional normal distributions have been found. The upper size limit should be between 5 to 10μm to enable safe clearance through the pulmonary vasculture. The size distribution can be sized and calculated by using electromagnetic impedance field calculation. As the particles pass through the 'coulter counter's' miniscule apertures located between two the complementary electrode the reading are produced. The detection and measurement of microbubble shadows cast as microbubble particles passage before a narrow uniform light source, is termed the 'light obscuration' method. The size range of this technique is has a maximum size tolerance of 400μm and requires low concentrations to be accurate. Laser diffraction approach can measure in the presence of high concentrations, is unaffected by buoyancy and can accurately assess any population distribution. Its particle size tolerance is 40-2 mm and requires calibration/optimisation for individual compounds. The dynamic light scattering method relies upon the 'Brownian motion of particles' is seldom utilised due its unreliability (Tinkov et al 2009).

9.2 Zeta potential

The zeta potential has important ramifications to both the intrinsic physiochemical characteristics of microbubble formulations and molecular pharmacokinetics. The active-product loading capacity, stability of the colloidal dispersion and the interaction with the endothelial cell wall properties can be usefully represented by the zeta potential behaviour. Preferable zeta potentials may significantly extend capillary retention time duration (Fisher et al., 2002).

The electrophoretic mobility can be determined by the degree of laser light scattering, which results from particle oscillation induced by an alternating electrical field. Using mathematical modelling the electrophoretic potential be translated into the zeta potential. The accuracy and precision of this presumptive calculation can be compromised by the inherent characteristics of microbubbles. The convection associated buoyancy compromises the reproducibility of original electrophoretic observations. Several environmental, physiochemical and electrochemical can affect the calculated zeta potential measurements (Tinkov et al 2009).

9.3 Mechanical properties

Critical to the modality of microbubbles is their characteristic acoustic behaviour on the application of an ultrasounic field. This behaviour is dictated by the shell characteristics. Advanced in the field of dynamic imaging has enabled the robust characterisation of ultrasound manipulated microbubbles.

The shell hardness plays a disproportionally large role in the determination of the ultrasonic threshold energy microbubble cracking threshold. Upon reaching their reaching their cracking threshold the encapsulating shell wall ruptures and the gaseous bubble is released instantaneously leaving a partially maintained shell. Harder shells require in the order of twice to twenty-fold the ultrasonic energy to rupture and liberate is payload, in comparison to softer shelled formulations. These necessarily high energy levels and more pronounced harder-shell microbubble oscillations may have a deleterious effect upon the adjacent cells (Postma et al., 2004; Tinkov et al., 2009).

Soft more pliable shells, such as phospholipid shell components oscillate through the process of expansion and contraction. With the expansions lipid aggregates are expelled and then reseals. Significantly lower ultrasonic energies are necessary to rupture softer shells (Sboros et al., 2006; Tinkov et al., 2009)

9.4 Lateral phase separation

Microbubble shells can be significantly compromised by the phenomenon of phase separation. The molecular findings demonstrated within the shell during lateral phase-shift are an uneven distribution of constituents. The differentials in inherent forces include heterogeneous electrostatic binding and steric-shielding binding forces. Quality assurance methods are essential to ensure microbubble stability. Fluorescent dye tagged microbubble laser scanning confocal microscopy can employed to characterise the quality of dispersion and drug-loading (Borden et al., 2006).

9.5 Chemical integrity

Novel and broad characterisation techniques are necessary to ensure quality control of payload compounds incorporated within microbubble formulations.

Reversed-phase and normal-phase liquid chromatography are useful tools for quality assurance analysis of smaller molecular therapeutic compounds and small phospholipid compounds respectively (Lentacker et al., 2006; Hvattum et al., 2006).

10. Safety

There is relatively little clinical experience of the microbubble agents as a result the extent and spectrum of adverse events has not yet completely revealed themselves. Thus far, the clinical use of these agents as largely contrast ultrasound agents has proven to be safe. The complicity of the therapeutic adjuncts will probably introduce further adverse effects.

Myocardial exposure to microbubble agents may induce a physiological response of premature cardiac contractions. Microvascular damage, pectechial haemorrhage, free radical damage and single strand DNA fractures are also postulated side-effect of the focused ultrasound microbubble destruction process. In clinical practice these complications have been seldom encountered.

In regards to the reasonable ultrasound energies feasible for focused therapeutic ultrasound microbubble destruction, 'the principle of as low as reasonably achievable' has been deemed appropriate by radiological committee guidelines. For clinical ultrasound of microbubble agents the accepted safe mechanical index parameters are 0.05 to 0.5. For harmonic and non-harmonic clinical imaging modalities exceeding the mechanical index of 0.5 would result in microbubble destruction. The American Food and Drug

Administration (FDA) have set the maximum accepted mechanical index tolerance of 1.9 (Tinkov et al., 2009).

10.1 Immunogenicity

Although microbubbles themselves are not immunogenic, the incorporation of targeted ligands to their shells does render them immunogenic. Immunoreactive components are desirable to target delivery, but, undesirable for ability to attract unwanted immunological attention whilst in transit to the target site.

Conventional exposed spacer-linked or directly linked ligands loaded upon the gaseous microbubble cores may be inappropriately degraded or trigger an aggressive detrimental immunology response. Buried ligands within an overbrushed grafted polymeric-PEG layer may enable the camouflage of these immunoreactive elements. This stealth layer can then be unveiled and the immunoreactive layer underneath revealed with the application of focused ultrasound beam. The ultrasonic waves push the microbubbles towards the vessel walls and the polymer sheath is unbrushed and enables ligand binding to the specific receptor targeting sites. The caveat of these stealth particle modifications is reduced target receptor site affinity (Kilbanov et al., 1999).

11. Gene-loading microbubbles

Gene-therapies have proven an intriguing, but nonetheless frustrating proposition. Their potentially enduring therapeutic promise has been slighted by some critical inherent deficiencies. The absence of any effective conventional carriers has hindered the progression of more widely clinically applicable nucleic acid therapeutic options. The essential criteria an advanced gene carrier must possess an exceptional safety profile. The carriage of their genetic payload must be hostile and protect it from the otherwise deleterious effects of native enzymes, immunological elements and pharmacokinetic elimination. The aim is the controlled and timely liberation of the active agent followed by appropriate transfection of the target cells. Transfection is mediated by attenuated viruses, thus rendered incapable of replication and instigating disease. Amongst critics of gene-therapy options typical concerns over safety exist as with non-microbubble preparations. These safety concerns pertain to the risks of genetic mutations, pathogenesis and immunological reactions. Work utilising non-viral vectors aimed at excluding these viron-vector related risks has not yielded practical or efficacious alternatives. Alternative delivery systems investigated include microbubble, lipoplexes, electroporation and microinjection.

Genetic therapy delivery mediated through microbubble formulations is widely considered to be a safe prospect. The enhancement in transfected genetic material quantities is significant. In vitro models have demonstrated transfection magnitiude enhancement rates of 10-3,000 fold in comparison to naked plasmid DNA constructs alone. Despite the observation under in vivo conditions, which have demonstrated 1,000 fold transfection improvement, true proven in vivo transfection enhancement remains elusive. Serum nucleases are pose a significant risk upon naked and viron loaded nucleic acid compounds. The shielding effect of microbubbles has been proposed in microbubble constructs loaded with adenovirus-associated microbubbles (Tinkov et al., 2009).

Unlike other therapeutic opportunities the response to therapy rarely correlates with the quantity delivered. It is essential however, to deliver a threshold quantity to the appropriate area. The limited surface absorption binding capacity of plasmid DNA to produce maximally loaded gene loaded microbubble is in the order of approximately 0.001-0.005 pg/μm^2. Adjuvants can be utilised to increase the gene binding and loading properties. These include the use dipalmitoylphosphatidylethanolamine (DPPE) containing phospholipid microbubbles, which have a higher binding capacity to plasmid DNA. Layer-by-layer technique with poly-L-lysine may enable pDNA packing to decimal points higher than conventional microbubble loading. The absorbed loading of genes on to albumin microbubbles has been found not to enhance pDNA loading. Utilising a formulation approach a greater than 200-fold loading to entire shell volume has been reported. More recent developments are secondary-carriers associated microbubbles and the active nucleic acid components. These constitute nanoparticles, polyplexes, liposomes and lipoplexes. They have multiple functions, including enhancement of the transfection and microbubble strong shell attachment to the microbubble shell with improved loading capacity. They function through the condensation of large DNA molecules, thus protecting them from the action of serum nucleases. Entrance into the intracellular compartment involves both endocytosis and pinocytosis (Frenkel et al., 2002; Tinkov et al., 2009).

In vivo microbubble transfection delivery studies have been undertaken in organ systems already validated for diagnostic contrast ultrasonography. Vascular paradigms are particular relevant to study therapeutic microbubble compounds. Numerous pathological vascular paradigms have indicated that the duration of action varies from condensed intensive therapeutic windows to the extremes of a month. No meaningful conclusions can be drawn from study of these gene-therapy microbubble therapies for atherosclerotic or intimal disease. Ultrasound induced sonoporation and of the microbubble containing genetic-material enables delivery into the cytoplastic compartment. Nanobubble delivery does not seem to enhance entry into the nuclear compartment resulting in its deactivation/destruction prior to activity (Newman & Bettinger 2007; Liu et al., 2006; Unger et al., 2006).

Elucidation of formulation, pharmacodynamics and pharmacokinetic principles is necessary to definitively ascertain the modality of this novel disease-specific therapeutic option. Establishing such studies are complicated by the absence of standardised models with comparable measurable parameters. Also more modern genetic therapy incorporated microbubble formulations have not been studied to the same extent as co-administration preparations. This leaves the door ajar for more modern formulations fulfilling their yet unrealised potential (Tinkov et al., 2009)

12. Conclusion

Safe *in vivo* utilisation of microbubble contrast enhanced ultrasound agents is well established and validated. The idea of directly translating microbubble into routine used therapeutic treatments is oversimplification of the challenge ahead. The premise of diagnostic microbubble is with low administrated doses to produce a sustained and clearly visible echogenic signal. Whereas, with therapeutic microbubbles the all the associated paradigm variables still remain too uncertain. The quantities of both ultrasonographic energy and microbubble dosage are many magnitudes greater, which imparts numerous complications. The most serious of these is significant unintentional tissue damage related to cavitation-inducing ultrasound beams. A pattern of cellular damage has been demonstrated on myocardial tissue between the upper safe ultrasonic energy limit and 2.5 times of this limit. Such a degree of irreversible cellular destruction will prove detrimental to organ function and subsequently result in significant morbidity. Despite these concerns, it has been proven that efficacious microbubble therapeutics used at the effective dosages with safely ultrasonography energies.

Further studies on all aspects of formulation, production and trials of therapeutic microbubbles are necessary as a precursor to establishing these agents as realistic therapeutic options. The promise in their therapeutic role may lie as co-administration agents to optimise selective uptake of the active agents. Therapeutic microbubbles may prove the 'missing-link' in fixing the inherent shortfalls of promising therapies, particularly in reference to gene therapy.

13. References

Barnett S. (1998). Nonthermal Issues: Cavitation-Its nature, detection and measurement. *Ultrasound in Medicine and Biology*, Vol 24, Suppl 1 pp 11-21. PMID:9841460

Bloch S.H., Dayton P.A., Ferrara K.W. (September-October 2004). Targeted Imaging using Ultrasound Contrast Agents. *IEEE Engineering in Medicine and Biology Magazine.* Vol 23, No. 5, pp 18-29. PMID:15565796

Borden M.A., Kruse D.E., Caskey C.F., Zhao S., Dayton P.A. & Ferrara K.W. (November 2005). Influence of lipid shell physicochemical properties on ultrasound-induced microbubble destruction. *IEEE Transaction in Ultrasonics, Ferroelectrics, and Frequency Control*, Vol 52, No. 11, 1992-2002. PMID:16422411

Borden M.A., Martinez G.V., Ricker J., Tsvetkova N., Longo M., Gillies R.J., Dayton P.A. & Ferrara K.W. (April 2006). Lateral Phase Separation in Lipid-coated Microbubbles. *Langmuir. the ACS Journal of Surfaces and Colloids*, Vol. 22, No. 9, pp 4291-4297. PMID:16618177

Bouakaz A., Tran T.A., Roger S., Leguennec J.Y. & Tranquart F (2006). On the Mechanisms of Cell Membrane Permeabilization with Ultrasound an Microbubbles. *Ultrasound in Medicine and Biology*, Vol 32, pp 90.

Dayton P., Kilbanov A., Brandenburger G. & Ferrara K. (October 1999). Acoustic Radiation Force In-vivo: A Mechanism to Assist Targeting of Microbubbles. *Ultrasound in Medicine and Biology.* Vol 25, No. 8, pp 1195-1201. PMID:10576262

Duncan P.B. & Needham D. (March 2004). Test of the Epstein-Plesset Model for Gas Microparticle Dissolution in Aqueous Media: Effect of Surface Tension and Gas Undersaturation in Solution. *Langmuir: the ACS Journal of Surfaces and Colloids*, Vol 20, No. 7, pp 2567-2578. PMID:15835125

Duvshani-Eshet M., Baruch L., Kesselman E., Shimoni E. & Machluf M. (January 2006). Therapeutic Ultrasound-mediated DNA to Cell and Nucleus: Bioeffects Revealed by Confocal and Atomic-force Microscopy. *Gene Therapy*, Vol 13, No. 2, pp 163-172. PMID16177822

El-Sherif D.M. & Wheatley M.A. (August 2003). Development of a Novel Method for Synthesis of a polymeric Ultrasound Contrast Agent. *Journal of Biomedical Materials Research. Part A*, Vol 66, No. 2, pp 347-355.

Fang J.Y., Hung C.F., Liao M.H & Chien C.C. (August 2007). A Study of the Formulation Design of Acoustically Active Lipospheres as Carriers for Drug Delivery. *European Journal of Pharmaceutical and Biopharmaceutics*, Vol 67, No. 1, pp 67-75. PMID:17320362

Fisher N.G., Christiansen J.P., Kilbanov A.L., Taylor R.P., Kaul S., Linder J.R. (August 2002). Influence of Microbubble Surface Charge on Capillary Transit and Myocardial Contrast Enhancement. *Journal of the American College of Cardiology*, Vol 40, No. 4, pp 811-819. PMID:12204515

Forsberg F., Lathia J.D., Merton D.A., Liu J.B., Le N.T. & Goldberg B.B., Wheatley M.A. (October 2004). Effect of Shell Type on the In-vivo Backscatter from Polymer-encapsulated Microbubbles. *Ultrasound in Medicine and Biology*, Vol 30, No. 10. pp 1281-1287. PMID:15582227

Frenkel P.A., Chen S., Thai T., Shohet R.V. & Grayburn P.A. (June 2002). DNA-loaded Albumin Microbubbles Enhance Ultrasound-mediated Transfection In-vitro. *Ultrasound in Medicine and Biology*, Vol 28, No. 6, pp 817-822.

Gramiak R., & Shah P.M. (September-October 1968). Echocardiography of the aortic root. *Investigative Radiology*, Vol 3, No. 3, pp 356-366. PMID:5688346

Guzman H.R., McNamara A.J., Nguyen D.X. & Prausnitz M.R. (August 2003). Bioeffects caused by changes in Acoustic Cavitation Bubble Density and Cell Concentration: A unified Explanation Based on Cell-to-bubble Ratio and Blast Radius. *Ultrasound in Medicine and Biology*, Vol 29, No. 8, pp 1211-1222. PMID:12946524

Hvattum E., Uran S., Sandeek A.G., Karlsson A.A. & Skotland T. (October 2006). Quantification of phosphatidylerine, phosphatidic acid and free fatty acids in an Ultrasound Contrast Agent by Normal-phase High-performance Liquid Chromatography with Evaporation Light Scattering Detection. *Journal of Pharmaceutical and Biomedical Analysis*, Vol 42, No. 4, pp 506-512.

Kaul S. (October 2010). New developments in ultrasound systems for contrast echocardiography. *Clinical Cardiology*, Vol 20. No. 10 Suppl 1 pp I27-30. PMID:9383599

Kilbanov A.L., Gu H., Wokdyla J.K., Wible J.H., Kim D.H., Needham D., Villanueva F.S. & Brandenburger G.H. (1999). Attachment of Ligands to Gas-filled microbubbles via PEG-spacer and Lipid Residues Anchored at the Interface. *Proc International Symposium Control Rel Bioact Mater.* 26.

Koike H., Tomita N., Azuma H., Taniyama Y., Yamasaki K., Kunugiza Y., Tachibana K., Ogihara T. & Morishita R. (January 2005). An Efficient Gene Transfer Method Mediated by Ultrasound and Microbubbles into the Kidney. *The journal of Gene Medicine*. Vol 7, No. 1, pp108-116. PMID:15515148

Pan H., Zhou Y., Izadnegahdar O., Cui J., Deng C.X. (June 2005). Study of Sonoporation Dynmaics Affected by Ultrasound Duty Cycle. *Ultrasound in Medicine and Biology*. Vol 31, No. 6, pp 849-856. PMID:15936500

Lawrie A., Brisken A.F., Francis S.E., Wyllie D., Kiss-Toth E., Qwarnstrom E.E., Dower S.K., Crossman D.C. & Newman C.M. (October 2003). Ultrasound-enhanced Transgene Expression in Vascular Cells is not Dependent upon Cavitation-induced Free Radicals. *Ultrasound in Medicine and Biology.* Vol 29, No. 10, pp 1453-1461. PMID:14597342

Lentacker I., De Smedt S.C., Demeester J., Van Marck V., Bracke M., Sanders N.N. (2007). Lipoplex-loaded Microbubbles for Gene Delivery: A Trojan Horse Controlled by Ultrasound. *Advanced Functional Material,* Vol. 17, pp 1910-1916.

Liu Y., Miyoshi H. & Nakamura M. (August 2006). Encapsulated Ultrasound Microbubles: Therapeutic Application in Drug /Gene Delivery. Journal of Control Release, Vol 114, No. 1, pp 89-99. PMID:16824637

Marmottant P., Hilgenfeldt S. (2003). Controlled Vesicle Deformation and Lysis by Single Oscillating Bubbles. *Nature,* Vol 423, No. 6936, pp152-156. PMID:12736680

Miller D.L. & Quddus J. (May 2000). Sonoporation of Monolayer Cells by Diagnostic Ultrasound Activation of Constrast-agent Gas Bodies. *Ultrasound in Medicine and Biology,* Vol 26, No. 4, pp 661-667 (7). PMID:10856630

Miller D.L., Pislaru S.V. & Greenleaf J.F. (November 2002). Mechanical DNA Delivery by Ultrasonic Cavitation. *Somatic and Cell Molecular Genetics,* Vol 27 No. 1-6, pp 115-134. PMID:12774945

Miller D.L. & Thomas R.M. (1993). A comparison of haemolytic and Sonochemical Activity of Ultrasonic Cavitation in a Rotating Tube. *Ultrasonic in Medicine and Biology,* Vol 19, No. 1, pp 83-90. PMID:8456532

Newman C.M. & Bettinger T. (March 2007). Gene Therapy Progress and Prospects: Ultrasound for Gene Transfer. Gene Therapy. Vol 14, No. 6, pp 465-475. PMID:17339881

O'Brien W.D. (January-April 2007). Ultrasound-Biophysics Mechanisms. Progress in *Biophysics and Molecular Biology.* Vol 93, No. 1-3, pp 823-829. PMID:16934858

Pitt W.G., Husseini G.A. & Staples B.J. (November 2004). Expert Opinion on Ultrasound Drug Delivery _ A General Review. *Expert Opinion on Drug Delivery,* Vol 1, No. 1, pp 37-56. PMID:16296719

Porter T.R. (1997). Perfluorobutane Ultrasound Contrast Agents and Methods for its Manufacture and Use. Board of Regents of the University of Nebraska. *United States of America Patents,* US 5,567,415.

Postma M., Van Wamel A., Lancee C.T. & De Jong N (June 2004). Ultrasound-induced Encapsulation Microbubble Phenomena. *Ultrasound in Medicine and Biology,* Vol 30, No. 6, pp 827-840. PMID:15219962

Rapoport N., Gao Z., Kennedy A. (July 2007). Multifunctional Nanoparticles for Combining Ultrasonic Tumor Imaging and Targeted Chemotherapy. *Journal of the National Cancer Institution,* Vol 99, No. 14 pp 1095-1106.

Sboros V., Glynos E., Pye S.D., Moran C.M., Butler M., Ross J., Short R., McDicken W.N. & Koutsos V. (April 2006). Nanointerrogation of Ultrasonic Contrast Agent Microbubbles Using Atomic Force Microscopy. *Ultrasound in Medicine and Biology,* Vol 32, No. 4, pp 579-585.

Schlicher R.K., Radhakrishna H., Tolentino T.P., Apkarian R.P., Zarnitsyn V. & Prausnitz M.R. (June 2006). Mechanism of Intracellular Delivery by Acoustic Cavitation. *Ultrasound Medicine and Biology*, Vol 32, No. 6, pp 915-924. PMID:16785013

Schneider M., Bichon D., Dussat P. & Puginier J. & Hybl E. (1991). Ultrasound Contrast agents and Methods of Making and Using them. *Patent* WO 91/15244.

Shi M., Wosnik J.H., Ho K., Keating A. & Shoichet M.S. (2007). Immuno-polymeric nanoparticles by Diels-Alder Chemistry. *Angewandte Chemie International Edition*, Vol 46,No. 32, pp 6126-6131

Short R.E. (2005). Method of Preparing Gas Filled Polymer Matrix Microparticles Useful for Echographic Imaging. *United States of America Patents Office*. Patent No. US 6,919,068 B2.

Siwak D.R., Tary A.M. & Lopez-Berestein G. (April 2002). The potential of drug-carrying immunoliposomes as anticancer agents. Commentary re: J. W. Park et al., Anti-HER2 immunoliposomes: enhanced efficacy due to targeted delivery. *Clinical Cancer Research.*, 8: 1172-1181, 2002. Clinical Cancer Research Vol 8, No. 4, pp 955-956.

Shohet R.V., Chen S., Zhou Y.T., Wang Z., Meidell R.S., Unger R.H. & Grayburn P.A. (June 2000). Echocardiographic Destruction of Albumin Microbubbles Directs Gene Delivery to The Myocardium. *Circulation*, Vol 101, pp 2554-2556 PMID:10840004

Skyba D., Price R., Linka A., Skalak T., & Kaul S. (July 1998) .Direct In-vivo Visualisation of Intravascular Destruction of Microbubbles by Ultrasound and Its Local Effects on Tissue. *Circulation*, Vol 98, Nov 4, pp 290-293. PMID:9711932

Tinkov S., Berkeredjian R., Winter G., & Coester C. (June 2009). Microbubbles as Ultrasound Triggered Drug Carriers. *Journal of Pharmaceutical Sciences*, Vol 98, No. 6, pp 1935-1961. PMID:18979536

Unger E.C., McCreey T.P., Sweitzer R.H., Shen D. & Wu G. (June 1998). In-Vitro Studies of a new Thrombus-specific Ultrasound Contrast Agent. *American Journal of Cardiology*, Vol 81, No. 12A, pp 58-61G. PMID:9662229

Vannan M., McCreery T., Li T., Han Z., Unger E., Kuersten B., Nabel E. & Rajagopalan S. (March 2002). Ultrasound-mediated Transfection of Canine Myocardium by Intravenous Administration of Cationic Microbubble-linked Plasmid DNA. *Journal of the American Society of Echocardiography*. Vol 15, pp 214-218. PMID:11875383

Voci P., Bilotta F., Merialdo P. & Agati L. (July-August 1994). Myocardial contrast enhancement after intravenous injection of sonicated albumin microbubbles: a transesophageal echocardiography dipyridamole study. *Journal of the American Society of Echocardiography*. Vol 7, No. 4. Pp 337-46. PMID:7917341

Wang Z.Q., Pecha R., Gompf B. & Eisenmenger W. (1999). Single Bubble Sonoluminescence: Investigations of the Emitted Pressure Wave with Fiberoptic Probe Hydrophone. Physical Review E 59:1777-1780.

Wei W., Zheng-Zhong B., Yong-Jie W., Qing-Wu Z., Ya-Lin M. (December 2004). Journal of Ultrasonic Gene Delivery and Safety on Cell Membrane Permeability Control. *Journal of Ultrasound in Medicine.* Vol 23, No. 12, pp 1569-1582.

11

Advanced Core-Shell Composite Nanoparticles Through Pickering Emulsion Polymerization

Lenore L. Dai
Arizona State University
USA

1. Introduction

Solid particles have been identified as a new type of emulsifying agent in addition to surfactants and amphiphilic polymers since the pioneer studies by Ramsden in 1903 (Ramsden, 1903) and Pickering in 1907 (Pickering, 1907). Such emulsions are later on named as Pickering emulsions. In Pickering emulsions, solid particles of intermediate wettability in the size range from several nanometers to several micrometers attach to liquid-liquid interfaces and provide emulsion stability. Recently, there has been growing interest in Pickering emulsions because they open new avenues of emulsion stabilization and have numerous practical applications. For instance, we have studied the fundamentals of particle assembly in Pickering emulsions (Dai et al., 2005; Tarimala and Dai, 2004) , utilized them as templates to investigate the dynamics of particles (Tarimala et al., 2004; Tarimala et al., 2006), and developed microrheology at liquid-liquid interfaces (Wu and Dai, 2006; Wu and Dai, 2007; Wu et al., 2009). In this chapter, we further apply the concept of Pickering emulsions to synthesize core-shell composite nanoparticles.

Organic-inorganic composites are vital in biological, medical, and chemical applications. Among them, core-shell composite nanoparticles are a unique class of materials which are attractive for wide applications. It is worthwhile to note that the composite nanoparticle structure in this study is opposite to the often reported core-shell structure in which inorganic particles serve as the core and polymer serves as the shell; here the polymer serves as the core and the inorganic particles serve as the shell. Such materials provide a new class of supramolecular building blocks and can "exhibit unusual, possibly unique, properties which cannot be obtained simply by co-mixing polymer and inorganic particles."(Barthet et al., 1999) In comparison with the recently reported methods to synthesize core-shell composite particles, for example, post-surface-reactio (Ding et al., 2004; Lynch et al., 2005), electrostatic deposition (Dokoutchaev et al., 1999), and layer-by-layer self-assembly (Caruso, 2001; Caruso et al., 1999; Caruso et al., 2001), we synthesize core-shell composite nanoparticles through a novel Pickering emulsion polymerization route (Ma and Dai, 2009; Ma et al., 2010). Figure 1 illustrates the polymerization route and its comparison with the conventional emulsion polymerization.

Pickering emulsion polymerization is superior in several aspects: (1) no sophisticated instrumentation is needed; (2) a commercialized nanoparticle powder or solution can be used without further treatment; (3) the synthesis can be completed in one-step; and (4) the produced particle dispersion is surfactant-free. Despite of these advantages, efforts to explore and utilize this approach are limited, although pioneer explorations have been initiated some approaches including miniemulsion polymerization (Bon and Colver, 2007; Cauvin et al., 2005), dispersion polymerization (Schmid et al., 2006; Schmid et al., 2007; Yang et al., 2008) inverse suspension polymerization (Duan et al., 2009; Gao et al., 2009), and inverse emulsion polymerization (Voorn et al., 2006) stabilized by fine solid particles.

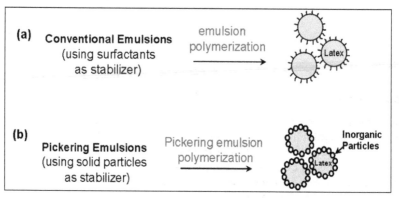

Fig. 1. Comparison between (a) conventional emulsion polymerization vs. (b) Pickering emulsion polymerization.

2. Pickering emulsion polymerization and its possible mechanisms

The general scope of Pickering emulsion polymerization is similar to that of emulsion polymerization, with the exception of using solid nanoparticles as emulsion stabilizers. Here we employ polystyrene (PS)-silica as a model system to illustrate its synthesis and explore possible mechanisms.

2.1 Synthesis and characterization of polystyrene-silica core-shell composite particles through Pickering emulsion polymerization

2.1.1 Materials and synthesis

IPA-ST silica solution, obtained from Nissan Chemicals, is 10–15 nm silica nanoparticles dispersed in 2-isopropanol. The silica concentration is 30–31% by weight. Nonionic azo initiator VA-086 (98%, 2,2-azobis(2-methyl-N-(2-hydroxyethyl)propionamide)), styrene monomer (99.9%), and HPLC grade water were purchased from Wako Chemicals, Fisher Scientific, and Acro Organics, respectively. All materials were used as received.

First, water, IPA-ST and styrene were agitated mechanically at 600 rpm for 8 min using Arrow 6000 (Arrow Engineering) in an ice bath to emulsify. Second, the emulsion was degassed with nitrogen and kept in nitrogen atmosphere under magnetic stir. When the temperature was raised to 70 °C, the initiator solution was added to start the polymerization. The composite particles were sampled at different time intervals ranging from 3 h to 24 h. Before characterization, samples were washed twice by centrifuging - redispersing cycles using an Eppendorf 5810R centrifuge. In each cycle, the sample was centrifuged at 7000 rpm for 5 min, the supernatant was replaced with water and the sediment was redispersed by shaking manually.

2.1.2 Characterization of the polystyrene-silica core-shell composite particles

The synthesized polystyrene-silica core-shell composite particles through Pickering emulsion polymerization were characterized by various techniques. Figure 2(a) is a representative scanning electron microscope (SEM) image of the composite particles sampled at 5 hour reaction time. The roughness of the composite particle surfaces suggests that the particles are covered by silica nanoparticles. The core-shell structure can be clearly observed in the transmission electron miscroscope (TEM) image presented in Figure 2 (b). In many regions, the thickness of the shell is close to the size of one silica nanoparticle (10-15 nm), which may suggest a monolayer coverage. Furthermore, we employed hydrofluoric acid (HF) to dissolve the silica shell which led to the smooth PS core in Figure 2(c). The PS-silica composparticle size and its distribution agree well with the dynamic light scattering (DLS) measurement, as shown in Figure 2(d). Finally, we performed energy dispersive x-ray (EDX) spectrum to confirm that a substantial amount of Si and O exist, as shown in Figure 2(e). Note that the relative intensity of the peak does not necessarily correspond to the true atom ratio in the sample. One of the main reasons is that the penetration depth of the electron beam is unknown. The penetration depth depends on various factors, such as the electron beam voltage, the nature of the sample, and the Au/Pd coating thickness during the sample peparation. The EDX result only provides qualitative information regarding the existence of silica, which composes of Si and O, in the composite particles.

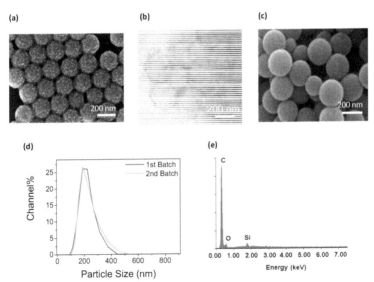

Fig. 2. Polystyrene-silica core-shell composite particles synthesized by Pickering emulsion polymerization: (a) a SEM image; (b) a TEM image of the cross-sectioned composite particles; (c) a SEM image of HF etched composite particles; (c) an overlay of DLS measurements of two batches; (d) an EDX spectrum.

It is worthwhile to note that we carefully selected VA-086 as the initiator. VA-086 is a water-soluble nonionic initiator and no success has been reported in surfactant-free emulsion polymerization of styrene (Tauer et al., 1999). In order to verify the sole stabilizing effect of silica nanoparticles, emulsifier-free emulsion polymerization using VA-086 as the initiator in the absence of nanoparticles was performed. No polystyrene particle formation was observed in the product, evidenced by SEM experiments. These experiments show that the initiator VA-086 has little effect on stabilizing the system in emulsion polymerization and therefore silica nanoparticles are the only source of stabilizer when present. In addition, VA-086 is neutral in charge thus is expected to minimize any electrostatic interactions with the negatively-charged silica nanoparticle surfaces which may complicate identifying silica nanoparticles as the sole stabilizer in emulsion polymerization.

The silica content is quantitatively determined by thermalgravimetric analysis (TGA), as shown in Figure 3. Two samples were measured: the composite particles (solid line) and the composite particles after removal of the silica component by hydrofluoric acid etching, which is essentially polystyrene cores (dashed line). The polystyrene cores show a residual weight of approximately zero at 800 °C. Thus it is reasonable to assume that the major weight loss during heating is associated with the thermo-oxidative degradation of polystyrene and the residue close to 800 °C is solely silica. The silica content of the composite particles is approximately 20 wt%. Although some silica nanoparticles remain in the continuous phase and are washed off by centrifuging-redispersing cycles, the silica content of particles prepared via solid-stabilized emulsion polymerization using nonionic initiator VA-086 is significantly higher than that of particles (1.1 wt%) prepared via dispersion polymerization using nonionic initiator AIBN (Schmid et al., 2007). The improvement is likely due to the distinct polymerization mechanisms. In contrast to the dispersion polymerization in which the polystyrene monomers are dissolved in alcohols, the emulsion polymerization here contains distinguished liquid-liquid interfaces due to the immiscibility between the monomers and the aqueous continuous phase. Therefore the nanoparticles, even in the absence of electrostatic interactions, are thermodynamically favorable to self-assemble and remain at the liquid-liquid interfaces, following the same argument in Pickering emulsions (Dai et al., 2005; Tarimala and Dai, 2004). At the initial

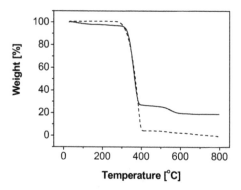

Fig. 3. Thermogravimetric analysis of the PS-silica core-shell composite particle prepared usinVA-086 as the initiator before (solid line) and after (dashed line) HF etching treatment.

stage of polymerization, the nanoparticles provide stability to the monomer droplets. During the nucleation stage, silica nanoparticles are at the interfaces between the monomer phase and continuous phase. It is worthwhile to note that the role of silica nanoparticles described here is not the same as that in the polymerization involving oppositely charged initiator and nanoparticles (Schmid et al., 2007). In the latter case, the initiator molecules or residues adsorb onto the silica nanoparticle surfaces after initiation (Schmid et al., 2007) thus the silica nanoparticles function as the surface-active initiator residue. The mechanism of the core-shell structure formation in Pickering emulsion polymerization will be detailed later on.

2.2 Possible mechanisms of Pickering emulsion polymerization

The mechanism of conventional emulsion polymerization stabilized by surfactants has been under active discussion for over half a century and some consensus has been reached. Harkins proposed three loci of particle nucleation in 1947 (Harkins, 1947), which are later developed into at least three different nucleation mechanisms (Chern, 2006): the micellar nucleation, the homogeneous coagulative nucleation, and the droplet nucleation. Upon initiator addition and decomposition, free radicals form in the aqueous phase. The micellar nucleation (Chern, 2006; Tauer et al., 2008) begins with the capture of free radicals by micelles, proceeds with the continuous swelling and polymerization of monomers in the monomer-swollen particles, and finally terminates with the exhaustion of monomers. While some researchers believe that the micellar nucleation mechanism dominates at a surfactant concentration above the critical micelle concentration, doubts have also been raised (Tauer et al., 2008). In the absence of micelles, the homogeneous coagulative nucleation mechanism is likely dominant. In homogeneous coagulative nucleation (Chern, 2006; Feeney et al., 1984, 1987; Yamamoto et al., 2004), monomers dissolve in water and undergo radical polymerization to form oligomers. The oligomers coagulate to form embryos, nuclei, and primary particles sequentially. These primary particles, stabilized by the adsorption of surfactant molecules, could grow either *via* swelling of particles by monomers or deposition of oligomers onto their surfaces (Yamamoto et al., 2006). Finally, droplet nucleation is another possible mechanism in conventional emulsion polymerization. Here the monomer droplets may be subjected to the oligomeric radical entry and solidify into particles, following the droplet nucleation mechanism. Droplet nucleation is usually minor in emulsion polymerization, except in miniemulsion polymerization when hydrophobic initiators are used.

Based on the fundamental understandings in conventional emulsion polymerization, we propose possible Pickering emulsion polymerization mechanisms, taking into account the differences between fine solid particles and surfactant molecules. Since the nanoparticles do not form micelles like surfactant molecules, micellar nucleation is excluded. Thus, there are two possible nucleation mechanisms involved in the initial stage of Pickering emulsion polymerization. Homogeneous coagulative nucleation is likely to be the dominating mechanism here, which yields the sub-micron-sized particles. The droplet nucleation might also occur, which yields micron-sized particles. The two mechanisms are illustrated in Figure 4. Upon initiator addition, monomers dissolved in the aqueous phase react with decomposed initiators and form oligomers with radicals. In homogeneous coagulative nucleation, the oligomers coagulate into nuclei, which subsequently become monomer swollen particles. Nanoparticles self-assemble at the interfaces between monomer and the continuous phase to provide stability. With the continuous supply of monomer molecules from the monomer droplets through diffusion, the particle size growth is mainly achieved by monomer swelling followed by polymerization within the core. In contrast, in droplet nucleation, initiated oligomers with radicals enter monomer droplets and subsequently polymerize into solid cores without significant size growth.

We use the hypothesized mechanisms to interpret the formation of polystyrene-silica nanocomposite particles prepared using VA-086 as the initiator. Figure 5 shows the dependence of particle size and surface coverage on reaction time and initiator concentration [Ma et al., 2010]. The composite particles are sampled from 3 h to 24 h reaction time and the initiator concentration relative to monomer is selected to be 0.83, 2.5, and 4.2 wt % respectively. At 3 h reaction time, well after the nucleation stage, composite particles with dense silica coverage are obtained. Since VA-086 initiator residues cannot provide sufficient stabilization to the monomer-swollen particles, silica nanoparticles would self-assemble at interfaces to provide stabilization and thus lead to high silica coverage. At initiator concentration 0.83 wt %, the silica coverage decreases with the particle size growth and the silica nanoparticles form patches on the nanocomposite particle surface with a low coverage. This might be an indication that the surface area of the polystyrene core increases with the particle growth without a significant increase of silica continuously attaching onto the polystyrene core. The particle growth mechanism is likely due to swelling of particles by monomers in the continuous phase. The same mechanism explains the surface coverage decrease in the system containing 2.5 wt % of initiator (images not shown) and from 3 h to 11 h in the system containing 4.2 wt % of initiator. These observations suggest that the Pickering emulsion polymerization using VA-086 as the initiator mainly follows the homogeneous coagulative nucleation mechanism.

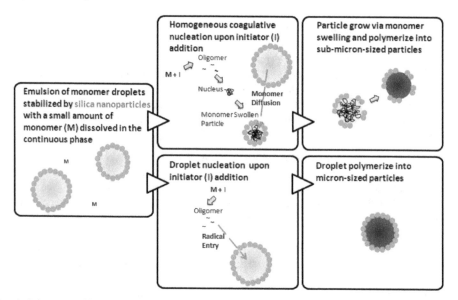

Fig. 4. Schematic illustration for possible mechanisms of Pickering emulsion polymerization.

One remaining mystery is the unexpected silica coverage from 11 h to 24 h in the system with 4.2 wt % initiator. Although the origin is unclear, we tentatively attribute the unusual silica coverage increase to the deposition of oligomers on the polystyrene core [Yamamoto et al., 2006], which adsorbed onto silica nanoparticles in the continuous phase [Yamamoto et al., 2006]. Excess initiator molecules might generate a large number of oligomers in the continuous phase, which could possibly adsorb onto silica nanoparticles. Thus when the oligomers on silica nanoparticles attach to preformed polystyrene surfaces, the silica nanoparticles are anchored there. It is also possible that the surface coverage increase might be due to the adsorption of depleted or close to depleted monomer droplets with a size below that of particles. It is worthwhile to note that the continuous phase contains approximately 21% isopropanol. The existence of isopropanol might increase the solubility of the monomer and the degree of polymerization required for an oligomer to be insoluble in the continuous phase, however, the solubility of monomer in the continuous phase is still low enough to enable emulsification and subsequent emulsion polymerization.

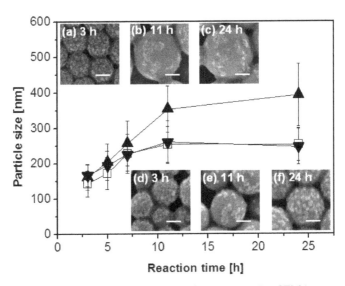

Fig. 5. Plot of particle size versus reaction time and representative SEM images with different initiator VA-086 concentrations: 0.83 wt % (▲, inset images a, b and c), 2.5 wt % (□) and 4.2 wt % (▼, inset images d, e and f). The error bars indicate the width of the particle size distribution and the scale bars represent 100 nm.

3. Environmentally responsive core-shell composite nanoparticles from Pickering emulsion polymerization

The Pickering emulsion polymerization opens a new and convenient way to synthesize core-shell composite nanoparticles. The simplicity enables further design and development advanced functional core-shell composite nanoparticles. One particular example to be illustrated here is the development of environmentally responsive core-shell composite nanoparticles from Pickering emulsion polymerization. By encapsulating drugs or chemicals, such nanoparticles enable controlled release upon environmental changes, as shown in Figure 6. For example, other than the styrene monomer, we have incorporated NIPAAm (N-isopropylacrylamide) as co-monomer into the Pickering polymerization and synthesized temperature responsive PS/PNIPAAm-silica composite nanopariticles. PNIPAAm is a well-understood temperature sensitive gel, which undergoes volume shrinkage at a transition temperature of approximately 32 °C in pure water (Schild, 1992). The mechanism of this change is based on different solubility below and above the lower critical solution temperature (LCST) in aqueous media. Below the LCST, the polymer chain is hydrophilic as the hydrogen bonding between the hydrophilic groups and water molecules dominates; above the LCST, the polymer chain becomes hydrophobic due to the weakened hydrogen bonding at elevated temperature and the hydrophobic interactions among hydrophobic groups (Qiu and Park, 2001). Figure 7(a) is a representative SEM image of the composite particles sampled at 5-hour reaction time which shows that the particles tend to be spherical. The roughness of the composite nanoparticle surfaces suggests that the nanoparticles are covered by silica nanoparticles; this is contrasted by the smooth surface of the hydrofluoric acid (HF)-treated particles in Figure 7(b). HF dissolves the silica layer and leaves behind the smooth polymer surface. It is also evidenced by the blue line in the Fourier transform infrared (FTIR) spectrum in Figure 7(c) which shows that the composite nanoparticles have a characteristic strong peak at 1104 cm^{-1}, corresponding to the asymmetrical vibration of the Si-O-Si bond. Such a peak is absent in the red line in Figure 7(c) which represents the HF-treated composite particles. FTIR is a strong analytical tool which gives information about specific chemical bonds simply by interpreting the infrared absorption spectrum; here it is used to identify the presence of silica.

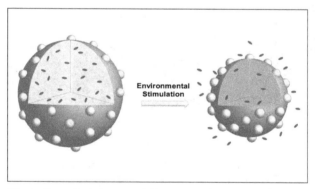

Fig. 6. Schematic illustration of environmentally responsive composite nanoparticles responding to an environmental change and releasing encapsulated materials such as a drug.

During our experiments, we also incorporated different ratios of monomer/comonomer in the formulation of the PS/PNIPAAm-silica core-shell composite nanoparticles. It is found that when the concentration of the NIPAAm monomer is high, the volume change of the nanoparticles is significantly greater with change in temperature, as shown in Figure 8. Control experiments of polystyrene-silica nanoparticles did not show a size transition over a temperature range of 25-45°C (data not shown). The transition temperature is not shifted by the silica nanoparticle encapsulation. This is consistent with the recently reported composite microspheres with a PNIPAAm core and a silica shell which also show a volume transition starting at 32°C (Qiu and Park, 2001). It is likely due to the fact that silica particles are physically adsorbed on the surfaces of PNIPAAm microspheres thus no chemical bond formation with silica occurs which might change the transition temperature. Moreover, the copolymerization with styrene has no significant effect on the transition temperature. One hypothesis is the relative phase separation of PNIPAAm and polystyrene within the core. Duracher et al. studied PNIPAAm-polystyrene particles and suggested a PNIPAAm-rich shell and a polystyrene-rich core structure (Duracher et al., 1998). Such phase separation may also occur in the core of the composite particles here although detailed morphology is unknown.

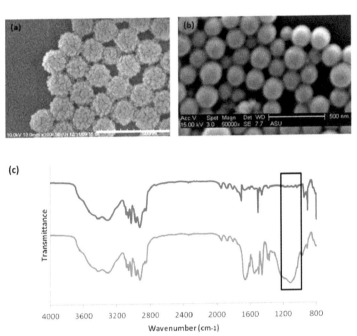

Fig. 7. (a) An SEM image of the composite particles; (b) SEM image taken after HF etching process (the scale bar represents 500 nm); (c) An FTIR spectrum of the composite nanoparticles where the blue line represents the composite particles and the red line is a sample of composite particles treated with HF. The box highlights the difference between the two spectra near 1104 cm⁻¹, which corresponding to the asymmetrical vibration of the Si-O-Si bond the two spectra near 1104 cm⁻¹ which corresponding to the asymmetrical vibration of the Si-O-Si bond.

Figure 9(a) shows the dependence of average diameter of the composite particles on temperature with 15% NIPAAm. The average particle size at 25°C is approximately 92 nm. The size decreases sharply as the temperature reaches 32°C, around the LCST for homopolymer PNIPAAm and size change is nearly reversible upon cooling. In addition, we have encapsulated a drug, 17-(Allylamino)-17-demethoxygeldanamycin (17AAG), during the Pickering emulsion polymerization and performed cumulative drug release measurements. Figure 9(b) depicts the cumulative fractional drug release at 25 °C and 40 °C of 17-AAG from the drug-loaded nanoparticles. No significant release of the drug was observed at room temperature (25°C). However, at a higher temperature of 40°C, the drug

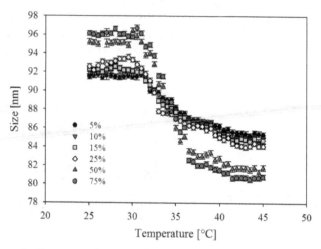

Fig. 8. Hydrodynamic diameters of PS/PNIPAAm-silica composite core-shell nanoparticles measured by DLS, decrease significantly near the transition temperature of PNIPAAm (32°C). The legend shows the various concentrations of NIPAAm, it is observed for higher concentrations of NIPAAm there is greater size change of nanoparticles.

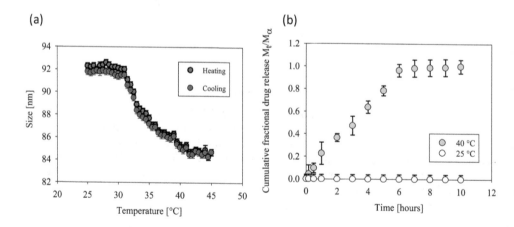

Fig. 9. (a) The dependence of average diameter of composite nanoparticles on temperature. The error bars show standard deviations of particles made in three different batches. The transition temperature is around 32°C. There size transition is nearly reversible; (b) Cumulative fractional drug release versus time curve indicating release at room temperature and at 40 °C. There is minimum release at 25 °C which is below the transition temperature of the nanoparticles.

releases from the nanoparticles reached a maximum after 7 h. The cumulative fractional drug release is calculated as M_t/M_∞, where t is the release time, M_t is the amount of drug released at a time t and M_∞ is the amount of drug released at time infinity. Infinity is taken to be when the maximum amount of drug gets released and there is no subsequent release after infinity. The concentration of the drug in the sample solution was read from the calibration curve as the concentration corresponding to the absorbance of the solution. To determine the release mechanisms of the composite nanoparticle system an equation proposed by Yasuda et al. was used (Yasuda et al., 1968), which analyses the release behavior of a solute from a polymer matrix, $\dfrac{M(t)}{M(\infty)} = kt^n$ where k is a constant related to the physical properties of the system, and the index, n, is the diffusional component that depends on the release mechanism. When n<0.5, the solute is released by Fickian diffusion; when 0.5<n<1.0, the solute is released by non-Fickian diffusion and when n=1, there is zero order release [Yasuda et al., 1968]. The calculated n value is 0.73 which indicates the non-Fickian diffusion. The mathematical model indicates that the drug diffusion behavior is non-Fickian and the rate of drug release is due to the combined effect of drug diffusion and polymer response due to increase in temperature.

4. Conclusion

In summary, polystyrene-silica core-shell composite particles were successfully synthesized by a novel one-step Pickering emulsion polymerization. The sole stabilizing effect of silica nanoparticles in the emulsion polymerization was verified. In addition, possible mechanisms of Pickering emulsion polymerization were explored and suggested that homogeneous coagulative nucleation is likely the dominating mechanism here. Finally, the temperature responsiveness of core-shell composite nanoparticles and drug release were validated by incorporating NIPAAM as a co-monomer into the Pickering emulsion polymerization.

5. Acknowledgments

We gratefully acknowledge Professor Paul Westerhoff at Arizona State University for the DLS usage and the Imaging Center at Texas Tech University/Solid State Center at Arizona State University for electron microscope usage. We are also grateful for the financial support provided by the National Science Foundation (CBET-0918282, CBET-0922277, and CBET-1050045) and the Texas Higher Education Coordinating Board.

6. References

Barthet, C.; Hickey, A. J.; Cairns, D. B.; Armes, S. P. (1999). Synthesis of novel polymer-silica colloidal nanocomposites via free-radical polymerization of vinyl monomers. *Adv. Mater., 11*, 408-410.

Bon, S. A. F.; Colver, P. J. (2007). Pickering miniemulsion polymerization using laponite clay as a stabilizer. *Langmuir, 23*, 8316-8322.

Caruso, F. (2001). Nanoengineering of particle surfaces. *Adv. Mater., 13*, 11-22.

Caruso, F.; Susha, A. S.; Giersig, M.; Möhwald, H. (1999). Magnetic core-shell particles: Preparation of magnetite multilayers on polymer latex microspheres. *Adv. Mater.,* *11,* 950-953.

Caruso, R. A.; Susha, A.; Caruso, F. (2001). Multilayered titania, silica, and laponite nanoparticle coatings on polystyrene colloidal templates and resulting inorganic hollow spheres. *Chem. Mater., 13,* 400-409.

Cauvin, S.; Colver, P. J.; Bon, S. A. F. (2005). Pickering stabilized miniemulsion polymerization: Preparation of clay armored latexes. *Macromolecules, 38,* 7887-7889.

Chern, C. S. (2006). Emulsion polymerization mechanisms and kinetics. *Prog. Polym. Sci., 31,* 443-486.

Dai, L. L.; Sharma, R.; Wu, C. Y. (2005). Self-assembled structure of nanoparticles at a liquid-liquid interface. *Langmuir, 21,* 2641-2643.

Ding, X. F.; Jiang, Y. Q.; Yu, K. F.; Hari-Bala; Tao, N. N.; Zhao, J. Z.; Wang, Z. C. (2004). Silicon dioxide as coating on polystyrene nanoparticles in situ emulsion polymerization. *Mater. Lett., 58,* 1722-1725.

Dokoutchaev, A.; James, J. T.; Koene, S. C.; Pathak, S.; Prakash, G. K. S.; Thompson, M. E. (1999). Colloidal metal deposition onto functionalized polystyrene microspheres. *Chem. Mater., 11,* 2389-2399.

Duan, L.; Chen, M.; Zhou, S.; Wu, L. (2009). Synthesis and characterization of poly(N-isopropylacrylamide)/Silica composite microspheres via inverse pickering suspension polymerization. *Langmuir, 25,* 3467-3472.

Duracher, D.; Sauzedde, F.; Elaissari, A.; Perrin, C. (1998). Pichot, *Colloid Polym. Sci., 276,* 219.

Feeney, P. J.; Napper, D. H.; Gilbert, R. G. (1987). Surfactant-Free Emulsion Polymerizations: Predictions of the Coagulative Nucleation Theory. *Macromolecules,* 20, 2922-2930.

Feeney, P. J.; Napper, D. H.; Gilbert, R. G. (1984). Coagulative Nucleation and Particle Size Distributions in Emulsion Polymerization. *Macromolecules,* 17, 2520-2529.

Gao, Q.; Wang, C.; Liu, H.; Wang, C.; Liu, X.; Tong, Z. (2009). Suspension polymerization based on inverse pickering emulsion droplets for thermo-sensitive hybrid microcapsules with tunable supracolloidal structures. *Polymer, 50,* 2587-2594.

Harkins, W. D. (1947). A general theory of the mechanism of emulsion Polymerization 1. *J. Am. Chem. Soc., 69,* 1428-1444.

Lynch, D. E.; Nawaz, Y.; Bostrom, T. (2005). Preparation of sub-micrometer silica shells using poly(1-methylpyrrol-2-ylsquaraine). *Langmuir, 21,* 6572-6575.

Ma, H.; Dai, L. L. (2009). Synthesis of polystyrene-silica composite particles via one-step nanoparticle-stabilized emulsion polymerization. *Journal of Colloid and Interface Science. 333,* 807-811.

Ma, H.; Luo, M. X.; Sanyal, S.; Rege, K.; Dai, L. L. (2010). The one-step Pickering emulsion polymerization route for synthesizing organic-inorganic nanocomposite particles. *Materials, 3,* 1186-1202.

Pickering, S. U. (1907). Emulsions. *J. Chem. Soc., 91,* 2001-2021.

Qiu, Y.; Park, K. (2001). Environment-sensitive hydrogels for drug delivery. *Adv. Drug Deliv. Rev.*, *53*, 321-339.

Ramsden, W. (1903). Separation of solids in the surface-layers of solutions and "suspensions" (observations on surface-membranes, bubbles, emulsions, and mechanical coagulation). preliminary account.W. *Proc. R. Soc. London, 72*, 156-164.

Schild, H. G. (1992). Poly(N-isopropylacrylamide): Experiment, theory and application. *Prog. Polym. Sci.*, *17*, 163-249.

Schmid, A.; Fujii, S.; Armes, S. P. (2006). Polystyrene-silica nanocomposite particles via alcoholic dispersion polymerization using a cationic azo initiator. *Langmuir*, *22*, 4923-4927.

Schmid, A.; Fujii, S.; Armes, S. P.; Leite, C. A. P.; Galembeck, F.; Minami, H.; Saito, N.; Okubo, M. (2007). Polystyrene-silica colloidal nanocomposite particles prepared by alcoholic dispersion polymerization. *Chem. Mater.*, *19*, 2435-2445.

Tarimala, S.; Dai, L. L. (2004). Structure of microparticles in solid-stabilized emulsions. *Langmuir, 20*, 3492-3494.

Tarimala, S.; Ranabothu, S. R.; Vernetti, J. P.; Dai, L. L. (2004). Mobility and in situ aggregation of charged microparticles at oil-water interfaces. *Langmuir, 20*, 5171-5173.

Tarimala, S.; Wu, C.; Dai, L. L. (2006). Dynamics and collapse of two-dimensional colloidal lattices. *Langmuir, 22*, 7458-7461.

Tauer, K.; Deckwer, R.; Kuhn, I.; Schellenberg, C. (1999). A comprehensive experimental study of surfactant-free emulsion polymerization of styrene. *Colloid Polym. Sci.*, *277*, 607-626.

Tauer, K.; Hernandez, H.; Kozempel, S.; Lazareva, O.; Nazaran, P. (2008). Towards a consistent mechanism of emulsion polymerization - new experimental details. *Colloid Polym. Sci.*, *286*, 499-515.

Voorn, D. J.; Ming, W.; van Herk, A. M. (2006). Polymer-clay nanocomposite latex particles by inverse pickering emulsion polymerization stabilized with hydrophobic montmorillonite platelets. *Macromolecules, 39*, 2137-2143.

Wu, J.; Dai, L. L. (2006). One-particle microrheology at liquid-liquid interfaces. *Appl. Phys. Lett.*, *89*, 094107.

Wu, J.; Dai, L. L. (2007). Apparent microrheology of oil-water interfaces by single-particle tracking. *Langmuir, 23*, 4324-4331.

Wu, C.; Song, Y.; Dai, L. L. (2009). Two-particle microrheology at oil-water interfaces. *Appl. Phys. Lett.*, *95*, 144104.

Yamamoto, T.; Kanda, Y.; Higashitani, K. (2006). Initial growth process of polystyrene particle investigated by AFM. *J. Colloid Interface Sci.*, *299*, 493-496.

Yamamoto, T.; Kanda, Y.; Higashitani, K. (2004). Molecular-scale observation of formation of nuclei in soap-free polymerization of styrene. *Langmuir, 20*, 4400-4405.

Yamamoto, T.; Nakayama, M.; Kanda, Y.; Higashitani, K. (2006). Growth mechanism of soap-free polymerization of styrene investigated by AFM. *J. Colloid Interface Sci.*, *297*, 112-121.

Yang, J.; Hasell, T.; Wang, W. X.; Li, J.; Brown, P. D.; Poliakoff, M.; Lester, E.; Howdle, S. M. (2008). Preparation of hybrid polymer nanocomposite microparticles by a nanoparticle stabilised dispersion polymerisation. *J. Mater. Chem.*, *18*, 998-1001.

Yasuda, H.; Lamaze, C. E.; Ikenberry, L. D. (1968). *Die Makromolekulare Chemie, 118,* 19.

12

Electrospray Production of Nanoparticles for Drug/Nucleic Acid Delivery

Yun Wu[1], Anthony Duong[1,2], L. James Lee[1,2] and Barbara E. Wyslouzil[1,2,3]
[1]NSF Nanoscale Science and Engineering Center,
The Ohio State University, Columbus, Ohio
[2]William G. Lowrie Department of Chemical and Biomolecular Engineering,
The Ohio State University, Columbus, Ohio
[3]Department of Chemistry, The Ohio State University, Columbus, Ohio
USA

1. Introduction

Nanomedicine – the application of nanotechnology to medicine – shows great potential to positively impact healthcare. Nanoparticles, including solid lipid nanoparticles, lipoplexes and polyplexes, can act as carriers to deliver the drugs or nucleic acid-based therapeutics that are particularly promising for advancing molecular and genetic medicine.

Many techniques have been developed to produce nanoparticles. Among them, electrospray has attracted recent research interest because it is an elegant and versatile way to make a broad array of nanoparticles. Electrospray is a technique that uses an electric field to disperse or break up a liquid. Compared with current technologies, such as bulk mixing, high pressure homogenization and double emulsion techniques, electrospray has three potential advantages. First, electrospray can generate monodisperse droplets whose size can vary from tens of nanometer to hundreds of micrometers, depending on the processing parameters. Secondly, it is a very gentle method. The free charge, induced by the electric field, only concentrates at the surface of the liquid, and does not significantly affect sensitive biomolecules such as DNA. Finally electrospray has the ability to generate structured micro/nanoparticles in a more controlled way with high drug/nucleic acid encapsulation efficiency.

Electrospray is a critical element of electrospray ionization mass-spectrometry, an analytical technique used to detect macromolecules that was developed by the 2002 chemistry Nobel Prize winner, Dr. John B. Fenn. (Fenn et al. 1989) Since then research has focused both on developing a fundamental understanding of the process as well as exploring potential applications of electrospray in fields ranging from the semiconductor industry to life science. During the past two decades, electrospray has been used to assist pyrolysis reactions and chemical vapor deposition processes, to produce inorganic particles including fine metal powder (Sn, Ag, Au, etc.), metal oxide particles (ZrO2, TiO2, etc.), ceramic particles (Si, SiO2), and semiconductor quantum dots (CdSe, GaAs, etc.) (Jaworek 2007; Salata 2005). Various structured particles have been produced via electrospray, including poly(methyl methacrylate)-pigment nanoparticles (Widiyandari et al. 2007), cocoa butter microcapsules containing a sugar solution or an oil-in-water emulsion (Loscertales et al. 2002; Bocanegra et al. 2005) and microbubble suspensions (Farook et al. 2007), to name a few. Electrospray has been used to deposit particle suspensions to form thin film (Jaworek & Sobczyk 2008, Jaworek 2010) or on-demand patterns, such as silica particle coatings on a quartz glass (Jaworek 2007; Salata 2005). Electrospray has been successfully applied in tissue engineering, for example, polymer materials including poly(lactide-co-glycolide) or poly(ethylene glycol) were electrospray-coated on biomedical implants (Kumbar et al. 2007). Combined with electrospinning, electrospray was used to fabricate smooth muscle cell integrated blood vessel constructs. (Stankusa et al. 2007) In the area of drug/nucleic acid delivery, many biological materials, such as DNA, proteins, and lipids have been electrosprayed without changing their biological activity (Pareta et al. 2005; Jayasinghe et al. 2005; Davies et al. 2005; Wu et al. 2009a, 2009b, 2010, 2011). Proteins, such as bovine serum albumin, have been encapsulated in biodegradable polymeric microcapsules (Pareta and Edirisinghe, 2006; Xie & Wang, 2006), and small molecule drugs, such as taxol and griseofulvin, have been encapsulated in polymeric microparticles for systemic or oral delivery. (Xie et al. 2006; Zhang et al, 2011)

2. Principles of electrospray

As illustrated in Figure 1a, in the standard electrospray configuration, a conducting liquid is slowly injected through a needle by a syringe pump. An electrical potential is applied to the needle to introduce free charge at the liquid surface. The free charge generates electric stress that causes the liquid to accelerate away from the needle. When the electrical potential rises to several kilovolts, the liquid meniscus at the needle opening develops into a conical shape, commonly called the Taylor cone. At the cone apex, where the free charge is highly concentrated, a liquid jet with high charge density is observed. Monodisperse particles are formed when the jet breaks into fine particles due to varicose or kink instabilities. (Cloupeau et al. 1994; Jaworek 2007; Salata 2005; Loscertales et al. 2002). Typically, the initial micron size droplets contain both solvent and a non-volatile solute and nanoparticles are produced as solvent evaporates from the high surface area aerosol.

Other electrospray configurations are also possible. Figure 1(b) shows a coaxial electrospray configuration, where two liquids are fed through the inner needle and the outer needle, respectively. This configuration is widely used to produce structured nanoparticles. Finally, since the flow rate in a single electrospray setup is always low (uL/hr or mL/hr), multiplexed electrospray configurations have been developed to scale up production. (Deng et al, 2006) As shown in Figure 1(c), the single needle is replaced by a micro-nozzle array. All nozzles work simultaneously and produce nanoparticles with the liquid flow increased by orders of magnitude.

The electrospray modes and resulting droplet sizes d_D are controlled by the process parameters that include the applied voltage V, liquid flow rate Q, and liquid properties including electrical conductivity γ, surface tension σ, and liquid density ρ. (Cloupeau et al. 1994; Jaworek 2007; Salata 2005; Loscertales et al. 2002, Ganan-Calvo AM. 2004; Hartman et al. 1999; Basak et al. 2007) As illustrated in Figure 2, different electrospray modes are obtained as the applied voltage increases. (Cloupeau et al. 1994; Jaworek and Krupa 1999; Chen et al. 2005; Yurteri et al. 2010) These include dripping, micro-dripping, spindle,

Taylor cone-jet and multi-jet mode. Taylor cone-jet mode is the most common used electrospray mode, because it can produce highly monodisperse particles in a stable manner.

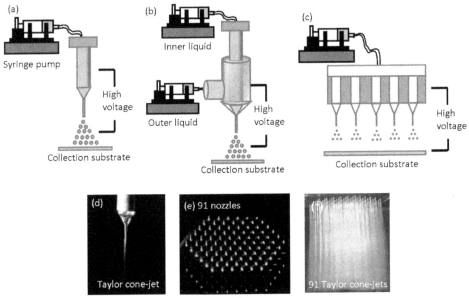

Fig. 1. Schematic diagrams of (a) standard electrospray (b) coaxial electrospray and (c) multiplexed electrospray. (d) Taylor cone-jet from standard electrospray (e) 91 nozzles used in multiplexed electrospray and (f) 91 Taylor cone-jets formed simultaneously. (Figures (e) and (f) are courtesy of Dr. Weiwei Deng).

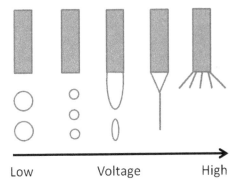

Fig. 2. The electrospray modes change as the voltage increases. From left to right: dripping, microdripping, spindle, Taylor cone-jet and multi-jet.

In cone-jet mode, the droplet size scales with the liquid flow rate, and is inversely proportional to the liquid conductivity. Theoretically droplet sizes can be determined from the following scaling law, confirmed by many experiments (Jaworek 2007).

$$d_D = \alpha \left(\frac{Q^3 \varepsilon_0 \rho}{\pi^4 \sigma \gamma} \right)^{1/6} \quad (\alpha = 2.9)$$

where ε_0 is the vacuum permittivity. In general, droplet size decreases with decreasing flow rate and increasing electrical conductivity of the liquid. By adjusting the flow rate and the liquid properties, droplets with the desired size can be produced in a well-controlled manner. Frequently, the initial solution contains both solvent and solute, and the final particle size (d_p) is related to the initial droplet size d_D by the following equation:

$$d_p = d_D \left(\frac{c\rho}{\rho_p} \right)^{1/3}$$

where c is the concentration of solute and ρ_p is the density of the particle.

3. Electrospray production of nanoparticles for drug/nucleic acid delivery

Our research group has been the first to explore electrospray as a means to produce solid lipid nanoparticles, lipoplexes and polyplexes for drug/nucleic acid delivery. Our results demonstrate the great potential of electrospray to produce nanoparticles in a well-controlled manner, and opens new avenues for the development of nanomedicine.

3.1 Solid lipid nanoparticles for hydrophobic drug delivery

3.1.1 Introduction

Solid lipid nanoparticles (SLNs) have been proposed as an alternative drug delivery system to traditional colloidal carriers, such as emulsions, liposomes and polymeric nanoparticles. As drug delivery vehicles, emulsions and liposomes are often limited by physical instability and low drug loading capacity. (Müller et al. 2000, 2004) Polymeric nanoparticles can exhibit cytotoxicity, and there are no industrialized production methods. SLNs may provide a drug delivery system that avoids many of these problems, and high pressure homogenization (HPH), the most popular preparation technique, is a reasonable method for large-scale production. Unfortunately, HPH requires molten lipids and high energy input. Thus, it is often unsuitable for use with temperature and shear sensitive biomolecules. Furthermore, as the mixtures cool, some drugs do not incorporate into the lipids but stick on the surface, resulting in burst drug release kinetics and low drug loading capacity. (Müller et al. 2000, 2004)

In our work we used electrospray to produce solid lipid nanoparticles. (Wu et al. 2009a, 2011) Cholesterol was chosen as a model since it is not only an important lipid but also a good model for lipophilic drugs. Similar to most lipophilic drugs, cholesterol has very low solubility in aqueous solutions, on the order of 1.8µg/mL or 4.7µM. (Haberland & Reynolds, 1973) Cholesterol has similar structure as the well-known anticancer drug Taxol®, and cholesterol (molecular weight = 387.6) is also a small molecule. In addition, the solubility of lipophilic drugs in lipids is much higher than in aqueous solutions, and, therefore, lipid nanoparticles, like the cholesterol nanoparticles discussed here, may also be useful as carriers to deliver lipophilic drugs.

3.1.2 Production of cholesterol nanoparticles

The standard electrospray configuration (Figure 1a) was used to produce cholesterol nanoparticles dispersed in aqueous media. Cholesterol powder (Sigma C3045) was dissolved in ethanol (Pharmco-Aaper E200GP) at a concentration of 2 mg/mL and then delivered to a 27 gauge needle by a syringe pump at a flow rate of 2 mL/h. A voltage of approximate 2.5 kV was applied to the needle and the observation of a stable Taylor cone-jet mode confirmed the proper conditions for generating fine droplets. The ethanol evaporated from the droplets, aided by heat input from a lamp, and the residual cholesterol particles were captured in 1X PBS supplemented with 1% Pluronic F68 placed in a grounded aluminum dish 5.5 cm below the needle tip. Figure 3 illustrates a typical size distribution measured for the cholesterol nanoparticle suspension. Highly monodisperse cholesterol nanoparticles were produced with sizes ~150 nm. The cholesterol concentration in the final solution was 180 ± 10 µg/mL prior to sterile filtration and 150 ± 8 µg/mL afterwards, i.e. up to 100 times higher than the solubility limit for cholesterol.

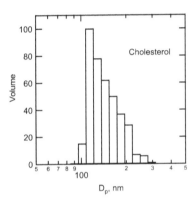

Fig. 3. A typical size distribution of cholesterol nanoparticles, in which Dp, mean, v was 140 nm and polydispersity was 0.107.

3.1.3 Bioavailability of cholesterol nanoparticles for NS0 cells

In this study, cholesterol itself was the bio-active ingredient, and NS0 cells were used to investigate the bioavailability of the cholesterol nanoparticles. While NS0 cells normally require serum as a cholesterol source, several research groups have demonstrated the growth of NS0 cells in protein free and chemically defined medium, using chemically defined cholesterol supplements. (Zhang & Robinson, 2005; Talley et al. 2005; Ojito et al. 2001) The auxotrophic nature of NS0 cells with respect to cholesterol makes this cell line a natural choice for characterizing the bioavailability of cholesterol nanoparticles produced by electrospray.

NS0 cells were continuously cultured in 250 mL shake flasks with 50 mL working volume in the basal medium supplemented with cholesterol nanoparticles or SyntheChol NS0 supplement (Sigma S5442, SyntheChol for short) at a cholesterol concentration level of 3.5 µg/mL. SyntheChol is a proprietary cholesterol supplement in liquid form and acts as the positive control. Figure 4 illustrates the viable cell density and cell viability as a function of time. NS0 cells grew better with cholesterol nanoparticles than with SyntheChol reaching much higher peak cell densities: 5.28e6 cells/mL vs. 3.35e6 cells/mL.

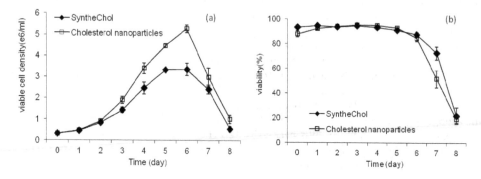

Fig. 4. Growth of NS0 cells in media supplemented with 3.5 µg/mL of cholesterol nanoparticles or SyntheChol: viable cell density **(a)** and cell viability **(b)**. The results are averages of triplicate runs.

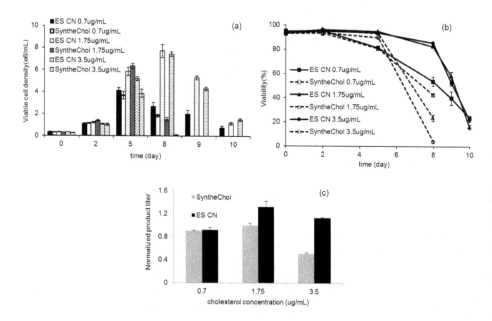

Fig. 5. The effect of cholesterol concentration on **(a)** the growth, **(b)** the viability, and **(c)** the product titer of NS0 cells during fed-batch culture. Cholesterol nanoparticles or SyntheChol supplement were fed on day 0 and daily from day 2 till the end of culture at cholesterol concentration levels of 0.7 µg/mL, 1.75 µg/mL or 3.5 µg/mL. Error bars indicate one standard deviation of the results of triplicate runs. (ES CN: electrosprayed cholesterol nanoparticles)

The NS0 cell line is one of the important cell lines widely used by the pharmaceutical industry to produce therapeutic antibodies. We therefore investigated the effect of the cholesterol source on product titer by conducting a fed-batch culture of NS0 cells using electrosprayed cholesterol nanoparticles and the SyntheChol supplement at feed rates of 0.7, 1.75 and 3.5 µg/mL/day. Figure 5 summarizes the viable cell density, cell viability and normalized product titer (relative to the 1.75 µg/mL/day SyntheChol supplement). At the lowest feed rate of 0.7 µg/mL/day, neither cholesterol nanoparticles nor SyntheChol supplement provided sufficient cholesterol to the cells and so cell growth was limited and the corresponding product titer was low (Figure 5c). When the feed rate of SyntheChol was increased, cell growth improved up to a feed rate of 1.75 µg/mL/day but was compromised at feed rate of 3.5 µg/mL/day. The inhibitory effect of 3.5 µg/mL/day feeding was presumably due to the compounds in SyntheChol that solubilize the cholesterol rather than to the presence of cholesterol. For cholesterol nanoparticles, both cell growth and viability improved at feed rates higher than 0.7 µg/mL/day. Compared to SyntheChol, culturing was extended from 8 days to 10 days. In general, the performance of cholesterol nanoparticles was better than that of SyntheChol supplementation at the same cholesterol feed rates. For both cholesterol nanoparticles and SyntheChol supplements, the highest product titer was achieved at a feed rate of 1.75 µg/mL/day and the product titer with cholesterol nanoparticles was ~32% higher than that with SyntheChol.

3.1.4 Conclusions

Compared with conventional technologies, we have demonstrated that electrospray is an efficient way to produce solid lipid nanoparticles for drug delivery. In our application we used electrospray to produce highly monodisperse cholesterol nanoparticles, and investigated the bioavailability of the nanoparticles using the cholesterol auxotrophic NS0 cell line with SyntheChol as the positive control. We found that the cholesterol nanoparticles not only supported NS0 cell growth but also improved the titer of therapeutic antibody by 32% when compared with SyntheChol.

3.2 DNA/polycation polyplexes for gene delivery

3.2.1 Introduction

Viral and non-viral vectors have been developed to deliver nucleic acids to treat genetic and acquired disease. (Pack et al., 2005; Mastrobattista et al., 2006) Compared to viral vectors, non-vrial vectors show lower immunogenicity, lower toxicity, better stability and lower cost. (Laporte et al., 2006; Kircheis, et al., 2001; Gebhart and Kabanov, 2001). Among the nonviral vectors, polyplexes and lipoplexes represent two major carrier systems. In this section, we focus on polyplexes for gene delivery, while in the following section we discuss the use of lipoplexes for oligonucleotides delivery.

Polyethylenimines (PEI) is a cationic polymer widely used to condense DNA and form polyplexes. Although PEI shows great promise for gene delivery both *in vitro* and *in vivo* (Godbey, et al., 1999; Neu et al., 2005), the delivery efficiency highly depends on the N/P ratio (the molar ratio of nitrogen in PEI to phosphate in DNA) and the preparation method. Bulk mixing (BM), the most commonly used method, is a simple and straightforward way to prepare DNA/PEI polyplexes. Unfortunately, the formation of the polyplexes is not well controlled in bulk mixing, significantly affecting the particle size, structure, and, thus, the delivery efficiency. (Kircheis, et al., 2001; Gebhart and Kabanov, 2001) Alternative methods are needed to overcome this challenge.

Because electrospray is a gentle method that does not damage biomolecules it has attracted research interest in the realm of gene delivery. In some of the earliest work, enhanced green fluorescent protein (eGFP) plasmid was successfully delivered into African Green Monkey fibroblast cells (COS-1) using coaxial electrospray. (Chen et al. 2000) Since electrospray generates an aerosol, Davies et al. (2005) investigated gene delivery by exposing mice to naked pCIKLux aerosol generated via electrospray. Unfortunately, the *in vivo* delivery efficiency, 0.075%, was much lower than the 0.2% delivery efficiency achieved by Koshkina et al. (2003) when they nebulized pDNA/PEI polyplexes for pulmonary gene delivery. One possible reason for the low delivery efficiency of electrosprayed pCIKLux might be that the naked pDNA used is not stable in biological fluids and does not interact with cell membrane due to its negative charges. Our hypothesis was that pDNA/PEI polyplexes produced via electrospray might overcome limitations in the earlier studies. Thus, we used coaxial electrospray to achieve better control over the mixing of the DNA solution and the PEI solution, and thus to produce DNA/PEI polyplexes in a well-controlled manner. (Wu et al. 2010)

3.2.2 Coaxial electrospray DNA/PEI polyplexes

Coaxial electrospray was explored as a method to produce pDNA/PEI polyplexes using pGFP and pSEAP for qualitative and quantitative studies, respectively. The pDNAs were dispersed in OPTI-MEM medium at a concentration of 20 μg/mL. Branched 25 kDa PEI was dissolved in OPTI-MEM medium at concentrations of 17.2 μg/mL. In the coaxial electrospray setup, the pDNA solution flowed through the inner 27 gauge needle and the PEI solution flowed through the outer 20 gauge needle. The flow rates of the pDNA and PEI solutions were both set at 6mL/hr to yield an N/P ratio of 6.7. A positive voltage, typically 5~6 kV, was applied between the inner needle and a grounded copper ring electrode. The polyplexes were produced in the stable Taylor cone jet mode, captured in a grounded aluminum dish 5 cm below the needle tip, and used within 10 minutes.

The conventional bulk mixing method was also used to prepare pDNA/PEI polyplexes. Here PEI solution (17.2 μg/mL) was added into equal volume pDNA (20 μg/mL) solution to achieve an N/P ratio of 6.7. The resulting mixture was vortexed for a few seconds, incubated at room temperature for 10 minutes and used immediately.

3.2.3 pDNA damage analysis

To investigate the effects of coaxial electrospray on pDNA integrity, pGFP and pSEAP solutions were coaxial electrosprayed with OPTI-MEM medium containing no PEI. The flow rate for both pDNAs was set at 6mL/h, and the voltages were set at 5.0 kV for pGFP and 5.6 kV for pSEAP. Agarose gel electrophoresis was used to detect if the pDNA was damaged by coaxial electrospray. As illustrated in Figure 6, compared to the non-electrosprayed pDNA controls, little degradation of pDNA is observed. In addition, the ratio of supercoiled to open circular pDNA of the sprayed samples appears unchanged relative to the controls. Our observations are consistent with those reported by Davies et al. (2005) and confirm that coaxial electrospray is a gentle but effective way to spray delicate bio-molecules.

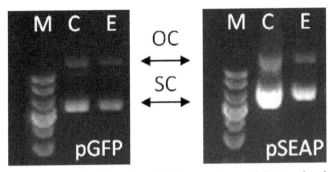

Fig. 6. Gel electrophoresis shows that the pDNA is not damaged during the electrospray process. M: marker lane. C: control. E: electrosprayed pDNAs. Electrospray conditions: pGFP: flowrate of 6 mL/h and voltage of 5.0 kV; pSEAP: flowrate of 6 mL/h and voltage of 5.6 kV. OC: open circular. SC: supercoiled plasmid.

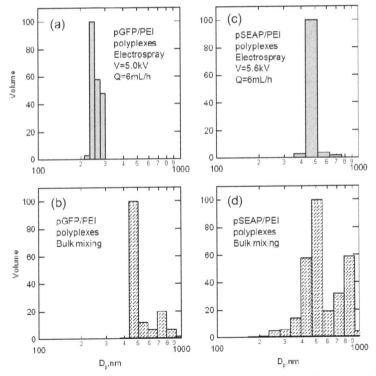

Fig. 7. Typical size distributions of the pDNA/PEI polyplexes prepared by coaxial electrospray and bulk mixing. The flow rate was 6 mL/h and the voltage was 5.0 kV for pGFP/PEI polyplexes and 5.6 kV for pSEAP/PEI polyplexes.

3.2.4 Size distribution of pDNA/PEI polyplexes

Dynamic light scattering was used to measure the size distribution of pGFP/PEI and pSEAP/PEI polyplexes prepared by coaxial electrospray, and bulk mixing. (Figure 7) The pDNA/PEI polyplexes prepared by either method are relatively monodisperse. The pGFP/PEI polyplexes prepared by coaxial electrospray had smaller particle size compared to those prepared by bulk mixing, while the sizes of the pSEAP/PEI polyplexes prepared by coaxial electrospray and bulk mixing were comparable. We note that the polyplexes prepared in this work were almost always larger than 400 nm because OPTI-MEM medium rather than NaCl solutions were used. OPTI-MEM medium was chosen to prepare polyplexes because it better mimics the *in vivo* situation than NaCl or PBS, and because its physical properties (electric conductivity and surface tension) are appropriate for the electrospray process. The downside is that OPTI-MEM medium contains reduced serum, and the adsorption of serum albumin and other negatively charged proteins enhances aggregation of polyplexes (Pack et al., 2005). Although the size distribution of each sample was measured within 10 minutes after preparation, the polyplexes were always larger than 300 nm suggesting that aggregates already formed. The adsorption of proteins to the polyplexes was also demonstrated by zeta potential measurements. The zeta potentials of all polyplexes samples were either close to zero or only slightly positive. Approaches such as grafting hydrophilic polymers, including polyethylene glycol (PEG) or polysaccharides, onto the PEI may help improve the stability of the polyplexes in serum (Pack et al., 2005; Laporte et al., 2006; Neu et al., 2001) and is one approach that could reduce aggregation.

3.2.5 pDNA/PEI polyplex transfection, DNA expression, and cell viability

The delivery efficiency of pGFP/PEI polyplexes produced by coaxial electrospray and bulk mixing was evaluated in NIH 3T3 cells and compared to Lipofectamine™ 2000, the positive control. As illustrated in Figure 8(a), at N/P ratio of 6.7, cells transfected with polyplexes produced by coaxial electrospray showed similar GFP expression as those transfected with Lipofectamine™ 2000, while cells transfected with polyplexes produced by bulk mixing showed much less GFP expression. No significant difference in cell viability was found among coaxial electrospray, BM and Lipofectamine™ 2000 (p<0.05).

Since the evaluation of GFP expression is only qualitative, pSEAP was used to compare the delivery efficiency of polyplexes in a more quantitative way. As shown in Figure 9 (a), at N/P ratio of 6.7, cells transfected with the pSEAP/PEI polyplexes produced by coaxial electrospray gave 2.6 times higher SEAP expression than those produced by bulk mixing. Although cells transfected with Lipofectamine™ 2000 showed the highest SEAP expression, 3.3 times higher than bulk mixing, the cell viability was low because of the toxicity of this material. (Figure 9(b))

3.2.6 Conclusions

Both pGFP/PEI and pSEAP/PEI polyplexes were successfully produced using coaxial electrospray. The delivery efficiency of pDNA/PEI polyplexes was evalulated in NIH 3T3 cells. At N/P ratio of 6.7, polyplexes produced by coaxial electrospray were more effective at delivering genes to NIH 3T3 cells than those prepared by bulk mixing. Since coaxial electrospray is an aerosol technique, it should be a useful way to deliver pDNA/PEI polyplexes directly to the lungs and achieve higher gene delivery efficiency than either electrospraying naked pDNA or nebulizing pDNA/PEI polyplexes produced by conventional methods.

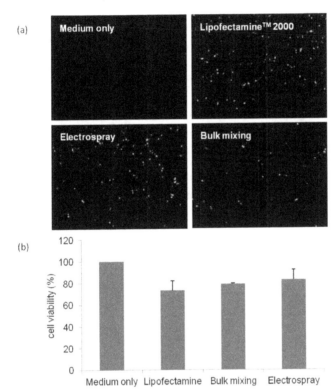

Fig. 8. **(a)** The GFP expression in NIH 3T3 cells and **(b)** the cell viability measured 2 days post transfection. All polyplexes were produced at the N/P ratio of 6.7. For coaxial electrospray, the flow rate was 6mL/h and voltage was 5.0kV. (n=4)

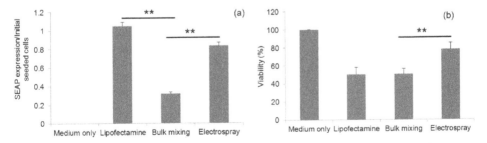

Fig. 9. The pSEAP expression in NIH 3T3 **(a)** and the cell viability **(b)** measured 2 days post transfection. All polyplexes were produced at the N/P ratio of 6.7. For coaxial electrospray, the flow rate was 6 mL/h and voltage was 5.6 kV. (n=5, **: p<0.005)

3.3 Coaxial electrospray lipoplexes for oligodeoxynucleotides (ODN) delivery

3.3.1 Introduction

Lipoplexes, a major non-viral carrier system, have attracted a lot of research interest in the drug and nucleic acid delivery field. The phospholipid bilayer structure of lipoplexes provides great flexibility in encapsulating various drugs or nucleic acids, either within the hydrophobic bilayers or in the hydrophilic core. Many techniques, such as bulk mixing (Chrai et al. 2001; Elouahabi et al. 2005), the film method (Bangham method) (Fan et al. 2007; Otake et al. 2006) and alcohol dilution (Stano et al. 2004), have been developed to produce lipoplexes. However, non-uniform particle size, low encapsulation efficiency and multistep production process remain as major challenges. (Müller et al. 2000, 2004)

Antisense oligodeoxynucleotides (ODN) are short pieces of specially designed DNA. ODN has been widely investigated as a potential therapeutic agent against viral infections, cardiovascular inflammation, hematological malignancies, pulmonary diseases and cancer. (Chiu et al. 2006; Yang et al. 2004) ODN inhibits gene expression by hybridizing to specific mRNA sequences, which interferes with the target protein expression. For the preparation of ODN encapsulated lipoplexes, the ethanol dilution method is very popular (Stano et al. 2004; Jeffs et al. 2005). In this approach, the lipid mixture is first dissolved in ethanol and then mixed with the aqueous ODN solution. The final mixture is dialyzed against buffer solutions to remove ethanol and as the ethanol is removed ODN encapsulated lipoplexes are formed. The ethanol dilution method is clearly a batch production method, and batch to batch variation of the product can be an issue. In addition, multiple steps that may take days to complete are involved, and contamination of the final product can be an issue.

Our alternative approach was to use coaxial electrospray to produce ODN encapsulated lipoplexes, where the lipids/ethanol mixture initially surrounded the aqueous-ODN core. (Wu et al. 2009b) As the liquids leave the needles, the electrical field breaks the compound liquid stream into fine droplets. The large surface area of the droplets results in rapid ethanol evaporation and the ODN encapsulated lipoplexes are formed immediately. Compared to ethanol dilution, coaxial electrospray is a simple, one step and continuous production process. In addition, lipoplexes produced by coaxial electrospray can be either collected in buffer solutions for intravenous injection or used directly in pulmonary delivery.

3.3.2 Electrospray G3139 encapsulated lipoplexes

In our study, G3139 (Genasense or oblimerson sodium) was chosen as the model ODN. G3139 is an 18-mer ODN (5'-TCT CCC AGC GTG CGC CAT-3'). It is specially designed to bind the first six codons of the human Bcl-2 mRNA and thus inhibit Bcl-2 expression, and may provide a way to decrease the resistance of tumor cells to chemotherapy. (Chiu et al. 2006) To prepare G3139 encapsulated lipoplexes, G3139 was dispersed in 1X PBS solution fed through the inner 27 gauge needle. The lipid mixture (DC-Cholesterol: EggPC:DSPE-PEG=30:68:2 molar ratio) was dissolved in ethanol and fed through the outer 20G needle. A positive voltage, typically ~3 kV, was applied between the inner needle and a grounded copper ring electrode to break the liquid into fine droplets. As the ethanol evaporates the lipoplexes are formed and then captured in a grounded aluminum dish 10 cm below the needle tip containing 15 mL PBS solution.

3.3.3 Size distribution and zeta potential of G3139 encapsulated lipoplexes

The size distribution and surface charge of G3139 encapsulated lipoplexes prepared by coaxial electrospray depend on the concentrations and flow rates of G3139 solution and lipid mixtures. By adjusting operation parameters, the size of lipoplexes can vary between 100 nm and 2500 nm and the surface charge of lipoplexes can vary between -16 mV and +20 mV. (Wu et al. 2009b) The best balance between lipoplex productivity, size and surface charge, was achieved by setting the flowrates of the G3139 solution and the lipid mixture to 1.2 mL/h, the concentration of the G3139 solution to 0.5 mg/mL and the concentration of the lipid mixture to 10 mg/mL. Under these conditions Figure 10 shows that the lipoplex average mean diameter by volume was 190 ± 39 nm. The corresponding zeta potential was +4.5 ± 0.43 mV and the final lipid/G3139 ratio was 20.

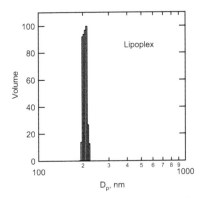

Fig. 10. A typical size distribution of G3139 encapsulated lipoplexes. The average mean diameter by volume was 190 ± 39 nm.

3.3.4 Structure of G3139 encapsulated lipoplexes

Small angle neutron scattering (SANS) and cryo-transmission electron microscopy (cryo-TEM) were used to characterize the structures of G3139 encapsulated lipoplexes. Figure 11 illustrates typical SANS spectra for lipoplexes made by coaxial electrospray and by an ethanol dilution method.

For the sample prepared by the ethanol dilution method, a Bragg Peak was observed, indicating the multi-lamellar structure of the lipoplexes, while the Bragg Peak was absent in lipoplexes produced by coaxial electrospray. From the position of the Bragg Peak we determined the inter-lamellar spacing d using the relation $d = 2\pi/q$, where $q = (4\pi/\lambda)\sin(\theta/2)$ is the momentum transfer vector, λ is the neutron wavelength, and θ is the scattering angle. When we fit the peak for the lipoplexes using a Gaussian function, we found a center-to-center lamellar spacing of 6.6 +/- 0.2 nm. The cryo-TEM images support the SANS results by showing that most electrosprayed lipoplexes have a uni-lamellar structure, while lipoplexes prepared by ethanol dilution have onion like multi-lamellar structure with center to center inter-lamella distance measured from the picture of about 7-10 nm. The difference in the structure of lipoplexes produced by these two methods may reflect the removal rate of ethanol from the solution. In coaxial electrospray, the liquids break into tiny droplets, and the ethanol evaporates in less than 1 s due to the large surface area of the fine droplets. Thus, the lipids and G3139 may be "locked" and do not have time to rearrange and form the multi-lamellar structure. In contrast, the dialysis step in the ethanol dilution method is a very slow process, usually requiring about 24 h to remove ethanol from the solution, and therefore the lipids and G3139 have enough time to form the more complex multi-lamellar structure.

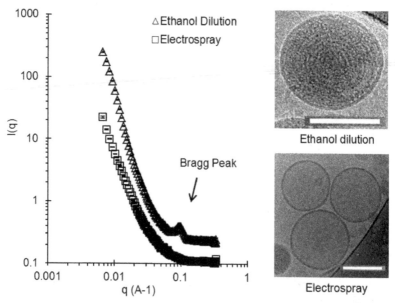

Fig. 11. Typical SANS spectra and cryo-TEM images show that the electrosprayed lipoplexes have unilamellar structure, while those prepared by ethanol dilution have a multi-lamellar structure. The error bars on the SANS data represent ± one standard deviation and are generally smaller than the symbol size. (Scale bar: 100nm)

3.3.5 Bcl-2 down regulation

The bioactivity of G3139 encapsulated lipoplexes was evaluated in K562 cells (chronic myelogenous leukemia cell line). In addition to non-targeted G3139 encapsulated lipoplexes, transferrin (Tf) conjugated G3139 encapsulated lipoplexes were also used in this study. Chiu et al. (2006) reported that transferrin (Tf) conjugated lipoplexes had targeting ability because they could binding to the transferrin receptor (TfR), a transmembrane glycoprotein over expressed on cancer and leukemia cells. Thus, transferrin conjugated lipoplexes provided better down regulation of Bcl-2 in K562 cells and were more effective for *in vivo* applications. K562 cells were treated with both non-targeted and Tf-targeted G3139 encapsulated lipoplexes at G3139 concentration of 1 µM. Bcl-2 expression were measured by western blotting 48 h post transfection. Figure 12 shows that the Bcl-2 expression was decreased by up to ~55% when cells were transfected with Tf-targeted G3139 encapsulated lipoplexes, compared to ~40% by non-targeted G3139 encapsulated lipoplexes and ~15% by free G3139.

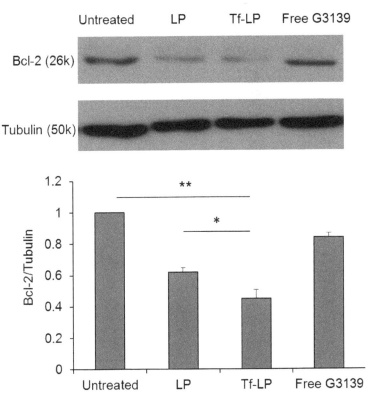

Fig. 12. A representative western blotting of Bcl-2 expression in K562 cells. LP: non-targeted G3139 encapsulated lipoplexes; Tf-LP: Tf-targeted G3139 encapsulated lipoplexes. The concentration of G3139 was 1 μM. (n=3, *: p < 0.05, **: p< 0.01.)

3.3.6 Cellular uptake of ODN encapsulated lipoplexes

K562 cells were transfected by free FAM-ODN, non-targeted FAM-ODN encapsulated lipoplexes and Tf-targeted FAM-ODN encapsulated lipoplexes at FAM-ODN concentration level of 1μM. Flow cytometry and confocal microscopy were used to investigate the cellular uptake of FAM-ODN encapsulated lipoplexes 4 h post transfection. As shown on Figure 13 the cellular uptake of FAM-ODN delivered by lipoplexes, particularly Tf-targeted lipoplexes, was much more efficient compared to cells treated with free FAM-ODN. Compared to cells transfected with non-targeted FAM-ODN encapsulated lipoplexes, those treated with Tf-targeted FAM-ODN encapsulated lipoplexes had ~40% higher fluorescence signal, indicating that transferrin improved the interaction between the lipoplexes and the K562 cells and facilitated the cellular uptake of lipoplexes, and thus more efficient Bcl-2 down-regulation was observed. For future *in vivo* applications Tf-targeted lipoplexes might also be more effective targeting cancer cells and minimize the rapid clearance by the reticuloendothelial system, or side effects, such as nonspecific cytokine production.

(a)

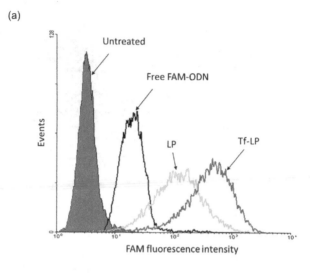

(b)

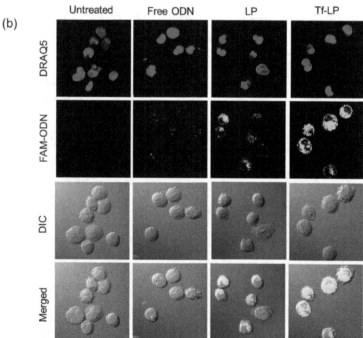

Fig. 13. **(a)** Flow cytometery and **(b)** confocal microscopy images showed the cellular uptake of FAM-ODN by K562 cells. K562 cells were treated with free FAM-ODN, non-targeted FAM-ODN encapsulated lipoplexes (LP) and Tf-targeted FAM-ODN encapsulated lipoplexes (Tf-LP) at FAM-ODN concentration of 1 μM. Control: cells cultured in medium with no transfection. DIC: differential interference contrast.

3.3.7 Conclusions

We developed a coaxial electrospray process to produce oligonucleotide encapsulated lipoplexes for nucleic acid delivery. The lipoplexes produced by coaxial electrospray can either be collected for intravenous injection or delivered as aerosol for inhalation therapy. This method allows for better control over the way in which the lipid and aqueous phases are mixed. Compared with the standard ethanol dilution technique, coaxial electrospray is a simple, one step, continuous process that significantly reduces the time and effort required to produce the lipoplexes. By adjusting operating parameters, such as flow rates and liquid concentrations, monodisperse lipoplexes with different size and surface charge can be easily produced to meet various application needs. In this work, G3139 encapsulated lipoplexes were successfully produced via coaxial electrospray with diameter of ~190nm and zeta potential of ~+4.5 mV. Due to fast ethanol removal, the lipoplexes produced by coaxial electrospray showed unilamellar structure compared to the multi-lammellar structure of lipopelxes produced by ethanol dilution method. Tranferrin was successfully conjugated to the G3139 encapsulated lipoplexes. Flow cytometry and confocal microscopy analysis showed that transferrin provided targeting ability for the lipoplexes, which greatly improved the cellular uptake of lipoplexes. Compared to ~40% Bcl-2 protein down-regulation observed by non-targeted G3139 encapsulated lipoplexes, Tf-targeted lipoplexes was more efficiently delivered to K562 cells and down regulated the Bcl-2 protein expression by ~55%.

4. Summary

To summarize, we have demonstrated the great potential of electrospray to produce nanoparticles for a variety of drug/nucleic acid delivery applications, including solid lipid nanoparticles for hydrophobic drug delivery, as well as polyplexes and lipoplexes for nucleic acid delivery. We hope this review can stimulate further development and utilization of electrospray in nanobiotechnology.

5. Acknowledgement

This work was supported by the National Science Foundation under Grant No. EEC-0425626. We acknowledge and thank Dr. Weiwei Deng at the University of Central Florida for providing the multiplexed electrospray pictures.

6. References

Bocanegra, R.; Gaonkar, A. G.; Barrero, A.; Loscertales, I. G.; Pechack, D. & Marquez, M. (2005). Production of Cocoa Butter Microcapsules Using an Electrospray Process. *Journal of Food Science*, Vol. 70, No. 8, pp. E492-E497, ISSN 1750-3841

Chen, D. R.; Wendt, C. H. & Pui, D. Y. H. (2000). A Novel Approach for Introducing Bio-materials into Cells. *Journal of Nanoparticle Research*, Vol. 2, No. 2, pp. 133-139, ISSN 1388-0764

Chen, X.; Jia, L.; Yin, X.; Cheng, J. & Lu, J. (2005). Spraying Modes in Coaxial Jet Electrospray with Outer Driving Liquid. *Physics of Fluids*, Vol. 17, No. 3, pp. 032101. ISSN 1070-6631

Chiu, S.; Liu, S.; Perrotti, D.; Marcucci, G. & Lee, R. J. (2006). Efficient delivery of a Bcl-2-Specific Antisense Oligodeoxyribonucleotide (G3139) via Transferrin Receptor-Targeted Liposomes. *Journal of Controlled Release*, Vol. 112, No. 2, pp. 199–207, ISSN 0168-3659

Chrai, S. S.; Murari, R. & Ahmad, I. (2001). Liposomes (a Review) Part One: Manufacturing Issues. *Biopharm International*, Vol. 14, No. 11, pp. 10-14, ISSN 1542-166X

Cloupeau, M. & Prunet-Foch, B. (1994) Electrohydrodynamic Spraying Functioning Modes: a Critical Review. *Journal of Aerosol Science*, Vol. 25, No. 6, pp. 1021-1036, ISSN 0021-8502

Davies, L. A.; Hannavy, K.; Davies, N.; Pirrie, A.; Coffee, R. A.; Hyde, S. C. & Gill, D. R. (2005). Electrohydrodynamic Comminution: a Novel Technique for the Aerosolisation of Plasmid DNA. *Pharmaceutical Research*, Vol. 22, No. 8, pp. 1294-1304, ISSN 0724-8741

Deng, W.; Klemic, J. F.; Li, X.; Reed, M. A. & Gomez, A. (2006). Increase of Electrospray Throughput Using Multiplexed Microfabricated Sources for the Scalable Generation of Monodisperse Droplets. *Journal of Aerosol Science*, Vol. 37, pp. 696-714, ISSN 0021-8502

Elouahabi, A. & Ruysschaert, J. (2005). Formation and Intracellular Trafficking of Lipoplexes and Polyplexes. *Molecular Therapy*, Vol. 11, pp. 336-347, ISSN 1525-0016

Fan, M.; Xu, S.; Xia, S. & Zhang, X. (2007). Effect of Different Preparation Methods on Physicochemical Properties of Salidroside Liposomes. *Journal of Agricultural and Food Chemistry*, Vol. 55, No. 8, pp. 3089-3095, ISSN 0021-8561

Farook, U.; Zhang, H.B.; Edirisinghe, M. J.; Stride, E. & Saffari, N. (2007). Preparation of Microbubble Suspensions by Co-axial Electrohydrodynamic Atomization. *Medical Engineering and Physics*. Vol. 29, No. 7, pp. 749-754, ISSN 1350-4533

Fenn, J. B.; Mann, M.; Meng, C. K.; Wong, S. F. & Whitehouse, C. M. (1989). Electrospray Ionization for Mass Spectrometry of Large Biomolecules. *Science*, Vol. 246, No. 4926, pp. 64-71, ISSN 0036-8075

Gebhart, C. L. & Kabanov, A. V. (2001). Evaluation of Polyplexes as Gene Transfer Agents. *Journal of Controlled Release*, Vol. 73, No. 2-3, pp. 401-416, ISSN 0168-3659

Godbey, W. T.; Wu, K. K. & Mikos, A. G. (1999). Poly(ethylenimine) and Its Role in Gene Delivery. *Journal of Controlled Release*, Vol. 60, No. 2-3, pp. 149-160, ISSN 0168-3659

Haberland, M. E. & Reynolds, J. A. (1973). Self-association of Cholesterol in Aqueous Solution. *Proceedings of the National Academy of Sciences*, Vol. 70, No. 8, pp. 2313-2316, ISSN 1091-6490

Jaworek, A. (2010). *Electrospray Technology for Thin-Film Deposition*, Nova Science Publishers Inc, ISBN 9781617612015, Hauppaug, New York, USA

Jaworek, A. (2007). Micro- and Nanoparticle Production by Electrospraying. *Powder Technology*, Vol. 176, No. 1, pp. 18-35, ISSN 0032-5910

Jaworek, A. & Krupa, A. (1999). Classificantion of the Modes of EHD Spraying. *Journal of Aerosol Science*, Vol. 30, No. 7, pp. 873-893, ISSN 0021-8502

Jaworek, A. & Sobczyk, A. T. (2008). Electrospraying Route to Nanotechnology: an Overview. *Journal of Electrostatics*. Vol. 66, No. 3-4, pp. 197-219, ISSN 0304-3886

Jayasinghe, S. N.; Qureshi, A. N. & Eagles, P. A. M. (2006). Electrohydrodynamic Jet Processing: an Advanced Electric-Field-Driven Jetting Phenomenon for Processing Living Cells. *Small*, Vol. 2, No. 2, pp. 216-219, ISSN 1613-6829

Jeffs, L. B.; Palmer, L. R.; Ambegia, E. G.; Giesbrecht, C.; Ewanick, S. & MacLachlan, I. (2005) A Scalable Extrusion-Free Method for Efficient Liposomal Encapsulation of Plasmid DNA. *Pharmaceutical Research*. Vol. 22, No. 3, pp. 363-372, ISSN 0724-8741

Kircheis, R.; Wightman, L. & Wagner E. (2001). Design and Gene delivery Activity of Modified Polyethylenimines. *Advanced Drug Delivery Reviews*, Vol. 53, No. 3, pp. 341-358, 0169-409X

Koshkina, N. V.; Agoulink, I. Y.; Melton, S. L.; Densmore, C. L. & Knight V. (2003). Biodistribution and Pharmacokinetics of Aerosol and Intravenously Administered DNA-Polyethylenimine Complexes: Optimization of Pulmonary Delivery and Retention. *Molecular Therapy*, Vol. 8, No. 2, pp. 249-254, ISSN 1525-0016

Kumbar, S. G.; Bhattacharyya, S.; Sethuraman, S. & Laurencin, C. T. (2007). A Preliminary Report on a Novel Electrospray Technique for Nanoparticle Based Biomedical Implants Coating: Precision Electrospraying. *Journal of Biomedical Materials Research Part B: Applied Biomaterials*, Vol. 81B, No. 1, pp. 91-103, ISSN 1552-4973

Laporte, L. D.; Rea, J. C. & Shea, L. D. (2006). Design of Modular Non-Viral Gene Therapy Vectors. Biomaterials, Vol. 27, No. 7, pp. 947-954, ISSN 0142-9612

Loscertales, I. G.; Barrero, A.; Guerrero, I.; Cortijo, R.; Marquez, M. & Ganan-Calvo, A. M. (2002). Micro/Nano Encapsulation Via Electrified Coaxial Liquid Jets. *Science*, Vol. 295, No. 5560, pp. 1695-1698, ISSN 0036-8075

Mastrobattista, E.; Van Der Aa, M. A. E. M.; Hennink, W. E. & Crommelin, D. J. A. (2006). Artificial Viruses: a Nanotechnological Approach to Gene Delivery. *Nature Reviews Drug Discovery*, Vol. 5, No. 2, pp. 115-121, ISSN 1474-1776

Müller, R. H. & Keck, C. M. (2004). Challenges and Solutions for the Delivery of Biotech Drugs – a Review of Drug Nanocrystal Technology and Lipid Nanoparticles. *Journal of Biotechnology*. Vol. 113, No. 1-3, pp. 151-170, ISSN 1684-5315

Müller, R. H.; Mader, K. & Gohla, S. (2000). Solid Lipid Nanoparticles (SLN) for Controlled Drug Delivery – a Review of the State of the Art. *European Journal of Pharmaceutics and Biopharmaceutics*. Vol. 50, No. 1, pp. 161-177, ISSN 0939-6411

Neu, M.; Fisher, D. & Kissel T. (2005). Recent Advances in Rational Gene Transfer Vector Design Based on Poly(ethyleneimine) and Its Derivatives. *Journal of Gene Medicine*, Vol. 7, No. 8, pp. 992-1009, ISSN 1521- 2254

Ojito, E.; Labrada, G.; Garcia, Z.; Garcia, N. & Chico, E. (2001). Study of NS0 Cell Line Metabolism in Lipid Supplemented Protein Free Media, *Animal Cell Technology: From Target to Market: Proceedings of the 17th ESACT Meeting*, pp. 179-182, ISBN 978-1-4020-0264-9, Tylösand, Sweden, June 10-14, 2001.

Otake, K.; Shimomura, T.; Goto, T.; Imura, T.; Furuya, T.; Yoda, S.; Takebayashi, Y.; Sakai, H. & Abe, M. (2006). Preparation of Liposomes Using an Improved Supercritical Reverse Phase Evaporation Method. *Langmuir*, Vol. 22, No. 6, pp. 2543-2550, ISSN 0743-7463

Pack, D. W.; Hoffman, A. S.; Pun, S. & Stayton, P. S. (2005). Design and Development of Polymers for Gene Delivery. *Nature Reviews Drug Discovery*, Vol. 4, No. 7, pp. 581-593, ISSN 1474-1776

Pareta, R. & Edirisinghe, M. J. (2006). A Novel Method for the Preparation of Biodegradable Microspheres for Protein Drug Delivery. *Journal of the Royal Society Interface*, Vol. 3, No. 9, pp. 573–582, ISSN 1742-5689

Pareta, R.; Brindley, A.; Edirisinghe, M. J.; Jayasinghe, S. N. & Lukinska, Z. B. (2005). Electrohydrodynamic Atomization of Protein (Bovine Serum Albumin). *Journal of Materials Science: Materials in Medicine*, Vol. 16, No. 10, pp. 919– 925, ISSN 0957-4530

Salata, O. V. (2005). Tools of Nanotechnology: Electrospray. *Current Nanoscience*, Vol. 1, No.1, pp. 25-33, ISSN 1573-4137

Stankusa, J. J.; Solettib, L.; Fujimoto, K.; Hong, Y.; Vorp, D. A. & Wagnera W. R. (2007). Fabrication of Cell Microintegrated Blood Vessel Constructs Through Electrohydrodynamic Atomization. *Biomaterials*, Vol. 28, No. 17, pp. 2738–2746, ISSN 0142-9612

Stano, P.; Bufali, S.; Pisano, C.; Bucci, F.; Barbarino, M.; Santaniello, M.; Carminati, P. & Luisi, P. L. (2004). Novel Camptothecin Analogue (Gimatecan)-Containing

Liposomes Prepared by the Ethanol Injection Method. *Journal of Liposome Research.* Vol. 14, No. 1-2, pp. 87–109, ISSN 0898-2104

Talley, D.; Cutak, B.; Rathbone, E.; Al-Kolla, T.; Allison, D.; Blasberg, J.; Kao, K. & Caple, M. (2003). SyntheChol™ Synthetic Cholesterol for Cholesterol Dependent Cell Culture-Development of Non-Animal Derived Chemically Defined NS0 Medium, *Animal Cell Technology Meets Genomics: Proceedings of the 18th ESACT Meeting,* pp. 577-580, ISBN 978-4020-2791-5, Granada, Spain, May 11-14, 2003.

Widiyandari, H.; Hogan, C. J. J.; Yun, K. M.; Iskandar, F.; Biswas, P. & Okuyama, K. (2007). Production of Narrow-Size-Distribution Polymer-Pigment-Nanoparticle Composites via Electrohydrodynamic Atomization. *Macromolecular Materials and Engineering,* Vol. 292, No. 4, pp. 495–502, ISSN 1438-7492

Wu, Y.; Chalmers, J. J.; Wyslouzil, B. E. (2009). The Use of Electrohydrodynamic Spraying to Disperse Hydrophobic Compounds in Aqueous Media. *Aerosol Science and Technology,* Vol. 43, pp. 902-910, ISSN 0278-6826

Wu, Y.; Yu, B.; Jackson, A.; Zha, W.; Lee L. J. & Wyslouzil, B. E. (2009). Coaxial Electrohydrodynamic Spraying: a Novel One-Step Technique to Prepare Oligodeoxynucleotide Encapsulated Lipoplex Nanoparticles. *Molecular Pharmaceutics,* Vol. 6, No. 5, pp. 1371–1379, ISSN 1543-8384

Wu. Y.; Fei, Z.; Lee, L. J. & Wyslouzil, B. E. (2010). Electrohydrodynamic Spraying of DNA/Polyethylenimine Polyplexes for Nonviral Gene Delivery. *Biotechnology and Bioengineering,* Vol. 105, No. 4, pp. 834-841, ISSN 0006-3592

Wu, Y.; Ma, N.; Chalmers, J.; Wyslouzil, B. E.; McCormick, E. L. & Casnocha, S. (2011). Enhanced Productivity of NS0 Cells in Fed-Batch Culture with Cholesterol Nanoparticle Supplementation. *Biotechnology Progress,* Vol. 27, No. 3, pp. 796–802, ISSN 1520-6033

Xie, J.; Lim, L. K.; Phua, Y.; Hua, J. & Wang, C. (2006). Electrohydrodynamic Atomization for Biodegradable Polymeric Particle Production. *Journal of Colloid and Interface Science,* Vol. 302, No. 1, pp. 103–112, ISSN 0021-9797

Xie, J. & Wang, C. (2007). Encapsulation of Proteins in Biodegradable Polymeric Microparticles Using Electrospray in the Taylor Cone-Jet Mode. *Biotechnology and Bioengineering,* Vol. 97, No. 5, pp. 1278–1290, ISSN 0006-3592

Yamane, I.; Kan, M.; Minamoto, Y. & Amatsuji, Y. (1981). α-Cyclodextin, a Novel Substitute for Bovine Albumin in Serum-Free Culture of Mammalian Cells. *Proceedings of the Japan Academy, Ser. B, Physical and Biological Sciences,* Vol. 57, No. 10, pp. 385-389 ISSN 0386-2208

Yang, L.; Li, J.; Zhou, W.; Yuan, X. & Li, S. (2004). Targeted delivery of antisense oligodeoxynucleotides to folate receptor-overexpressing tumor cells. *Journal of Controlled Release,* Vol. 95, No. 2, pp. 321– 331, ISSN 0168-3659

Yurteri, C. U.; Hartman, R. P. A. & Marijnissen J. C. M. (2010). Producing Pharmaceutical Particles via Electrospraying with an Emphasis on Nano and Nano Structured Particles - A Review. *KONA Powder and Particle Journal* No.28, pp. 91-115, ISSN 0288-4534

Zhang, J. & Robinson, D. (2005). Development of Aninal-free, Protein-free and Chemically-defined Media for NS0 Cell Culture. *Cytotechnology,* Vol. 48, No. 1-3, pp. 59-74, ISSN 0920-9069

Zhang, S.; Kawakami, K.; Yamamoto, M.; Masaoka, Y.; Kataoka, M.; Yamashita, S. & Sakuma, S. (2011). Coaxial Electrospray Formulations for Improving Oral Absorption. *Molecular Pharmaceutics,* Vol. 8, No. 3, pp. 907-813, ISSN 1543-8384

Permissions

The contributors of this book come from diverse backgrounds, making this book a truly international effort. This book will bring forth new frontiers with its revolutionizing research information and detailed analysis of the nascent developments around the world.

We would like to thank Dr. Abbass A. Hashim, for lending his expertise to make the book truly unique. He has played a crucial role in the development of this book. Without his invaluable contribution this book wouldn't have been possible. He has made vital efforts to compile up to date information on the varied aspects of this subject to make this book a valuable addition to the collection of many professionals and students.

This book was conceptualized with the vision of imparting up-to-date information and advanced data in this field. To ensure the same, a matchless editorial board was set up. Every individual on the board went through rigorous rounds of assessment to prove their worth. After which they invested a large part of their time researching and compiling the most relevant data for our readers. Conferences and sessions were held from time to time between the editorial board and the contributing authors to present the data in the most comprehensible form. The editorial team has worked tirelessly to provide valuable and valid information to help people across the globe.

Every chapter published in this book has been scrutinized by our experts. Their significance has been extensively debated. The topics covered herein carry significant findings which will fuel the growth of the discipline. They may even be implemented as practical applications or may be referred to as a beginning point for another development. Chapters in this book were first published by InTech; hereby published with permission under the Creative Commons Attribution License or equivalent.

The editorial board has been involved in producing this book since its inception. They have spent rigorous hours researching and exploring the diverse topics which have resulted in the successful publishing of this book. They have passed on their knowledge of decades through this book. To expedite this challenging task, the publisher supported the team at every step. A small team of assistant editors was also appointed to further simplify the editing procedure and attain best results for the readers.

Our editorial team has been hand-picked from every corner of the world. Their multi-ethnicity adds dynamic inputs to the discussions which result in innovative outcomes. These outcomes are then further discussed with the researchers and contributors who give their valuable feedback and opinion regarding the same. The feedback is then collaborated with the researches and they are edited in a comprehensive manner to aid the understanding of the subject.

Apart from the editorial board, the designing team has also invested a significant amount of their time in understanding the subject and creating the most relevant covers. They scrutinized every image to scout for the most suitable representation of the subject and create an appropriate cover for the book.

The publishing team has been involved in this book since its early stages. They were actively engaged in every process, be it collecting the data, connecting with the contributors or procuring relevant information. The team has been an ardent support to the editorial, designing and production team. Their endless efforts to recruit the best for this project, has resulted in the accomplishment of this book. They are a veteran in the field of academics and their pool of knowledge is as vast as their experience in printing. Their expertise and guidance has proved useful at every step. Their uncompromising quality standards have made this book an exceptional effort. Their encouragement from time to time has been an inspiration for everyone.

The publisher and the editorial board hope that this book will prove to be a valuable piece of knowledge for researchers, students, practitioners and scholars across the globe.

List of Contributors

Jelena Kolosnjaj-Tabi, Henri Szwarc and Fathi Moussa
UMR CNRS 8612 and LETIAM, EA 4041, University of Paris Sud, France

Hassan Korbekandi
Genetics and Molecular Biology Department, School of Medicine, Isfahan University of Medical Sciences, Iran

Siavash Iravani
School of Pharmacy and Pharmaceutical Science, Isfahan University of Medical Sciences, Iran

Roghayeh Abbasalipourkabir, Aref Salehzadeh and Rasedee Abdullah
Hamedan University of Medical Science, Iran
Universiti Putra Malaysia, Malaysia

Georg Garnweitner
Technische Universität Braunschweig, Germany

Vijaykumar Nekkanti, Venkateswarlu Vabalaboina and Raviraj Pillai
Dr. Reddy's Laboratories Limited, Hyderabad, India

Michał Piotr Marszałł
Department of Medicinal Chemistry, Collegium Medicum in Bydgoszcz, Nicolaus Copernicus University, Poland

Hassan Namazi
Research Center for Pharmaceutical Nanonotechnology, Tabriz University of Medical Science, Tabriz, Iran
Research Laboratory of Dendrimers and Nanopolymers, University of Tabriz, Tabriz, Iran

Farzaneh Fathi and Abolfazl Heydari
Research Laboratory of Dendrimers and Nanopolymers, University of Tabriz, Tabriz, Iran

A. Martínez
Departamento de Farmacología, Facultad de Farmacia, Universidad Complutense de Madrid, Spain

A. Fernández, J.M. Teijón and M.D. Blanco
Departamento de Bioquímica y Biología Molecular, Facultad de Medicina, Universidad Complutense de Madrid, Spain

M. Benito
Centro Universitario San Rafael-Nebrija, Ciencias de la Salud, Madrid, Spain

E. Pérez

Veli C. Özalp and Thomas Schäfer
University of the Basque Country, Spain

Ajay Sud
Whipps Cross University Hospital, London, United Kingdom
The Royal London Hospital, London, United Kingdom
The Royal Liverpool University Hospital, Liverpool, United Kingdom

Shiva Dindyal
Whipps Cross University Hospital, London, United Kingdom
The Royal London Hospital, London, United Kingdom

Lenore L. Dai
Arizona State University, USA

Barbara E. Wyslouzil
NSF Nanoscale Science and Engineering Center, The Ohio State University, Columbus, Ohio, USA
William G. Lowrie Department of Chemical and Biomolecular Engineering, The Ohio State University, Columbus, Ohio, USA
Department of Chemistry, The Ohio State University, Columbus, Ohio, USA

Anthony Duong and L. James Lee
NSF Nanoscale Science and Engineering Center, The Ohio State University, Columbus, Ohio, USA
William G. Lowrie Department of Chemical and Biomolecular Engineering, The Ohio State University, Columbus, Ohio, USA

Yun Wu
NSF Nanoscale Science and Engineering Center, The Ohio State University, Columbus, Ohio, USA

9 781632 383396